MANUAL OF GEOTECHNICAL LABORATORY SOIL TESTING

MANUAL OF GEOTECHNICAL LABORATORY SOIL TESTING

Bashir Ahmed Mir

CRC Press
Taylor & Francis Group
Boca Raton London New York

CRC Press is an imprint of the
Taylor & Francis Group, an **informa** business

First edition published 2022
by CRC Press
6000 Broken Sound Parkway NW, Suite 300, Boca Raton, FL 33487-2742

and by CRC Press
2 Park Square, Milton Park, Abingdon, Oxon, OX14 4RN

Library of Congress Cataloging-in-Publication Data
Names: Mir, Bashir Ahmed, author.
Title: Manual of geotechnical laboratory soil testing/Bashir Ahmed Mir.
Description: Boca Raton: CRC Press, 2022. | Includes bibliographical references and index.
Identifiers: LCCN 2021018349 (print) | LCCN 2021018350 (ebook) | ISBN 9781032060095 (hbk) | ISBN 9781032060125 (pbk) | ISBN 9781003200260 (ebk)
Subjects: LCSH: Soil mechanics--Laboratory manuals. | Soils--Testing--Laboratory manuals. | CYAC: Soils--Experiments--Laboratory manuals.
Classification: LCC TA710. M55 2022 (print) | LCC TA710 (ebook) | DDC 624.1/51360287--dc23
LC record available at https://lccn.loc.gov/2021018349
LC ebook record available at https://lccn.loc.gov/2021018350

ISBN: 978-1-032-06009-5 (hbk)
ISBN: 978-1-032-06012-5 (pbk)
ISBN: 978-1-003-20026-0 (ebk)

DOI: 10.1201/9781003200260

Typeset in Times
by MPS Limited, Dehradun

Contents

Systems of Units: Units Conversion Factors/Tables

It may be necessary to express the results of laboratory tests in a given system of units. At this time in the United States, both the English and the SI system of units are used. Conversion of units may be necessary in preparing reports. Some selected conversion factors are given in tabular form below:

Length

1 m	3.28 ft
1 cm	0.3937 in
1 ft	30.48 cm
1 in	2.54 cm
1 μm	1×10^{-6} m
	0.0001 cm
	0.001 mm
1 Å	1×10^{-10} m
1 Å	3.28×10^{-10} feet
1 yd	0.9144 m

Volume

1 L	1000 cm^3
1 L	61.02 in^3
1 m^3	35.32 ft^3
1 ft^3	0.0283 m^3
1 in^3	16.39 cm^3
1 ft^3	28317 cm^3
1cc	10^{-6} m^3

Force

1 N	102 g
1 N	0.2248 lb
1 lb	4.448 N

Area

1 m^2	10.76ft^2
1 ft^2	929 cm^2
1 cm^2	0.0001 m^2
1 in^2	0.007 ft^2
1 acre	4047 m^2
1 ha	2.471 acre
1 $mile^2$	640 acre

Mass/Weight

1 kg	2.202 lb
1 lb	0.454 kg
1 gram	10^{-5} kN
1 gram	0.01 N
1N	100 gram
1kg	10N

Power P

1 W	1 J/sec
1 W	1.1622 cal/hr
1 W	3.41 Btu/hr
1 W	1 W
1 hp	745.7 W
1 hp	0.7457 kW
1 kW	1.34 hp

Stress/Pressure

1 kPa	1 kN/m^2
1 psi	6.895 kPa
1 N/m^2	1 Pa
1 kPa	20.89 lb/ft^2
1 lb/ft^2	0.0479 kPa
1 kPa	0.1450 psi
1 kg/cm^2	100 kPa
1 t/ft^2	107.6 kPa

Density

1 g/cc	1 T/m^3 or 1 Mg/m^3
1 g/cc	62.43 p/ft^3
1 lb/ft^3	0.01602 Mg/m^3
1 T/m^3	10 kN/m^3
1k gf/m^3	0.00981 kN/m^3
1 $kN/\ m^3$	102 kgf/m^3
1 pci	271.3714 kN/m^3
1 pci	1728 pcf
1 pci	27.6799 T/m^3
1 pcf	0.01602 g/cc
1 pcf	0.01602 T/m^3
1 pcf	0.1602 $kN/\ m^3$
1 pcf	

Multiplication Factors

10^{18}	exa	E
10^{15}	peta	P
10^{12}	tera	T
10^{9}	giga	G
10^{6}	mega	M
10^{3}	kilo	k
10^{-3}	milli	m
10^{-6}	micro	μ
10^{-9}	nano	n
10^{-12}	pico	p
10^{-15}	femto	f
10^{-18}	atto	a

Gravitational acceleration

1 g	9.81 m/s^2
1 g	32.2 ft/sec^2

Temperature

°C	0.555 °F -17.778
°F	1.8°C+32

Angle

1 rad	57.296"
1°	0.017453 rad
π	180°

μm: micron, lb: pound, pci: pounds force per cubic inch, pcf: pounds force per cubic feet

1 gm-s/cm^2 = 981 poises (1000 poises); 1 pound-s/sq.ft = 478.69 poises; 1 MN = 10^6 N; 1μ = 10^{-3} mm = 10^{-6} m

1 poise = 1000 millipoises = 10^{-4} kN-s/m^2 = 10^{-3} gm-sec/cm^2 = 1 dyne-s/sq.cm; 1 N = 10^6 μN

1 g/cc = 1 t/m^3 = 10 kN/m^3; 1 kg/cm^2 = 10 N/cm^2 = 0.1 N/mm^2 = 10 t/m^2 = 10^5 N/m^2 = 0.1 MN/m^2 = 100 kN/m^2

Length Units

Millimeters mm	Centimeters cm	Meters m	Kilometers km	Inches in	Feet ft	Yards yd	Miles mi
1	0.1	0.001	0.000001	0.03937	0.003281	0.001094	6.21e-07
10	1	0.01	0.00001	0.393701	0.032808	0.010936	0.000006
1000	100	1	0.001	39.37008	3.28084	1.093613	0.000621
1000000	100000	1000	1	39370.08	3280.84	1093.613	0.621371
25.4	2.54	0.0254	0.000025	1	0.083333	0.027778	0.000016
304.8	30.48	0.3048	0.000305	12	1	0.333333	0.000189
914.4	91.44	0.9144	0.000914	36	3	1	0.000568
1609344	160934.4	1609.344	1.609344	63360	5280	1760	1

Area Units

Millimeter square mm^2	Centimeter Square cm^2	Meter Square m^2	Inch Square in^2	Foot Square ft^2	Yard Square yd^2
1	0.01	0.000001	0.00155	0.000011	0.000001
100	1	0.0001	0.155	0.001076	0. 00012
1,000,000	10,000	1	1,550.003	10.76391	1.19599
645.16	6.4516	0.000645	1	0.006944	0.000772
92,903	929.0304	0.092903	144	1	0.111111
836,127	8,361.274	0.836127	1296	9	1

Volume Units

Centimeter Cube cm^3	Meter Cube m^3	Liter ltr	Inch Cube in^3	Foot Cube ft^3	US Callons US gal	Imperial Gallons Imp. gal	US Barrel (oil) US brl
1	0.000001	0.001	0.061024	0.000035	0.000264	0.00022	0.000006
1,000,000	1	1,000	61,024	35	264	220	6.29
1,000	0.001	1	61	0.035	0.264201	0.22	0.00629
16.4	0.000016	0.016387	1	0.000579	0.004329	0.003605	0.000103

(*Continued*)

Centimeter Cube cm³	Meter Cube m³	Liter ltr	Inch Cube in³	Foot Cube ft³	US Callons US gal	Imperial Gallons Imp. gal	US Barrel (oil) US brl
28,317	0.028317	28.31685	1728	1	7.481333	6.229712	0.178127
3,785	0.003785	3.79	231	0.13	1	0.832701	0.02381
4,545	0.004545	4.55	277	0.16	1.20	1	0.028593
158,970	0.15897	159	9,701	6	42	35	1

Mass Units

Grams g	Kilograms kg	Metric tonnes tonne	Short Ton shton	Long Ton Lton	Pounds lb	Ounces oz
1	0.001	0.000001	0.000001	9.84e-07	0.002205	0.035273
1,000	1	0.001	0.001102	0.000984	2.204586	35.27337
1,000,000	1,000	1	1.102293	0.984252	2204.586	35273.37
907,200	907.2	0.9072	1	0.892913	2000	32000
1,016,000	1016	1.016	1.119929	1	2239.859	35837.74
453.6	0.4536	0.000454	0.0005	0.000446	1	16
28	0.02835	0.000028	0.000031	0.000028	0.0625	1

Density Units

Gram/milliliter g/ml	Kilogram/meter cube kg/m³	Pound/foot cube lb/ft³	Pound/inch cube lb/in³
1	1,000	62.42197	0.036127
0.001	1	0.062422	0.000036
0.01602	16.02	1	0.000579
27.68	27,680	1727.84	1

Volumetric Liquid Flow Units

Liter/ second L/sec	Liter/ minute L/min	Meter cube/hour m³/hr	Foot cube/ minute ft³/min	Foot cube/hour ft³/hr	US gallons/ minute gal/min	US barrels (oil)/day US brl/d
1	60	3.6	2.119093	127.1197	15.85037	543.4783
0.016666	1	0.06	0.035317	2.118577	0.264162	9.057609

(*Continued*)

Liter/ second L/sec	Liter/ minute L/min	Meter cube/hour m³/hr	Foot cube/ minute ft³/min	Foot cube/hour ft³/hr	US gallons/ minute gal/min	US barrels (oil)/day US brl/d
0.277778	16.6667	1	0.588637	35.31102	4.40288	150.9661
0.4719	28.31513	1.69884	1	60	7.479791	256.4674
0.007867	0.472015	0.02832	0.01667	1	0.124689	4.275326
0.06309	3.785551	0.227124	0.133694	8.019983	1	34.28804
0.00184	0.110404	0.006624	0.003899	0.2339	0.029165	1

Volumetric Gas Flow Units

Normal meter cube/hour Nm³/hr	Standard cubic feet/hour scfh	Standard cubic feet/minute scfm
1	35.31073	0.588582
0.02832	1	0.016669
1.699	59.99294	1

Mass Flow Units

Kilogram/hour kg/h	Pound/hour lb/h	Kilogram/second kg/s	Ton/hour t/h
1	2.204586	0.000278	0.001
0.4536	1	0.000126	0.000454
3,600	7,936.508	1	3.6
1,000	2,204.586	0.277778	1

High-Pressure Units

Bar bar	Pound/ square inch psi	Kilo- pascal kPa	Mega- pascal MPa	Kilogram force/ centimeter square kgf/cm²	Millimeter of mercury mm Hg	Atmospheres atm
1	14.50326	100	0.1	1.01968	750.0188	0.987167
0.06895	1	6.895	0.006895	0.070307	51.71379	0.068065
0.01	0.1450	1	0.001	0.01020	7.5002	0.00987
10	145.03	1,000	1	10.197	7500.2	9.8717
0.9807	14.22335	98.07	0.09807	1	735.5434	0.968115

(*Continued*)

Bar	Pound/ square inch	Kilo- pascal	Mega- pascal	Kilogram force/ centimeter square	Millimeter of mercury	Atmospheres
bar	psi	kPa	MPa	kgf/cm^2	mm Hg	atm
0.001333	0.019337	0.13333	0.000133	0.00136	1	0.001316
1.013	14.69181	101.3	0.1013	1.032936	759.769	1

Dynamic Viscosity Units

Centipoise*	Poise	Pound/foot·second	Remarks
cp	poise	lb/(ft·s)	*: centistokes x specific gravity = centipoise
1	0.01	0.000672	
100	1	0.067197	
1,488.16	14.8816	1	

Kinematic Viscosity Units

Centistoke*	Stoke	Foot square/ second	Meter square/ second	Remarks
cs	St	ft^2/s	m^2/s	*: centistokes x specific gravity = centipoise
1	0.01	0.000011	0.000001	
100	1	0.001076	0.0001	
92,903	929.03	1	0.092903	
1,000,000	10,000	10.76392	1	

Low-Pressure Units

Meter of water mH$_2$O	Foot of water ftH$_2$O	Centimeter of mercury cmHg	Inches of mercury inHg	Inches of water inH$_2$O	Pascal Pa
1	3.280696	7.356339	2.896043	39.36572	9,806
0.304813	1	2.242311	0.882753	11.9992	2,989
0.135937	0.445969	1	0.39368	5.351265	1,333
0.345299	1.13282	2.540135	1	13.59293	3,386
0.025403	0.083339	0.186872	0.073568	1	249.1
0.000102	0.000335	0.00075	0.000295	0.004014	1

Temperature Conversion Formulas

Degree Celsius (°C)	(°F - 32) x 5/9
	(K - 273.15)
Degree Fahrenheit (°F)	(°C x 9/5) + 32
	(1.8 x K) - 459.67
Kelvin (K)	(°C + 273.15)
	(°F + 459.67) ÷ 1.8

Speed Units

Meter/Second m/s	Meter/Minute m/min	Kilometer/Hour km/h	Foot/Second ft/s	Foot/Minute ft/min	Miles/Hour mi/h
1	59.988	3.599712	3.28084	196.8504	2.237136
0.01667	1	0.060007	0.054692	3.281496	0.037293
0.2778	16.66467	1	0.911417	54.68504	0.621477
0.3048	18.28434	1.097192	1	60	0.681879
0.00508	0.304739	0.018287	0.016667	1	0.011365
0.447	26.81464	1.609071	1.466535	87.99213	1

Torque Units

Newton meter Nm	Kilogram force meter kgfm	Foot pound ftlb	Inch pound inlb
1	0.101972	0.737561	8.850732
9.80665	1	7.233003	86.79603
1.35582	0.138255	1	12
0.112985	0.011521	0.083333	1

Glossary of Symbols

A	Activity
A_f	for saturated soils
$\overline{A}$	for partially saturated soils
A_J	Internal cross- sec. area of Jar containing hydrometer
a_v	Coefficient of compressibility
B	Skempton's Pore pressure parameter (which is a function of the degree of saturation of the soil)
B_f	Width of footing
C	Clay
C_u	Coefficient of uniformity
C_c	Coefficient of curvature
CH	Clay of high plasticity
CI	Clay of medium plasticity
CL	Inorganic clay of low plasticity
C_m	Meniscus correction
C_d	Dispersing agent correction
C_t	Temperature correction
C	Composite correction $= C_m - C_d \pm C_t$
CI	Consistency index
C_C	Compression index
C_r or C_s	Re-compression/swelling index
C_α	Secondary compression index
C_v	Coefficient of consolidation in vertical direction
C_r or C_h	Coefficient of consolidation in radial or horizontal direction
c_u	Undrained shear strength of the original clayey ground ($q_u/2$)
c'	Effective cohesion parameter of clayey soil
CN	Correction factor proposed by Peck et al. (1974)
CD	Consolidated drained
CU	Consolidated undrained
D_{10}	Diameter at 10% finer, also known as effective size
D_{30}	Diameter at 30% finer,
D_{60}	Diameter at 60% finer
D	Diameter of sphere (mm), which is given by the relation
D_f	Foundation depth from ground surface
DST	Direct shear test
e	Void ratio
e_o	Original/initial void ratio or void ration in in-situ state
e_{max}	Void ratio of the soil in the loosest state attained in the laboratory
e_{min}	Void ratio of the soil in the densest state attained in the laboratory
E	Elastic modulus of the soil = stress/strain
e_f	Final void ratio

GC   Clayey gravels
GM   Silty gravels
G_s Specific gravity of soil material
G   Gravel
GW-SW   Well-graded gravels with well-graded sand binder
GW-GC   Well-graded gravels with clay binder
GP   Poorly graded gravels
g   Acceleration due to gravity
g   Gravitational acceleration constant ($g = 9.81 \text{m/s}^2 = 981 \text{cm/s}^2$)
H   Distance from the lowest graduation (1.030) to the graduation mark (R_h) of the stem at the top surface of soil water mixture
h   Height of hydrometer bulb
H   Thickness of clay layer
h_n Distance between neck of hydrometer to the lowest graduation mark
H_e Effective depth from center of hydrometer bulb to the hydrometer reading
h   Hydraulic head difference across length L
I_f Influence factor, depending upon the rigidity of the loaded area
I_F Flow index (slope of flow curve of liquid limit test)
I_T Toughness index
i   Hydraulic gradient
k   Coefficient of permeability
L   Length or effective height of specimen
LL   Liquid limit
LL_{WSO} Liquid limit value for oven-dry soil sample
LL_{WSA} Liquid limit value for air-dry soil sample
LI   Liquidity index
L.C.   Least count
LOI   Loss on ignition
L   Length of footing
LE   Triaxial lateral extension test
LC   Triaxial lateral compression
M   Silt
MI   Silt of medium plasticity
MH   Silt of high plasticity
ML   Silt of low plasticity
MDD   Maximum dry density
MDU   Maximum dry unit weight
m_v Modulus of volume change
M_s Mass of dry soil solids
M Mass of wet soil sample
NP   Non-plastic
NC   Normally consolidated soil
N'   Observed values of N-blows during SPT test
N   Corrected value of N-values for effective overburden pressure

N_c	Corrected value of N-values for effective overburden pressure and dilatancy
N_{cd}	Cone penetration resistance (DCPT test)
n	Porosity
OMC	Optimum moisture content
OC	Overconsolidated soils
OCR	Over consolidation ratio
PI	Plasticity index
PI_{A-line}	Plasticity index of A-line (Plasticity chart)
PI_{U-line}	Plasticity index of U-line (Plasticity chart)
PL	Plastic limit
p'_c	Preconsolidation pressure
q_u	Unconfined compression strength of the original clayey ground
q_{ns}	Net safe footing pressure at the base of footing against shear
q_s	Safe footing pressure at the base of footing = $q_n + \gamma_* D_f$
$q_{n\rho}$	Safe bearing pressure at the base of footing against settlement
q_a	Allowable bearing capacity safe against both shear and settlement criterion
q_c	Cone resistance
R_o	Hydrometer reading at upper rim of the meniscus in suspension jar with hydrometer immersed in the jar
R_h	Hydrometer reading corrected for meniscus correction = $R'_h + C_m$
R'_h	$R_o - R_w$ = Reduced observed hydrometer reading after taking density of water
$R_{min}{}^m$	Minimum value = 0.995 on hydrometer stem
R_{max}	Lowest graduation mark (highest value) on hydrometer stem
RC	Relative compaction
R_D	Relative density
R_w	Water table correction
SPT	Standard penetration test
S	Degree of saturation
SI	Shrinkage index
SM	Silty sands
SC	Clayey sands
S	Sand
SP	Poorly graded sands
WG	Well-graded gravels
SW	Well-graded sands
S_N	Suitability number
SL	Shrinkage limit
S_T	Sensitivity
t_p	Time when primary consolidation is complete
t	Time in seconds
TC	Triaxial compression test
TE	Triaxial extension test
UCS	Unconfined compression test

UU	Unconsolidated undrained
μ	Viscosity of liquid in absolute units ($=\eta/g$)
u	Pore water pressure
μ	Poisson's ratio
η	Viscosity or dynamic viscosity of liquid in ($kN\text{-}s/m^2$)
V_h	Volume of hydrometer (cm^3)
V	Volume of soil sample
v or v_D	Darcian velocity or discharge velocity
v_s	Seepage velocity
W_s	Weight of dry soil solids
W	Weight of wet soil sample
w_n	Natural moisture content
w_f	Final water content at the end of consolidation test
γ_b	Bulk unit weight of soil material
γ_d	Dry unit weight of soil material or field dry unit weight of the soil in place, called "in-place dry density"
γ_{dmin}	Minimum dry unit weight of the soil in the loosest state
γ'	Effective unit weight of soil material ($=\gamma_{sat} - \gamma_w$)
γ_{sat}	Saturated unit weight of soil material
γ_w	Unit weight of water
γ_{dth}	Theoretical dry unit weight
γ_{dmax}	Dry unit weight of the soil in the densest state
Δh	Head difference to create flow
ε_a	Axial strain
σ_o	Effective overburden pressure
$\Delta\sigma$	Vertical stress increment or deviator stress
τ_f	Shear resistance or shear stress at failure
$\sigma_{n)f}$	Normal stress at failure
ϕ	Angle of internal friction also known as angle of shearing resistance
ϕ'	Effective angle of shearing resistance
σ	Total stress
σ'	Effective stress
σ_1	Major Principal Stress
σ_3	Minor principal stress
δa	Allowable/permissible settlement
δ_i	Immediate/elastic settlement
δ_c	Consolidation settlement
δ_t	Total settlement ($\delta_i + \delta_c$)

Preface

Soil exists as a naturally occurring material in its undisturbed state, or as a compacted material and is perhaps most common and probably the most complex construction material. Geotechnical engineering involves the understanding and prediction of the engineering behavior of soils. In addition to the physical and index properties, there are a number of other properties of soils, i.e. engineering properties, which play a vital role insofar as design of foundation system for various infrastructures (such as road structures, bridges, culverts, retaining walls, water tanks, high-rise buildings, fill embankment, reactor facilities) is concerned. Engineering properties of soils are those properties which provide a basis for determining the appropriate input parameters for design of foundation system for various types of infrastructures. Furthermore, the behavior of a structure resting upon a soil or rock foundation depends on the engineering characteristics of the foundation materials. The main engineering properties of soils are compaction, permeability, compressibility, and shear strength and there are a number of laboratory test methods which have been developed over the years to evaluate these engineering properties.

In this manual, the current practices in the Geotechnical Engineering Laboratory, National Institute of Technology Srinagar and feedback from the previous years' undergraduate and postgraduate students have been taken into account. Also, this soil laboratory testing manual is the outcome of the author's experience in teaching elementary and advanced courses in geotechnical engineering at the UG and PG level. Accordingly, this *Manual of Geotechnical Laboratory Soil Testing* has been prepared to present the basic essentials of the laboratory testing of soils for determination of various properties of in-situ and remolded soil samples and to assist technicians, training personnel, students, and geotechnical engineers in particular, and civil engineers in general in understanding the method of soil testing, testing equipment and testing materials, step-by-step test procedure, recording of observation data and analysis, test results, and discussions in a very lucid manner. A careful study of the laboratory tests and of the instructions for preparation of test reports will indicate the intent of the author to provide a maximum of understanding of the purpose of the test and the basic theory involved, as well as to teach the students the mechanics of performing a particular test. The material contained in this manual is specific to relevant standard codes and testing protocols and should be used as a basis for establishing a qualification program.

Finally, on a personal note, I have had a tremendous pleasure writing this testing manual with one important message that "One good test is not only far better than many poor tests, but is also less expensive and less likely to permit a misjudgment in design." I hope readers, students, practicing engineers, and consultants will enjoy reading it. I have also endeavored to make this testing manual enjoyable to read, but with respect to this subjective aspect, only your judgment matters, and so, do not hesitate to compare this manual with other manuals in terms of style, presentation, and content while remembering the words of Benjamin Disraeli "*... this shows how much easier it is to be critical than to be correct*" and what Sir Arthur Conan Doyle said "*it is a capital mistake lo Theories without data.*"

Further, it is my pleasant duty to acknowledge the support and help extended to me by my colleagues and Ph.D. scholars for completion of this book. At the last and not least, valuable suggestion from esteemed readers to improve this edition would greatly be appreciated. Kindly input any suggestion through e-mail to: p7mir@nitsri.net/bashiriisc@yahoo.com/bamiriitb@gmail.com

Acknowledgments

This testing manual entitled, *Manual of Geotechnical Laboratory Soil Testing*, is the outcome of the author's experience in teaching elementary and advanced courses in geotechnical engineering at the UG and PG level. In addition, the material presented has been drawn from many sources, persons, websites, and organizations and are accordingly acknowledged. References are listed at the end of each test, under the names of authors arranged alphabetically for further reading or reference purposes.

Further, I highly appreciate the support of my Ph.D. research scholars who have spent their quality time in preparation of the book.

Valuable suggestions to improve this edition would greatly be appreciated. Kindly submit suggestions to:

p7mir@nitsri.ac.in; bashiriisc@yahoo.com; bamiriitb@gmail.com

Prof. Bashir Ahmed Mir, Room No. 203, Department of Civil Engineering, National Institute of Technology Srinagar-190006, J&K

Acknowledgments

Overview and Main Goals of Manual Contents

The main objectives of a laboratory course in soil mechanics are to introduce soil mechanics laboratory techniques to civil engineering undergraduate students, and to familiarize the students with common geotechnical test methods, test standards, and terminology. The procedures for all of the tests described in this Geotechnical Laboratory Soil Testing Manual are written in accordance with applicable Indian Standards (IS) Codes, American Society for Testing and Materials (ASTM) standards, and British Standard (BS) Codes. It is important to be familiar with these standards to understand, interpret, and properly apply laboratory results obtained using a standardized method. In addition, some of the important references are also included at the end of each test so as to avoid any cause of difficulty or misunderstanding by the readers.

In this manual, each of the tests starts with an objective, brief introduction to the topic, followed by definitions and brief theory as applicable for better understanding so that the test is carried out without any difficulty. A section on theory presents sufficient theoretical background to enable the tests to be understood before conducting them. For better understanding and convenience of students, a step-by-step procedure is outlined as follows:

- Objective
- Definitions and Theory
- Method of Testing
- Soil Testing Material and preparation of test specimens (undisturbed and disturbed)
- Testing Equipment and Materials, Testing Procedure, Observation Data Sheet and Analysis
- Applications of Test Results, Sources of Error, Precautionary Measures
- References

The observation of test data, analysis, and calculation and plotting of graphs and presentation of results are described in a very lucid manner.

Finally, general comments, additional considerations, applications, sources of error, and precautionary measures have also been highlighted briefly at the end of each test. The main goals of writing this testing manual are to:

- Provide basic concepts, general understanding, and appreciation of the geotechnical principles for determination of physical, index, and engineering properties of soil materials.

- Provide a general understanding and appreciation of the geotechnical principles gearing towards a sound, safe, and cost-effective design and construction of buildings and structures.
- Serve as a consistent guidance for the practitioners involved in the geotechnical planning, design, and construction activities.
- Encourage the readers to follow through the topic of interest in one or more of the reference books mentioned in the references.
- Present state-of-the-art applications for geotechnical engineers in particular and civil engineers in general.

I have had a tremendous pleasure writing this soil testing manual with one important message that "the value of a book is not determined by its cost, but its contents and use."

Declaration

Civil and geotechnical engineering includes the conception, analysis, design, construction, operation, and maintenance of a diversity of structures, facilities, and systems. All are built on, in, or with soil or rock. The properties and behavior of these materials have major influences on the success, economy, and safety of the work. Soil exists throughout the world in a wide variety of types. Different types of soil exhibit diverse behavior properties. The engineering properties of soils are governed by their physical properties. Therefore, we need to describe and identify soils in terms that will convey this information clearly and accurately. The principal objective in geotechnical testing is to know the properties and engineering behavior of soil as an engineered construction material and as a foundation medium for development and construction of various engineering infrastructures. Therefore, it is necessary to know about various soil tests, which need to be scored in accordance with norms for the soil type, region, and anticipated use.

This *Manual of Geotechnical Laboratory Soil Testing* has been written to present the very basic essentials of the geotechnical laboratory testing of soils, including methods of data collection, analysis, computations, and the presentation of the test results in a very lucid manner for the benefit of readers/students. Every effort and care has been taken in selecting methods and recommendations that are appropriate to desired conditions wherever applicable. Notwithstanding these efforts, no warranty or guarantee, express, implied, or statutory is made as to the accuracy, reliability, suitability, or results of the methods or recommendations. Therefore, the use of this *Manual* requires professional interpretation and judgment. Appropriate design procedures and assessments must be applied to suit the particular circumstances under consideration.

The Author(s) shall have no liability or responsibility to the user or any other person or entity with respect to any liability, loss, or damage caused or alleged to be caused, directly or indirectly, by the adoption and use of the methods and recommendations of this Manual, including but not limited to, any interruption of service, loss of business or anticipatory profits, or consequential damages resulting from the use of this Manual. Also, this Manual cannot be used as evidence in any Judicial, Criminal or Quasi Judicial Matter or as a part of any Legal Matter.

Guidelines for Conducting Soil Tests

The students are advised to follow the following laboratory rules or guidelines for smooth conduct of soil tests in the laboratory

1. Attendance is required for all students for all lab classes. Students who do not attend lab will not receive credit for their group's report.
2. Ensure that you are aware of the test and its procedure before each lab class.
   - **You will NOT be allowed to attend the class if you are not prepared!**
3. Personal safety is top priority. Do not use equipment that is not assigned to you.
4. All accidents must be reported to your instructor or laboratory supervisor.
5. The surroundings near the equipment must be cleaned before leaving each lab class.
6. Ensure that the readings are checked by your TA for each lab period.

Dress Code

Students are expected to dress in a manner appropriate for laboratory and field work. Lab coats, ear plugs, safety glasses, dust masks, and gloves are available in the lab area. Open-toed shoes or sandals are not allowed in the laboratory or on field trips.

Laboratory Report

1. General

   - Each student has to submit the report for each experiment.
   - The test results and graphs are to be written and drawn by the students themselves only. The test report should be submitted to the TAs at the start of next lab. Test compulsorily.
   - **You will NOT be allowed to attend the class if you are not prepared with report!**
   - Please write your roll number, batch, and group number along with the name of the soil sample given to your batch on the first page of the report.

2. Report Organization

   - **Abstract**

This contains a brief summary of the experiment. Students should explain what data you will gather, the procedure for gathering the data, and the reason for the same.

The abstract should be very specific high lightening the main objective, purpose, procedure, and outcome of the test.

- **Measured Data**

Tabulate the measured and calculated test values, preferably in the format given in the lab manual. You should also show sample calculations. Attach reference figures and/or tables used if any in the evaluation of the data.

- **Summary**

Tabulate the results/findings from the experiment. Towards the end of the semester, you should generate a detailed summary of soil properties in a table during the submission of the manual after the last experiment.

- **Inference**

Write briefly on your inference from test observations and analysis of results.

- **Discussion**

Discuss the possible sources of error, accuracy of the test method, and anything noteworthy you observed during the test. State the specific Indian Standard number for test procedure and also, other standards like ASTM and BS Standards may be referred to during testing of soil samples.

Notes for Students

- Your report must be neat, well-organized and make a professional impact.
- Figures and tables should be presented immediately following their first mention in the text. Short tables and figures (say, less than half the writing area of the page) should be presented within the text, while large tables and figures may be presented on separate pages.
- The discussion shall logically lead to inferences and conclusions as well as scope for possible further future work.
- Report should be submitted within 7 days and before the beginning of the next lab class of your batch.
- Your reports will be graded on a scale from 0 to 5 as follows:

5 = Complete and excellent work

4 = Satisfactory, but with some minor errors

3 = Significant errors or omissions

2 = Very little correct or useful work

1 = Lab report handed in, but with minimal work

0 = Missed lab

- Two points will be deducted for late submission.
- **Anyone caught plagiarizing work in the laboratory report, from a current or past student's notebook, will receive 0 on the grading scale or XX grade in the course.**

==

Declaration

I have read and understood the above guidelines.

Signature of student: _____

Name   : _____

==

Prof. (Dr.) B A Mir (9419002500)
Faculty Incharge/
PG Head, Geotech Engg Divn.

******************************END***********************************

About the Author

Prof. (Dr.) Bashir Ahmed Mir, Ph.D., C. Engg (I), MASCE (USA), P.Engg (UK), is a Professor and PG Head Geotechnical Engineering Division in the Department of Civil Engineering, National Institute of Technology Srinagar, Kashmir (J&K), obtained his Ph.D. (Geotech. Engg) from the Indian Institute of Technology Bombay, M.E. (Geotech. Engg) from the Indian Institute of Science (IISc) Bangalore, and M.B.A. from the Institute of Chartered Managers, IAS Academy Chennai. Over the last 25 years, Dr. Mir, an established teacher, researcher, and active consultant has conducted extensive research on various aspects of geotechnical engineering. Dr. Mir has published more than 100 research papers in international journals and conferences and he is the recipient of many Best Paper Awards. Dr. Mir has traveled to many countries abroad including Australia, Bangkok (Thailand), China, Italy, Japan, Malaysia, Singapore, Taiwan, Tunisia, and United Kingdom (UK) and presented his research papers. Dr. Mir has also chaired technical sessions in international conferences abroad and in India. His research areas include prediction of soil behavior, constitutive modeling, ground improvement, foundation engineering, reinforced soils, critical state soil mechanics, environmental geotechnics, characterization of expansive soils and fly ash, characterization of highway materials, and engg. geology. Dr. Mir is also an active journal reviewer and advisory board member for various journals such as *Ground Improvement, ICE–UK, GEGE Springier, Geotechnique–UK, IJGE–*Ross Pub. USA, *Geomate–*Japan. Dr. Mir is also an active member of various scientific and professional societies and holds membership of Sr. M. IEEE (Delhi Section), MASCE (USA), MISSMGE (UK), MICE (UK), LMIGS, LMISTE, LMISCMS, LMISWE, LMICI, LMISRMTT, LMIE, C.Engg (I), UACSE (USA), and P. Engg (UK). Dr. Mir has provided consultancy services to more than 150 projects of national importance of various government and private agencies. Dr. Mir is also Hon. Secretary, IGS Srinagar Chapter and A/Dean Planning & Development Wing NIT Srinagar. Dr. Mir is also representing India as a technical committee (TC 216) member on "Frost Geotechnics" of the International Society for Soil Mechanics and Geotechnical Engineering (ISSMGE) for the years 2018–2021.

1 Natural Water Content of a Soil Sample

1.1 OBJECTIVES

Natural water content (w_n) is a parameter which has direct bearing on the shear strength of fine grained soils in soil engineering practice. Therefore, the main objective of determination of natural moisture content is to evaluate and understand the engineering behaviour of fine grained soils. This parameter has been very helpful in the development of correlations among various properties of soils so as to assess the consistency classes of fine grained soils in the field.

Since the soil is a user-friendly material abundantly available throughout the universe, it is only the water that makes it a problematic material for sustainable development. Therefore, it is mandatory to quantify the water content in the soil mass to assess the compressibility and stability problems in a soil mass due to presence of natural water content.

1.2 INTRODUCTION

Water on, above, and below the ground surface may exist in different forms such as liquid, gaseous, (e.g. vapors) or solid (e.g. frozen ice) states. Water below or underneath ground surface is termed as subsurface water. At varying depths under the ground surface, there is a zone of saturation in which water fills all the pores in the soils and the openings in the underlying rock. The water in the zone of saturation is commonly termed as "ground water," its upper surface being the "ground water table or water table" of phreatic surface, which has the atmospheric pressure.

The term "water" and "moisture" may be interchangeably used to describe all kinds of subsurface water, except gravitational water (ground water is free or gravitational and moves by obeying the law of gravity in contrast to the "attached or sorbed" water), to which the term "moisture" is not applied. The interstices between soil particles beyond ground water table contain some water which is termed "vadose" (e.g. water between the ground surface and the water table) by geologists. In engineering terms, it is known as moisture or simply "water" or moisture content. Thus, the water present in the soil voids or any other material is generally termed as the natural moisture content. Natural water content (w_n) is used to access the state of compactness of in situ soils. The natural moisture content can be determined either on volume basis (known as volumetric moisture content) or mass/weight basis (known as gravimetric moisture content). The water content is always expressed as percentage (%) defined as below:

DOI: 10.1201/9781003200260-1

$$w = \frac{W_w}{W_s} * 100 \,(\%) \,..based\ on\ Wt.\ of\ soil\ sample\ known\ as\ gravimetric\ water\ content$$

$$w = \frac{V_w}{V_s} * 100 \,(\%)... based\ on\ Vol.\ of\ soil\ mass\ known\ as\ volumetric\ water\ content$$

$$0 \le w \le \infty \hspace{5cm} [1.1]$$

Where: V_W = volume of water (cm^3) in a soil sample
　　V_S = volume of solids in a soil sample
　　W_W = weight of water in a soil sample, and
　　W_S = weight of soil solids or dry weight of soil sample
　　e.g. Range of water content in a soil sample is zero (in completely dry soil state) to infinity (in fully saturated state).

However, the above definition of water content is different from the one used in geology. The geologists define the natural water content as:

$$w = \left[\frac{W_w}{W} * 100 \,(\%) \,..based\ on\ Wt.\ of\ soil\ sample\ known\ as\ gravimetric\ water\ content \right]$$

Where: W_W = weight of water in a soil sample, and
　　W = total weight of the soil sample

The natural water content in soils varies with minimum percentage of about 10% for sandy soils to a maximum value of about 120% for clayey soils such as montmorrillionite clays and highly organic soils. It may be noted that the water content is expressed as a percentage with one decimal such as 10.3% or 110.5% (instead of 10.27% or 110.47%), respectively. Water in clayey soils exists in different forms as described briefly below:

1.2.1　Free Water Content or Moisture Content or Water Content

Free water content exists in the voids or pores of soil mass. This is also called pore water or natural moisture content of an in situ soil sample in the field, which can be removed by the oven drying method at 110°C. The free water remains outside the surface of adsorbed moisture films and determines ordinarily the physical properties of soils.

1.2.2　Adsorbed Water Content

Adsorbed water content encases the soil mass as a part of the mineral grains and cannot be removed by the oven drying method at 110°C (Figure 1.1). The adsorbed moisture layer may be divided into two parts: i. *Solidified water layer* very close to the soil grain and ii. *Cohesive water layer* surrounding solidified water layer. This water content is also sometimes termed as *bound water* or *hydrated water*. This layer is of very small thickness, perhaps of the order of 0.005 μm. It may be noted that the free water content and adsorbed water content have different physical properties.

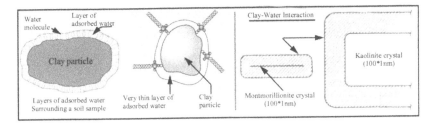

FIGURE 1.1   Adsorbed water content encompassing clay particle.

1.2.3   SOIL WATER

The water content present in the equilibrium solution in the soil below the ground surface is refereed to as soil water.

1.2.4   GRAVITATIONAL WATER

The natural water content which moves under the influence of gravity into or out of the soil is referred to as gravitational water content or free water content.

1.2.5   CAPILLARY WATER

The free water content held by the surface tension forces on the surface of soil grains and in the pore void spaces between the soil grains is referred to as capillary water content.

1.2.6   HYGROSCOPIC WATER

The natural water content adsorbed from the atmosphere by the soil grains is referred to as hygroscopic water content.

1.2.7   INTERLAYER WATER

Interlayer water is specific to expansive clay minerals, being water that is adsorbed between clay mineral layers. Considerable interlayer water remains in place after oven-drying at 110°C, which is sufficient to remove hygroscopic water. Temperatures in excess of 200°C may be required to remove all interlayer water.

1.2.8   SALINE WATER

Saline water content is the water content with a high concentration of dissolved salts such as sodium chloride, which is usually expressed in parts per million (ppm). This water content is generally found in off-shore soil deposits and as sea water in particular. The saline fluid content, denoted by w_{sf} (%), can be determined from the known water content w (%) expressed either by mass per unit mass, or mass per unit volume, of fluid as below:

i. For known salt content in terms of mass of salts per gram of fluid (i.e., p_{ppt} parts per thousand), the fluid content (w_{sf}) can be calculated as:

$$w_{sf} = \left[\frac{1000 * w}{100 - p_{ppt}\left(1 + \frac{w}{100}\right)} \right] (\%) \qquad [1.2]$$

ii. For known salt content in terms of mass of salts per liter of fluid (q_{gpl} g per liter) and the density of fluid content, ρ_{sf} (Mg/m^3), the fluid content (w_{sf}) can be calculated as:

$$w_{sf} = \left[\frac{1000 * w}{1000 - \frac{q_{gpl}}{\rho_{sf}}\left(1 + \frac{w}{100}\right)} \right] (\%) \qquad [1.3]$$

Where the sea water has a density of 1.024 g/cm^3 and contains 35 g per liter dissolved salts.

1.2.9 TYPICAL VALUES OF NATURAL WATER CONTENT OF SOILS

The natural water content in a soil specimen has a direct affect on its compressibility, stability, and consistency (Reddy, 2000). The natural water content plays a vital role for the determination of Atterberg limits of the soils, compaction of soils in the field, and for the evaluation of stability analysis of various earth structures and foundations. Therefore, before determination of water content test on soil samples, it is necessary to have an idea of typical values of natural water content of soils in different conditions. Typical values of natural moisture content for a saturated soil sample are given in Table 1.1.

1.3 DETERMINATION OF MOISTURE CONTENT BY OVEN-DRYING METHOD (ON GRAVIMETRIC BASIS)

References: IS: 2720 (Part 2)-1973; ASTM D2216-66; BS 1377: Part 2, 1990

1.3.1 DEFINITIONS AND THEORY

For a given soil specimen, the water content is defined as the ratio of the weight of water to the weight of dry soil in the soil specimen. In the oven-drying method, the soil is placed in Oven and dried with a maximum temperature of 105°–110°C for about 16–24 hours. However, peat and soil containing organic matter should be a drying temperature of 60°C is to be preferred to avoid oxidation of the organic content). For soils containing gypsum content, a temperature not exceeding 80°C should be used as the water of crystallization may be lost at temperatures above 100°C. After drying the soil specimen, the moisture content

TABLE 1.1

Typical Values of Natural Water Content of Saturated Soils

Description of Soil Type	Water Content w%	Unit Weight (kN/m³)	
		γ_{sat}	γ_d
Uniform loose sand	25–30	18.5	14.0
Uniform dense sand	10–16	20.5	17.2
Mixed-grained loose sand	25	19.5	15.6
Mixed-grained dense sand	16	21.2	18.2
Loose angular-grained silty sand	25	–	16
Dense angular-grained silty sand	15	–	19
Glacial till, mixed grained	8–10	22.7	20.8
Soft glacial clay	45	17.3	11.9
Stiff glacial clay	22	20.3	16.7
Soft slightly organic clay	70–110	15.6	9.1
Soft highly organic clay	80–130	14.0	6.8
Soft bentonite	194	12.4	4.2
Laterite soils	–	20.0	19.5
Loess	25	–	13.5

(w) is determined in the laboratory by the oven-dry method, expressed as (nearest 0.1%):

$$w = \left[\frac{(W_2 - W_1)}{(W_3 - W_1)} * 100 \right] (\%) \qquad [1.4]$$

Where W_1 = Weight of empty container with lid
W_2 = Weight of the container with moist soil and lid
W_3 = Weight of the container with dry soil and lid.

1.3.2 METHOD OF TESTING

The water content of a given natural soil sample is determined by the *oven-drying method*.

1.3.3 SOIL TESTING MATERIAL

The soil specimen is collected from the field at selected locations as disturbed sample in sealed plastic bags so that the field or natural water content is not lost during transportation to geotechnical testing laboratory. The minimum quantity of soil sample taken for water content determination should not be less than 20 g for the soil sample passing through the 0.25 mm IS sieve. The recommended quantity

TABLE 1.2

Recommended Quantity of Soil Sample for General Lab. use (IS: 2720 (Part II)

Size of Particles More Than 90% Passing	Minimum Quantity of Soil Sample to Be Taken for Test (g)	Remarks
0.425 mm IS Sieve (No. 40)	20	1.. For sieve sizes, refer IS:
2 mm IS Sieve (No. 10)	50	460 (Part-I)-1978.
4.75 mm IS Sieve (No. 4)	200	2.. Drier the soil, greater shall
9.50 mm IS Sieve	300	be the quantity of soil taken
19 mm IS Sieve	500	
37.5 mm IS Sieve	1,000	
75.0 mm IS Sieve	5,000	

of soil sample for general laboratory use as per IS: 2720 (Part II) is given in Table 1.2.

Further, the relationship between recommended minimum weight of moist soil sample for determination of moisture content with increasing maximum particle size is illustrated in Figure 1.2 (ASTM D2216). However, it may be noted that the minimum weight of moist soil sample for determination of moisture content should not be less than 20 g.

1.3.4 Testing Equipment and Accessories

The following equipment and accessories are required for determination of natural moisture content of a soil sample in the laboratory:

1. Oven (110°C ± 5°C)
2. Moisture content cans or containers
3. Weighing balance (with accuracy nearest 0.1%)
4. Tongs
5. Desiccators

1.3.5 Testing Program

1. Take weight of a properly cleaned container with lid (W_1).
2. Take the moist soil sample in the container and weigh it with lid (W_2).
3. Remove the lid from the container and keep it in the oven at a temperature of 110°C ± 5°C for 16–24 hours (or until the weight becomes constant).
4. After oven-drying is complete, remove the container with lid from the oven using tongs and cool it in a desiccator.
5. After cooling the soil sample in the desiccator, weigh the container with the lid and dry soil sample (W_3).

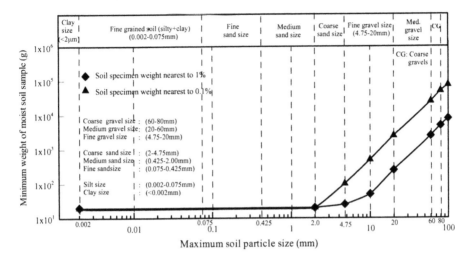

FIGURE 1.2 Recommended minimum weight soil specimen for determination of natural water content.

6. Repeat the above procedure and find the average value of natural water content (%) for each soil sample.

The maximum variation between any one determination and the average value should not exceed 1%.

1.3.6 OBSERVATION DATA SHEET AND ANALYSIS

Test data analysis, observations and calculations for determination of natural water content are given in Table 1.3.

1.3.7 RESULTS AND DISCUSSIONS

Based on test results, water content of the soil sample = _____%

Soil type: ....

1.4 DETERMINATION OF MOISTURE CONTENT BY PYCNOMETER METHOD

References: IS: 2720 (Part 2)-1973; ASTM D2216-66; BS 1377: Part 2, 1990

1.4.1 DEFINITIONS AND THEORY

In the case of coarse grained soils (sands), the moisture content is determined by using a Pycnometer fitted with a brass conical cap. The cap has a small hole of about 6 mm in diameter at its apex for expulsion of air if any entrapped in the pycnometer while filling with water and soil sample. The natural water content of

TABLE 1.3

Natural water content of a soil sample by the oven-dry method

Project/Site Name: **Date:**

Client Name: **Job. No.**   **Sample No.**

Sample Recovery Depth: **Sample Recovery Method:**

Sample Description:

Tested By:

Test Type: on Gravimetric basis/on Volumetric basis

Sl. No.	Observations and Calculations	Trial No.		
Observations		**1**	**2**	**3**
1	Container No.			
2	Weight of empty container, W_1 (g)			
3	Weight of container + soil, W_2 (g)			
4	Weight of container + dry soil, W_3 (g)			
Calculations				
5	Weight of water, $W_w = W_2 - W_3$ (g)			
6	Weight of solids, $W_s = W_3 - W_1$ (g)			
7	Water content, $w = (W_w)/(W_s) * 100$ (%)			
8	Average Water content, w (%)			

the oven-dried soil sample is determined by a pycnometer method and expressed as (nearest 0.1%):

$$w = \left[\frac{(W_2 - W_1)}{(W_3 - W_4)} \left(\frac{G - 1}{G} \right) - 1 \right] * 100 \ (\%) \qquad [1.5]$$

Where W_1 = Weight of empty Pycnometer

W_2 = Weight of the Pycnometer with wet soil

W_3 = Weight of the Pycnometer and soil, filled with water

W_4 = Weight of Pycnometer filled with water only

G = Specific gravity of solids.

1.4.2   METHOD OF TESTING

The water content of a given natural soil sample (sands) is determined by the *Pycnometer Method.*

1.4.3   SOIL TESTING MATERIAL

The soil specimen is collected from the field at selected locations as a disturbed sample in sealed plastic bags so that the field or natural water content is not lost during transportation to geotechnical testing laboratory. The minimum quantity of soil sample taken for water content determination should not be less than 20 g for the soil sample passing through a 0.25 mm IS sieve. The recommended quantity of soil sample for general laboratory use as per IS: 2720 (Part II) is given in Table 1.2.

1.4.4 TESTING EQUIPMENT AND ACCESSORIES

1. Pycnometer
2. Weighing balance (with accuracy nearest 0.1%)
3. Glass rod
4. Moisture cans or containers with lid
5. Tongs
6. Oven (110°C ± 5°C)
7. Desiccators with any suitable desiccating agent

1.4.5 TESTING PROGRAM

1. Take a Pycnometer with its cap and wash and dry it properly.
2. Weigh the Pycnometer with its cap and washer accurate to 1.0 g. (W_1).
3. Take 250 to 450 g of moist soil specimen in the Pycnometer and weigh it with its cap and washer accurate to 1.0 g. (W_2).
4. Fill distilled water in the Pycnometer containing the moist soil specimen to about half its height.
5. Mix the soil and water thoroughly with a glass road in the pycnometer. Fill the Pycnometer by adding more water, flush with the hole in the conical cap.
6. Clean and dry the Pycnometer from outside and take its weight (W_3).
7. Empty the Pycnometer. Clean it thoroughly. Fill it with water flush with the hole in the conical cap and weigh (W_4) as shown in Figure 1.3.
8. Repeat the above procedure and find the average value of natural water content (%) for each soil sample.
9. The maximum variation between any one determination and the average should not exceed 1%.

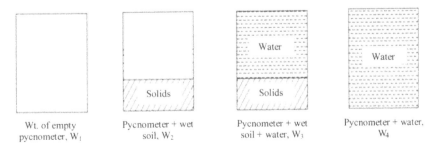

| Wt. of empty pycnometer, W_1 | Pycnometer + wet soil, W_2 | Pycnometer + wet soil + water, W_3 | Pycnometer + water, W_4 |

FIGURE 1.3 Determination of moisture content by Pycnometer Method.

TABLE 1.4

Moisture content determination by Pycnometer Method

Project Name: **Date:**

Client Name: **Job. No.   Sample No.**

Sample Description:

Tested By:

Test Type: on Gravimetric basis/on Volumetric basis

Sl. No.   Observations an Calculations   **Determination No.**

                                                           1   2   3

Observation

1   Weight of empty Pycnometer, W_1 (g)

2   Weight of pycnometer + wet soil, W_2 (g)

3   Weight of Pycnometer + soil + water, W_3 (g)

4   Weight of Pycnometer + water only, W_4 (g)

Calculations

5          $(G - 1) / G$

6          $W_2 - W_1$ (g)

7          $W_3 - W_4$ (g)

8   Water content, w (%):

$$w = \frac{W_s}{W_w} = \left[\frac{W_2 - W_1}{W_3 - W_4} \left(\frac{G-1}{G} \right) - 1 \right] * 100 \, (\%)$$

9   Average water content, w (%)

1.4.6   OBSERVATION DATA SHEET AND ANALYSIS

Test data analysis, observations, and calculations for determination of natural water content are presented in tabular form, as below (Table 1.4).

1.4.7   RESULTS AND DISCUSSIONS

Based on test results, water content of the soil sample = _____%.
     Soil type: ....

Note: For other methods, refer to respective standards (e.g. IS 2720-Part 2), IS: 2720 (Part-19)-1978. The other methods to determine moisture content in the laboratory are an infrared lamp with a torsion balance moisture meter. The rapid moisture meter method and sand bath method are quick and convenient field methods of moisture content determination.

1.5   GENERAL COMMENTS

1. The soil water content plays a vital role in soil engineering practice for the characterization of fine grained soils, which is used in analysis of compressibility and stability problems.

2. The fine grained soils possess natural moisture content about 75 to 120%. But it may be noted that highly organic and peaty soils can posses more than 600% water content, especially found in Malaysia.
3. The main reason for fine grained soils, especially clays for exhibiting higher water holding capacity, is due to their particle shape and ability of clay minerals to adsorb moisture content.
4. The main reason for coarse grained soils, especially sands for exhibiting lower water holding capacity as compared to fine grained soils is due to their particle shape, and higher void spaces between the soil particles and hence higher permeability or excellent drainage characteristics.
5. Organic soils should be oven-dried at 60°C as per standard codal procedures.
6. The oven-dried soil specimen should be cooled in a desiccator when removed from the oven.

1.6 APPLICATIONS/ROLE OF NATURAL WATER CONTENT IN SOIL ENGINEERING

The natural water content or natural moisture content or simply water content plays a vital role in characterization of fine grained soils. The natural water content not only controls the consistency of fine grained soils, but also has direct bearing on the compressibility and shear strength and of soils.

Natural water content (w_n) can be used to access the state of compactness of soils with known values of liquid limit (LL) and plastic limit (PL) whether soil is normally or heavily consolidated e.g.:

- If "w_n" is close to "LL," soil is normally consolidated (NC).
- If "w_n" is close to "PL," soil is some to heavily overconsolidated.
- If "w_n" is intermediate, soil is somewhat overconsolidated (OC).
- If "w_n" is greater than "LL," soil is on the verge of being a viscous liquid.

1.7 SOURCES OF ERROR

The major sources of error during the measurement of natural moisture content are:

1. The representative soil sample is not properly sealed after collecting from the selected site while transporting it to geotechnical testing laboratory.
2. The soil specimen may have not been dried as per standard codal procedures in the oven. The temperature variation (above- or below-standard temperature of 110°C for inorganic soils) and time line (more than or less than recommended time line of 16–24 hours) during drying of the soil specimen will definitely affect the quantification of water content.
3. Highly organic soils might have been dried at a standard temperature of 110°C for about 24 hours, which will definitely produce erratic measurement of moisture content. However, it may be noted that the highly organic soils should be dried at about 60°C as compared to 110°C for inorganic soils.

1.8   PRECAUTIONS

1. The soil sample collected in the field should be properly sealed in air-tight polythene bags to avoid any loss of moisture content during transporting it to soil testing laboratory.
2. Determined the water content of the soil sample immediately when it is received in the laboratory
3. The moist soil sample should be placed in a clean and dried moisture can with its lid underneath for drying in the oven.
4. Make sure that the oven temperature is constant at 105°C–110°C
5. The moisture can with lid and dry soil should be cooled in a desiccator.

REFERENCES

ASTM D 2216. "Method for Laboratory Determination of Water (Moisture) Content of Soil, Rock, and Soil-Aggregate Mixtures." *Annual Book of ASTM Standards*, Vol. 04–08. American Society for Testing and Materials, Philadelphia, United States, www.astm.org.

ASTM D 4643. 2008. "Test Method for Determination of Water (Moisture) Content of Soils by the Microwave Oven Method." *Annual Book of ASTM Standards*, Vol. 04–08. American Society for Testing and Materials, Philadelphia, United States, www.astm.org.

Bowels, J. E. 1992. *Engineering Properties of Soils and Their Measurements*. 4th ed. McGraw-Hill: England, UK.

BS 1377-2. 1990. "Methods of Test for Soils for Civil Engineering Purposes Part 2: Classification Tests for Determination of Water Content." British Standards, UK.

IS: 460-Part 1. 1985. "Indian Standard Specification for Test Sieves Part-1: Wire Cloth Test Sieves." Bureau of Indian Standards, New Delhi.

IS: 2720 (Part 1). 1980. "Indian Standard Code for Preparation of Soil Samples." Bureau of Indian Standards, New Delhi.

IS: 2720 (Part I). 1983. "Method of Test for Soils: Preparation of Dry Soil Specimen for Various Tests." Bureau of Indian Standards, New Delhi.

IS: 2720 (Part II). 1973. "(Reaffirmed 2002). Method of Test for Soils: Determination of Water Content." Bureau of Indian Standards, New Delhi.

Lambe, T. W. 1951. *Soil Testing for Engineers*. New York: John Wiley & Sons, Inc.

Reddy, Krishna R. 2000. "Engineering Properties of Soils Based on Laboratory Testing." Department of Civil and Materials Engineering University of Illinois at Chicago, USA.

2 Field/In-Place Dry Density of a Soil Sample

2.1 OBJECTIVES

1. The main objective of determining the in-place or field dry density is to access the compactness of in-situ state of soil deposit in the field. The field dry density is used as an important parameter for writing the specifications such as degree of compaction, relative density, and compactability factor, etc. for compaction control in the field.
2. Generally, density is indicative of strength of a material. The higher the value of density, the higher the strength of that material.

2.2 INTRODUCTION

Field dry density or field dry unit weight is one of the controlling parameters for ensuring that the filled soils are compacted as per established design criteria in the various construction works (Reddy, 2000). However, dry density is determined in the laboratory and not directly in the field. The deformation characteristics of fine grained soils are dependent on the natural water content and the density. Density parameter is determined for each soil specimen collected from the desired site under consideration and compared with minimum requirements of the engineering structure for sustainable development. The in situ or field density is determined for an undisturbed fine grained soil specimen obtained from the field using a core cutter of known volume. However, an undisturbed soil sample cannot be collected from the coarse grained soil stratum in the field. Therefore, the density of a coarse grained soil sample can be determined either by the sand or water displacement method as per standard codal procedures. Generally, two methods are widely specified for determination of the volume of the soil sample. The first method applies to the undisturbed soils collected in a core cutter (e.g. clays) of known volume. In the second method, the volume is measured by either the sand or water replacement method.

2.3 FIELD/IN-PLACE DRY DENSITY OF A SOIL SAMPLE BY CORE CUTTER METHOD

References: IS: 2720 (Part-29) - 1975 (Reaffirmed 2006); BS 1377: Part 2, 1990

2.3.1 DEFINITIONS AND THEORY

The in situ density (or unit weight) parameter is equal to the mass of the undisturbed soil specimen collected in the seamless cylindrical tube or core cutter divided by its volume (e.g. the volume of the equipment or item in which the undisturbed soil

DOI: 10.1201/9781003200260-2

sample is collected in the field). Generally, an undisturbed soil sample of fine grained soils (e.g. clays) is collected in a seamless cylindrical core cutter of known weight and volume. Then the net mass (or weight) of the undisturbed soil sample is obtained for the determination of its bulk density using the expression:

$$\rho_b = \left[\frac{M_s}{V}\right] (g/cc) \ or \ \gamma_b = \left[\frac{W_s}{V}\right] (kN/m^3)$$

Once the bulk density is computed, the in situ dry density is determined using the expression:

$$\rho_d = \left[\frac{\rho_b}{(1 + w/100)}\right] (g/cc) \ or \ \gamma_d \left[\frac{\gamma_b}{(1 + w/100)}\right] (kN/m^3) \qquad [2.1]$$

Where M_s = mass of the moist soil in the core cutter
  V = internal volume of the core cutter = $\pi(D)^2 * H/4$ (cm^3)
  D = Diameter of core cutter (generally 10 cm for standard laboratory tests)
  H = Height of core cutter (12.5 cm for standard laboratory tests)
  ρ_b = Bulk density or moist or total density of the soil sample (g/cm^3)
  W_s = Weight of the moist soil in the core cutter
  γ_b = Bulk unit weight or moist or total unit weight of the soil sample (kN/m^3)
  w = Natural water content (%) of the representative soil sample (refer Test 1).

2.3.2   Method of Testing

Field/in-place dry density of a natural in-situ soil is determined by the core cutter method. This method is suitable for the soils where the cutter can be easily driven.

2.3.3   Soil Testing Material

The soil specimen is collected from the field at selected locations in a core cutter and properly sealed and packed in plastic bags so that the field or natural water content is not lost during transportation to the geotechnical testing laboratory. A schematic of collection of undisturbed soil sample using a core cutter of known dimensions (e.g. of known volume) is shown in Figure 2.1.

2.3.4   Testing Equipment and Accessories

1. Cylindrical core cutter, 10 cm internal diameter and 13 cm long with wall thickness of 3 mm and beveled at one end to form a cutting edge.
2. Steel dolly, 25 mm high and 10 cm internal diameter and 11.5 cm outside diameter.
3. Steel hammer, mass 9 kg, having a tamping foot of 14 cm diameter and 7.5 cm height, fitted with a handle making a total height of about 1 m.
4. Weighing balance, knife, straight edge, steel rule etc.

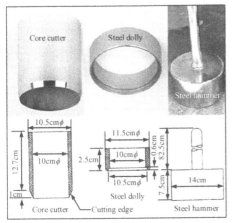
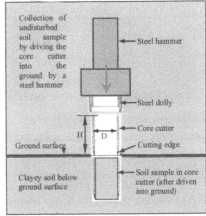

FIGURE 2.1 Schematic illustration for collection of undisturbed soil sample in a core cutter.

2.3.5 TESTING PROGRAM

1. Select seamless cylindrical core cutter of known internal diameter (D) and height (H) and compute its volume, V (cm^3).
2. Apply a thin layer of grease gently all around the inner surface of the core cutter and determine the mass of the empty cutter (M$_1$).
3. Go to the field at a selected site and clean and level the ground surface by removing the top loose soil layer and grass roots, etc.
4. Keep the core cutter on the leveled ground surface along with a dolly on top of the core cutter.
5. Drive the core cutter into the soil mass using the steel hammer. Stop the driving when the core cutter is completely driven into the ground and the dolly remains about 10–15 mm above the ground surface.
6. Excavate the soil around the core cutter carefully and take out the core cutter filled with a soil sample such that the soil should project from the lower end of the cutter.
7. Remove the dolly from the top of the core cutter and trim the top and bottom surfaces of the core cutter carefully using a cutting knife and a straight edge.
8. Clean the core cutter from outside and top and bottom surfaces with a neat cloth to remove any soil flushed with the core cutter.
9. Weigh the core cutter filled with the soil (M$_2$).
10. Extrude the soil sample from the cutter using a soil extruder. Remove about 1 inch thick soil from the top and bottom of the soil sample. Take a representative soil sample at three locations of this soil sample at the top, middle, and bottom for the water content determination.
11. Determine the field dry density using Equation [2.1]. Repeat the test for at least three soil core samples and determine the average field dry density for the selected site.

12. Always express density in terms of unit weight (e.g. by SI system of units) using the following relationship between density and unit weight:

$$\gamma = [\rho * g] \Rightarrow \gamma_d = [\rho_d * g] \, (g = 10 m/s^2)$$ [2.2]

2.3.6 OBSERVATION DATA SHEET AND ANALYSIS

Observations, test data analysis, and calculations for determination of field/in-place density are given in a data sheet presented in tabular form as below (Table 2.1).

2.3.7 RESULTS AND DISCUSSIONS

Based on test results, field dry unit weight of the soil = _____ kN/m^3.
  Soil type is: ....

2.4 FIELD/IN-SITU DRY DENSITY OF SOIL BY SAND REPLACEMENT METHOD

References: IS: 2720 (Part-28) - 1974 (Reaffirmed 2006); ASTM D 1556-90; BS 1377: Part 2, 1990

2.4.1 DEFINITIONS AND THEORY

The in situ dry density plays a vital role as a controlling parameter for ensuring that the required degree of compaction is achieved for the compacted soil fills in the field. However, in the case of cohesionless soils (e.g. sandy soils), it is very difficult to obtain undisturbed soil sample from the field for determination of in situ density. Therefore, the in situ density is determined indirectly by sand or water replacement method. IS 2720 (Part 28)-1974 gives the procedural details of the sand replacement method. Since an undisturbed soil sample can not be collected in case of sandy soils, the volume of an undisturbed soil sample is computed indirectly by the sand replacement method. In this method, the soil is excavated through a template tray and a hole as close as possible to the calibration container is prepared and filled with sand of known mass (M_{sand}) and density ($\rho_{d(sand)}$) as per standard codal procedure for determination of the volume of the hole. Then the volume of the soil sample excavated from the hole is determined using the expression:

$$V_{hole} = \left[\frac{M_{sand}}{\rho_{d\,(sand)}} \right] (cc)$$ [2.3]

Where: V_{hole} = total volume of representative soil sample excavated from the hole
  ρ_{sand} = density of sand filled into the hole, and
  M_{sand} = mass of sand filled into the hole.

TABLE 2.1
Field Dry Density of Soil by Core Cutter Method

			Date:
Project/Site Name:			
Test Site:	**Job. No.**		**Sample No.**
Sample Recovery	**Sample Recovery Method:**		
Depth:			
Sample Description:			
Tested By:			

Conversion factors used: 1 g/cc. = 62.43 p/ft^3; γ_w = 10kN/m^3, γ = ρg; g = 10 m/s^2

Sl. No.	Observations and Calculations	Trial (Soil core) No.		
		1	2	3
Observation				
1	Core cutter No.			
2	Internal diameter, D (cm)			
3	Internal height, H (cm)			
4	Volume of cutter, V (cm^3) = $\pi/4 * (D^2 * H)$			
5	Mass of empty core cutter (M_1)			
6	Mass of core cutter with moist soils (M_2)			
Calculations: Bulk density, ρ_b				
7	Mass of moist soil, $M_s = M_2 - M_1$ (gm)			
8	Bulk density, $\rho_b = M_s/V$ (g/cm^3)			
Calculations: Water content, w				
9	Container No.	1	2	3
10	Mass of empty container (M_1)			
11	Mass of container + moist soil (M_2)			
12	Mass of container + dry soil (M_3)			
13	Mass of water $M_w = M_2 - M_3$			
14	Mass of dry soil solids, $M_d = M_3 - M_1$			
15	Water content (fraction), w = $(M_w)/(M_d)$			
Calculations: Field dry density, ρ_d				
16	Field dry density, $\rho_d = \rho_b/(1 + w/100)$ (g/cm^3)			
17	Average dry density, ρ_d (g/cm^3)			
18	Average dry unit weight, $\gamma_d = \rho_d * 10$ (kN/m^3)			

Note: It may be ensured that the volume of the soil specimen is nearly close to the volume at which the sand was calibrated in the laboratory. Further, the sand passing through the 2.00 mm (No. 10) sieve must be free-flowing, dry and clean having constant water content. However, it should be ensured that sand material has less than 3% passing the 0.30 mm sieve.

Further, the volume of the soil sample depends on the maximum soil grain size. The other notable requirements to be considered while determining the volume of the hole are:

1. Soil sample of grain size of 10–12 mm requires a minimum hole volume of 1,500 cc.
2. Soil sample of grain size of about 20–25 mm requires a minimum hole volume of 2,000 cc.
3. Soil sample of grain size of about 40–50 mm requires a minimum hole volume of 3,000 cc.

The excavated soil is very carefully collected in a polythene bag and brought to the geotechnical laboratory and weighed (M_s). Once the volume of the representative soil is known, the bulk density of the excavated soil is determined as:

$$\rho_b = \left[\frac{M_s}{V_{hole}} \right] (g/cc)$$

Where M_s = Mass of the excavated soil from the hole, and
V_{hole} = Total volume of representative soil sample excavated from the hole.
Similarly, knowing the bulk density of the representative excavated soil sample, the in situ dry density (or dry unit weight) of the soil sample can be determined using the expression (same as Eq. 2.1 above):

$$\rho_d = \left[\frac{\rho_b}{(1 + w/100)} \right] (g/cc) \, or \, \gamma_d \left[\frac{\gamma_b}{(1 + w/100)} \right] (kN/m^3)$$

2.4.2   METHOD OF TESTING

The sand replacement method is adopted to determine the in situ dry density of natural or compacted soil in the field. The method has a wider application for granular soils and should be preferred wherever possible.

2.4.3   SOIL TESTING MATERIAL (SAND)

Dry, clean, uniform sand passing in the 2 mm IS sieve and retained on the 0.60 mm IS sieve in a sufficient quantity is taken for testing purposes.

2.4.4   TESTING EQUIPMENT AND ACCESSORIES

The following equipment and accessories are required for the determination of in situ dry density by the sand replacement method (sand displacement method apparatus is shown in Figure 2.2):

1. Sand pouring cylinder (3-liter capacity).
2. Calibrating container/compaction mold/cylinder of desired dimensions.

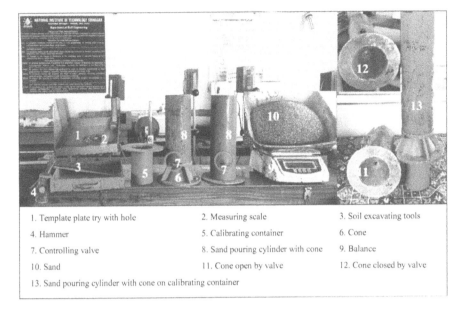

1. Template plate try with hole	2. Measuring scale	3. Soil excavating tools
4. Hammer	5. Calibrating container	6. Cone
7. Controlling valve	8. Sand pouring cylinder with cone	9. Balance
10. Sand	11. Cone open by valve	12. Cone closed by valve
13. Sand pouring cylinder with cone on calibrating container		

FIGURE 2.2   Apparatus for sand displacement method.

3. Template plate tray, 300 mm ∗ 300 mm ∗ 40 mm with a hole of about 10 cm in diameter at the center.
4. Soil excavating tools, such as scrapper tool, bent spoon, hammer, etc.
5. Good quality plastic bags for collection of excavated soil.
6. Weighing balance of 20-kg capacity (nearest to 1.0 g).
7. Weighing balance 500-g capacity (nearest to 0.01 g).
8. Laboratory accessories such as moisture content cans, oven, desiccators, etc.
9. Other accessories such as a ruler, a varnish brush, knife, nails, glass plate, etc. as required.

2.4.5   TEST PROGRAM, DATA SHEET, AND ANALYSIS FOR THE SAND DISPLACEMENT METHOD

Step 1: Density of sand/calibration of sand pouring cylinder

a. Take a clean sand pouring cylinder affixed with a cone at its bottom and operated by a controlling valve (Figure 2.2).
b. Close the controlling valve and the pouring cylinder with clean, dry, and free-flowing poorly graded sand.
c. Weigh the sand pouring cylinder filled with sand and fixed with funnel as M_1 (g), as shown in Table 2.2.

TABLE 2.2

Determination of Density of Sand Material

Field density determination of a soil sample by sand cone method

Project Name: **Date:**

Test Site: Job. No.   Sample No.

Addl. information, if any:

Conversion factors used: 1 cu.ft. = 62.4 lb; 1 cm^3 = 1 g, γ_w = 10 kN/m^3, γ = ρg; g = 10 m/s^2

Determine the density of sand material/calibration of sand (standard material)

Trial No.		1	2	3
1	Mass of sand pouring cylinder filled with dry sand, before pouring sand on template tray: M_1 (g)			
2	Mass of sand pouring cylinder with remaining dry sand after pouring sand on template tray: M_2 (g)			
3	Mass of sand on template tray: $M_{sand(T)}$ = M_2 − M_1 (g)			
4	Mass of sand pouring cylinder filled with dry sand, before pouring sand into the calibrating container: M_3 (g)			
5	Mass of sand pouring cylinder with remaining dry sand after pouring sand into the calibrating container: M_4 (g)			
6	Mass of sand in the calibrating container:$M_{sand(TC)}$ = M_3 − M_4 − $M_{sand(T)}$ (g)			
7	Volume of calibrating container: V_C (cm^3)			
8	Dry density of sand material: $\rho_{d(sand)}$ = $M_{sand(TC)}$ / V_C (g/cm^3)			
9	Dry unit weight of sand: $\gamma_{d(sand)}$ = ρ_{sand} * g (kN/m^3)			
10	Average unit weight of sand, $\gamma_{d(sand)av}$ = (γ_1 + γ_2 + γ_3)/3			

d. Level the ground surface carefully and place the template tray on it or alternatively, place the template tray on a leveled surface such as a table or counter in the laboratory.

e. Place the sand pouring cylinder on the template tray and open the valve of the cone to allow the sand to fall into the cone and the template tray.

f. Close the valve when the sand stops falling into the cone and the template tray.

g. Remove the sand pouring cylinder from the template tray and weigh it with the remaining sand as M_2 (g).

h. Determine the mass of sand filled in the cone and the template tray as M_{sand} $_{(T)}$ = M_1 − M_2 (g).

i. Now refill the sand pouring cylinder with dry sand and weigh it along with the cone as M_3 (g).

j. Take a calibrating container of known volume (V_C) and place the template tray on it.

k. Place the sand pouring cylinder on the template tray carefully and open the valve of the cone to allow the sand to fall into the cone, template tray, and the calibrating container.

l. Close the valve when the calibrating container, template tray, and the cone are completely filled with sand or when the sand stops falling into the cone, template tray, and the calibrating container.

m. Replace the sand pouring cylinder from the template tray and weigh it with the remaining sand as M_4 (g).

n. Determine the mass of the sand in the calibrating container as: $M_{sand(C)}$ = $M_3 - M_4 - M_{sand(T)}$ (g).

o. Repeat the above procedure and record the average of about three trials for determination of mass of sand ($M_{sand(C)}$).

p. Determine the dry density of the sand as:

$$\rho_{d\,(sand)} = \left[\frac{M_{sand\,(C)}}{V_C} \right] (g/cc)$$

Where: V_C = volume of the calibrating cylinder (cm^3).

q. Convert the density of the materials into unit weight and compute the average unit weight of the sand material as given in Table 2.2.

Step 2: Site preparation for the digging hole

a. Clean and level the ground surface at which in situ density is to be determined.

b. Now place the template tray on the leveled ground surface and fasten it in place with nails and seal the inside edges properly if so required.

Step 3: Digging the hole by excavating the soil through the template tray

a. Dig the hole through the template tray by excavating the soil, making about a 150 mm deep circular hole by using the required soil cutting and excavating tools.

Note: It may be noted that the minimum test hole depth is dependent upon the maximum soil grain size.

Step 4: Volume of the excavated hole using sand replacement method

a. Fill the sand pouring cylinder with the same sand as used for determination of density of sand/calibration of sand pouring cylinder in Step 1.

b. Weigh the sand pouring cylinder filled with sand and affixed with cone as M_5 (g).

c. Now align the sand pouring cylinder on the template tray placed on the excavated hole carefully and open the valve to allow the sand to fall into the cone, template tray, and the excavated hole.

d. Close the valve when the excavated hole, template, and the cone are completely filled with sand.

e. Remove the sand pouring cylinder from the template tray and weigh it with the remaining sand as M_6 (g).

f. Determine the mass of the sand filled in the cone, template tray, and the excavated hole as: $M_{sand(fth)} = M_5 - M_6$ (g).

g. Determine the mass of the sand filled into the excavated hole as: $M_{sand(hole)} = M_{sand(fth)} - M_{sand(T)}$ (g), where $M_{sand(T)}$ is calculated by Step 1(h).

h. Determine the volume of the excavated hole (cm^3) using the following formula:

$$V_{hole} = [(M_{sand(hole)} * \rho_{sav})]$$

Where: $M_{sand\ (hole)}$ = Mass of sand in the hole (g), and
ρ_{sav} = Average density of sand obtained from Table 2.2.

i. Collect the excavated soil specimen in sealed plastic bags so that there is no moisture loss during its transportation from the field to the geotechnical testing laboratory.

j. Weigh the excavated soil specimen (M_s g) (as representative sample of undisturbed soil sample) for determination of bulk density (g/cm^3):

$$\rho_b = \left[\frac{M_s}{V_{hole}} \right] (g/cc)$$

Where: V_{hole} is the volume of excavated soil in the field (sub-step- h above).

k. Take representative samples of the excavated soil specimen for the determination of water content (wn%).

l. Determine the dry density: $\rho_d = \left[\frac{\rho_b}{(1 + w/100)} \right] (g/cc)$.

m. Determine the dry unit weight: $\gamma_d = [\rho_d * g] (kN/m^3)$.

Where: g = gravitational constant (10 m/s^2)

n. Ensure that the shape of the hole is as close as the shape of the calibrating container.

2.4.6 OBSERVATION DATA SHEET AND ANALYSIS

Observations, test data analysis, and calculations for determination of field/in-place density are given in the data sheet presented in tabular form as below (Table 2.2).

2.4.7   RESULTS AND DISCUSSIONS

Based on test results, the average dry unit weight of the soil, γ_{dav} = _____
(kN/m^3).
   Soil type: ....

2.5   FIELD/IN-SITU DRY DENSITY OF A SOIL SAMPLE BY WATER DISPLACEMENT METHOD

2.5.1   DEFINITIONS AND THEORY

This method is suitable for cohesive soils only, where it is possible to have a lump soil sample. A small sample is trimmed to a regular shape from a large sample brought from the field. The soil pat so prepared for determination of its volume is covered with melted wax solution of known density (ρ_p) all around its surface allowed to harden. After hardening the waxed soil pat, the total volume of waxed soil (V_T) pat is determined by the volume of water displaced by the waxed soil pat using a graduated measuring water jar. The volume of unwaxed soil pat (V) is determined as:

$$V = \left[V_T - \frac{(M_T - M)}{\rho_p} \right] \qquad [2.4]$$

Where: M_T = Mass of waxed soil pat (e.g. mass of soil specimen plus mass of wax applied)
   M = Mass of the soil pat without wax
   ρ_p = Density of paraffin, and
   V_T = Volume of the waxed soil pat.
   Once the actual volume (V) is known, the dry density is computed as (same as Eq. 2.1 above):

$$\rho_d = \left[\frac{M/V}{(1 + w/100)} \right]$$

2.5.2   METHOD OF TESTING

The water replacement method is adopted to determine the in situ dry density of natural or compacted soil fills in the field. The method has a wider application for cohesive soils of large lumped sizes and should be preferred wherever possible.

2.5.3   SOIL TESTING MATERIAL

Soil pat coated with paraffin wax.

2.5.4 Testing Equipment and Accessories

1. Soil material required for testing purpose
2. Measuring jar
3. Weighing balance, accuracy 1 g
4. Paraffin wax
5. Straight edge
6. Heater
7. Oven
8. Brush
9. Measuring jar
10. Water content container

2.5.5 Testing Program

a. Prepare a soil pat of regular shape and weigh it (M_s g).
b. Prepare paraffin wax solution and apply a thin coat of melted wax solution all-around the surface of the soil pat and leave it for some time to harden it.
c. Apply second coat of wax solution and weigh the waxed soil specimen (M_T g).
d. Fill the measuring jar with water and place it in a tray carefully.
e. Now immerse the waxed soil pat slowly into the water jar completely.
f. Collect the overflowed water due to immersion of waxed soil pat in the tray and determine the volume of waxed soil specimen (V_T).
g. Determine the volume of wax, V_{wax} (cm^3) = (M_T − M_s)/ ρ_p. Where: ρ_p = Density of paraffin wax
h. Determine the volume of unwaxed soil specimen: $V = V_T - V_{wax}$.
i. Determine the bulk density of the soil specimen: $\rho_b = \left[\dfrac{M_s}{V}\right]$ (g/cc).
j. Take out the waxed specimen from the water jar. Dry it from the outside and remove the paraffin wax by peeling it off.
k. Cut the specimen into two pieces. Take a representative sample for the water content determination.
l. Determine the dry density of the soil specimen as:

$$\rho_d = \left[\frac{\rho_b}{(1 + w/100)}\right] (g/cc)$$

m. Determine the dry unit weight: $\gamma_d = [\rho_d * g] (kN/m^3)$.

Where: g = gravitational constant (10 m/s^2)

2.5.6 Observation Data Sheet and Analysis

Test data analysis, observations, and calculations for determination of field/in-place density are given in the data sheet presented in tabular form as below (Table 2.3).

TABLE 2.3
In-Situ Density Determination of Soil Sample by Water Displacement Method

Project Name: Date:

Test Site: Job. No.   Sample No.

Density of paraffin (ρ_p) = 0.91 g/ml.

Conversion factors used: 1 cu. ft.= 62.4 lb; 1 cm^3 = 1 g, γ_w = 10kN/m^3, γ = ρg; g = 10 m/s^2

Sl. No.	Observations and Calculations	Trial No.		
		1	2	3
Observation				
1	Mass of excavated soil specimen, M_s (g)			
2	Mass of waxed soil pat, M_T (g)			
3	Volume of waxed specimen by weight displacement, V_t (cm^3)			
Calculations				
4	Mass of wax, $M_{wx} = M_T - M$ (g)			
5	Volume of wax, V_{wax} (cm^3) = $(M_T - M_s)/ \rho_p$			
6	Volume of specimen, $V = V_t - V_{wax}$			
7	Bulk/wet density, $\rho_b = M_s/V$ (g/cm^3)			
8	Water content, w (%) as per procedure			
9	Dry density, $\rho_d = \rho_b/(1 + w/100)$ (g/cm^3)			
10	Dry unit weight, $\gamma_d = \rho_d * g$ (kN/m^3)			
11	Average Dry unit weight, γ_{dav} (kN/m^3)			

2.5.7 Results and Discussions

Based on test results, the average dry unit weight, γ_{dav} = _____(kN/m^3).

   Soil type: ....

2.6 GENERAL COMMENTS

1. The density or unit weight of the in situ sample can be determined on an undisturbed soil specimen presuming that there has been no moisture loss and the structure of in situ soil sample remains in tact during collection of undisturbed soil specimen.

2. The field dry unit weight of soils may vary over a wide range between 12–18 kN/m^3, whereas maximum dry density (MDD) for compacted soils lies in the range of 14.5–20.0 kN/m^3. However, it may be noted that the degree of compaction for compacted soil is always less than 100% in the field. Therefore, it is recommended that the remolded soils may be prepared at 90% of MDD (e.g. 0.9 * MDD) for determination of shear strength parameters.

3. Water replacement test can also be adopted for cohesionless soils wherein the volume of the excavated soil is determined by water replacement method using a graduated water measuring jar. In such a case, the soil is excavated

from the test site and weighed (M_s). After the hole is prepared, a good quality polythene sheet is inserted very carefully into the hole and filled with water slowly such that while filling the hole with water, the polythene sheets sit closely with the soil surface in the hole. The volume of the excavated soil from the hole is equivalent to the volume of water filled into the hole. Once the volume of the hole (V_{hole}) is known, the in situ density of the excavated soil sample is determined as $\rho_b = [M_s/V_{hole}, g/cc]$.

2.7   APPLICATIONS/ROLE OF DRY DENSITY IN SOIL ENGINEERING

1. Density is indicative of strength of a material. The higher the value of density, the higher the strength of that material.
2. For construction of pavements and other compacted fills, the density is used as a controlling parameter to ensure that the in situ density is achieved adequately as per standard codal procedures.
3. The density is used as a design parameter for the computation of degree of compaction, relative density, and compactability factor, etc. in the construction of various engineering structures.
4. This test helps in justifying that the construction works in the field are compacted in accordance with the established construction specifications.

2.8   SOURCES OF ERROR

The following sources of error may be considered during the sand replacement test:

1. Sand may densify due to unknown vibrations if any during the test.
2. The error may occur if the ground surface is not properly leveled at the test site. In such a case, volume measurement of the hole may be incorrect.

2.9   PRECAUTIONS

1. Steel dolly or a wooden block should be placed on the top of the core cutter before hammering it down to avoid any damage to core cutter top.
2. Core cutter should be carefully removed from the ground surface and no soil should drop down.
3. A hole of exact size as that of the calibrating cylinder should be excavated for the sand replacement method.
4. Make sure that no loose material is left in the hole before pouring sand into it.
5. Make sure that there are no vibrations during the test.
6. Make sure that a regular shape soil specimen is chosen for the water displacement method.
7. Make sure that the soil specimen is properly coated with paraffin wax to make it impervious to water.

REFERENCES

ASTM D 1556-2007. "Method for Laboratory Determination of In-Place Density and Unit Weight of Soils by Sand Cone Method." *Annual Book of ASTM standards*, Vol. 04–08. American Society for Testing and Materials, Philadelphia, United States, www.astm.org.

BS 1377-2. 1990. "Methods of Test for Soils for Civil Engineering Purposes – Part 2: Classification Tests for Determination of Water Content." British Standards, UK.

IS: 2720 (Part 29). 1975. "(Reaffirmed 2006). Method of Test for Soils: Determination of Dry Density of Soils in Place by Core Cutter Method." Bureau of Indian Standards, New Delhi.

IS: 2720 (Part 28). 1974. "(Reaffirmed 2006). Method of Test for Soils: Determination of Dry Density of Soils in Place by the Sand Replacement Method." Bureau of Indian Standards, New Delhi.

IS: 2720 (Part 33). 1971. "(Reaffirmed 2006). Methods of Test for Soils: Determination of Density of Soils In-Place by the Ring and Water Replacement Method." Bureau of Indian Standards, New Delhi.

Reddy, Krishna R. 2000. "Engineering Properties of Soils Based on Laboratory Testing." Department of Civil and Materials Engineering University of Illinois at Chicago, USA.

3 Specific Gravity of Soil Solids

3.1 OBJECTIVES

1. Specific gravity of soil solids is used for the classification of soils for various applications in geotechnical engineering applications and the design of hydraulic structures.
2. Specific gravity is used for grading fine grained soils in the hydrometer/ sedimentation analysis passing the 0.075 mm sieve.
3. Specific gravity is often used for the computation of various parameters such as hydraulic gradient, degree of saturation, theoretical dry density, void ratio, and concentration of substances in aqueous solutions indirectly.

3.2 INTRODUCTION

Specific gravity (G) is a dimensionless quantity, defined as the ratio of the weight of dry soil solids in air of a given volume to the weight of equal volume of distilled water at that temperature (20°C or 27°C). Specific gravity of the soil solid is a function of the packing of its constituents or the minerals. The specific gravity indicates the degree of packing of constituents of a solid mineral in a given volume whether a mineral is heavier or lighter compared to water. The specific gravity of any desired material is determined either in distilled water or kerosene oil at standard or as desired temperatures. The temperature has a direct bearing on specific gravity measurements. The suspension solution will be lighter if the test temperature is higher than the standard temperature, having a lower value of specific gravity of the material in suspension and vice versa. Therefore, temperature correction has to be applied if the test temperature is different than the standard temperature (20°C or 27°C). The specific gravity of soil solids is used for the classification of soils for various applications in geotechnical engineering applications and design of hydraulic structures. The specific gravity also plays an important role in the derivation of interrelationship of mass-weight, weight-volume, and volumetric relations for developing various empirical correlations between the soil properties. In soil engineering practice, the specific gravity of a soil sample passing through the 4.75 mm sieve size is determined. Generally, the specific gravity of granular soils varies in the narrow range between 2.61 to 2.67, with a standard value of 2.65, which is the specific gravity of quartz mineral. However, granular soils exhibit specific gravity on lower side compared to cohesive soils, which varies over a wide range of 2.67 to 2.90. Specific gravity (G) is a dimensionless quantity, as it is the ratio of the weight in air of a given volume of dry soil solids to the weight of equal volume of distilled water at that temperature

DOI: 10.1201/9781003200260-3

(20°C or 27°C). The specific gravity of soils depends on the packing of the minerals building the individual soil particles (Reddy, 2000). Generally, the typical values of various soil types are given as below:

1. Specific gravity of granular soils: 2.60–2.87.
2. Specific gravity of inorganic clays: 2.67–2.84.
3. Specific gravity of soils with large amounts of organic matter: 2.20–2.40.
4. Specific gravity of soils with porous particles such fly ash: 2.10–2.64.
5. Specific gravity of highly organic soils such peaty soils: < 2.00.
6. Specific gravity of iron-rich soils: 2.75–3.20.

Note: It may be noted that the soil specimen may be oven-dried before the specific gravity test. However, organic soils or low compressible soils are tested at their natural moisture content.

The specific gravity of a soil material used in various applications in geotechnical engineering is usually expressed in two forms as given below:

1. The specific gravity of soil solids, G_s, for soils finer than a 4.75 mm (No. 4) IS sieve.
2. The apparent or bulk specific gravity, G_a, for soils coarser than a 4.75 mm (No. 4) IS sieve.

However, the specific gravity of coarse grained soils is always lower than the fine grained soils due to the fact that coarse grained particles have void pores filled with air and hence are lighter compared to fine grained particles. Therefore, soils coarser than a 4.75 mm sieve exhibit apparent or bulk specific gravity rather than specific gravity of solids.

3.3   SPECIFIC GRAVITY OF SOIL SOLIDS BY DENSITY BOTTLE METHOD

References: IS: 2720-Part 3(1); IS: 2720-Part 3(2); ASTM D854-14; ASTM D5550-14; BS 1377: Part 2, 1990

3.3.1   DEFINITIONS AND THEORY

Specific gravity (G) is a dimensionless quantity, defined as the ratio of the weight of dry soil solids in air of a given volume to the weight of equal volume of water at standard temperature (20°C or 27°C). Thus, the specific gravity of soil solids is defined as:

$$G_s = \left[\frac{\rho_s}{\rho_w} \right]$$

[3.1]

Where ρ_w is the density of water at 20°C (≈ 1 g/cm^3).

Since the soil is a heterogeneous material and the specific gravity of different soil grains in a given soil sample may not be same, the specific gravity is referred to as the average values of all the soil solid particles present in the given soil sample (G_s). The G_s value of some inorganic clay may be as high as 2.92, whereas the specific gravity of organic soils is very variable, may fall below 2.00.

The theory involved in the determination of G_s is very simple. A soil sample passing a 2 mm sieve is first oven-dried and taken in a neat and clean dried density bottle and weighed. Then some distilled water or kerosene oil is added to entrap air bubbles if any. After removing the air bubbles, this bottle is filled with distilled water or kerosene oil up to the top and weighed. Lastly, the bottle is cleaned and filled either with distilled water or kerosene oil and weighed again. Then G_s is computed by the following equation:

$$G_s = \left[\frac{(W_2 - W_1)}{(W_2 - W_1) - (W_3 - W_4)} \right]$$ [3.2]

Where: W_1 = Weight of empty bottle
   W_2 = Weight of the bottle and dry soil
   W_3 = Weight of bottle, soil and water, and
   W_4 = Weight of bottle filled with water only.

3.3.2 Method of Testing

The specific gravity of a soil specimen is determined by using a density bottle method.

3.3.3 Soil Testing Material

The soil specimen is collected from the field at selected locations and transported to a soil testing laboratory. The soil sample is air-dried, pulverized, and sieved through desired sieves required for testing. About 50 g of an oven-dried *soil sample passing a 2 mm IS sieve* is taken for determination of specific gravity of a given soil sample by the density bottle method.

Note: Soil specimen containing natural moisture (e.g. air-dried) may also be taken, but oven-dry weight of the soil must be determined at the end of the test.

Soil sample passing a 2 mm IS sieve is used if the value of specific gravity is used in connection with hydrometer analysis.

3.3.4 Testing Equipment and Accessories

The following equipment and accessories are required to determine the specific gravity of a soil specimen (Figure 3.1).

1. Oven-dried soil sample passing a 2 mm IS sieve
2. Density bottle (Borosil) of 50 to 100 ml capacity with stopper having capillary hole at its center
3. Constant temp water bath (20°C)
4. Balance-sensitivity 0.001 g
5. Vacuum pump, funnel, and spoon
6. Wash bottles filled with distilled water
7. Drying oven and hot plate
8. Desiccator, vacuum desiccator
9. IS Sieves (4.75 mm, 2 mm, 0.425 mm)
10. Thermometer
11. Glass road 150 mm long and 3 mm diameter
12. Porcelain dish, spatula, and funnel
13. Soil sample passing a 2.00 mm IS sieve

3.3.5 Testing Program

1. Clean the 50 ml capacity density bottle thoroughly and dry it in a micro-oven for about half an hour and then cool it.
2. Weigh the density bottle along with stopper to the nearest 0.001 g (W_1).
3. Take about 5 to 10 g of the oven-dried soil sample and transfer it into the density bottle carefully and then weigh the density bottle plus dry soil with stopper to the nearest 0.001 g (W_2).
4. Fill de-aired distilled water up to half of the bottle, just enough to cover the soil, shake gently to mix water and soil and after removing stopper, place it in a water bath (@ 30 to 35°C) or in the vacuum desiccator, and connect to the vacuum pump (electrically operated) to remove the entrapped air if any. The entrapped air may also be removed by boiling the density bottle suspension in a sand bath for at least 10 min. It may be noted that suspension is boiled gently to avoid any loss of material out of suspension during the boiling process. However, in addition to boiling the suspension, the density bottle should be stirred to assist in the removal of air.

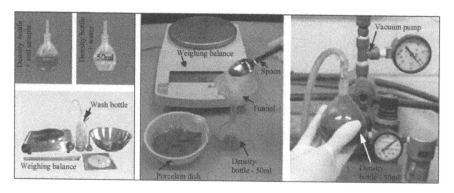

FIGURE 3.1 Equipments and accessories required for specific gravity test.

Note: It may be noted that entrapped air in highly plastic and organic soils cannot be removed by vacuum. Therefore, the entrapped air may be removed by boiling the suspension in the density bottle for about 30 min gently. However, in such a case, distilled water may be added from time to time so that the suspension may not dry during prolonged boiling. Also, vigorous boiling may be avoided to avoid loss of soil material out of the density bottle.

5. Be careful to release the vacuum slowly and observe the level of suspension in the density bottle. If the suspension level drops by about 2 to 3 mm, then the entrapped air has been removed and the suspension is presumed as de-aired.
6. When no further air movement is noticed, take the density bottle out and stir gently and allow the density bottle and contents to cool.
7. Fill the density bottle with deaired distilled water up to the top stopper with free flow of water. Then weigh the density bottle filled with soil and water to the nearest 0.001 g (W_3).
8. Clean the density bottle thoroughly and fill with de-aired distilled water fully and weigh it to the nearest 0.001 g (W_4).
9. Make sure that during the experiment that the temperature is about 20°C, and fluctuation if any, may be recorded (ToC).
10. The specific gravity based on the above observations is computed as (Figure 3.2):

$$G_s = \left[\frac{(W_2 - W_1)}{(W_2 - W_1) - (W_3 - W_4)} \right]$$

11. Repeat the specific gravity test for additional values as per codal procedure until two values (largest value of G_s and smaller G_s) are obtained, which are within 2% of each other using the largest value as the base value: 0.02 (largest G_s)+smaller G_s ≥ largest G_s.
12. Average the results of these two tests and report the value, rounded off to the nearest 0.01, as the specific gravity, G_s, of the soil. For an average value of G_s, three trials of the test are performed and it is ensured that the difference between any two values does not exceed 0.03, else the test is repeated.

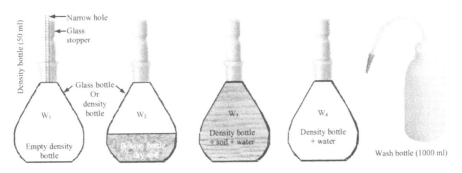

FIGURE 3.2   Schematic presentation for specific gravity test by density bottle method.

3.3.6 OBSERVATION DATA SHEET AND ANALYSIS

Test observations, data analysis, and calculations for determination of specific gravity of soil solids by the density bottle method are given in Table 3.1. Density of water (g/cm³) at temperatures from 0°C to 30.9°C by 0.1°C increments is given in Table 3.2.

If room temperature is different than standard ($T_{std} = 20°C$), then the corrected specific gravity is obtained as:

$$G_{s(T_{std})} = \left[G_{s(Lab.Temp)} * \frac{\text{Density of water at lab.Temperature } (T°C)}{\text{Density of water at standard Temperature } (20°C)} \right]$$

[3.3]

3.3.7 RESULTS AND DISCUSSIONS: BASED ON TEST RESULTS, SPECIFIC GRAVITY OF SOLIDS AT 24°C = 2.72

Soil type: Inorganic clayey soil

TABLE 3.1
Specific Gravity of Soil Solids by Density Bottle Method

Project/Site Name:		Date:	
Test Site:		**Job. No.**	**Sample No.**
Sample Recovery Depth:		**Sample Recovery Method:**	
Sample Description:			
Tested By:			
Lab. Temp.: $\rho_w = 1.00$ g/cm³ 24°C		**Density of water (20°C): 0.9982 ≅ 1 g/cm³**	
Density of water at test lab. temperature		**0.997296 g/cm³** $\gamma_w = 10$ kN/m³	

Sr. No.	Observations and calculations	Trial No./Sample No.		
		1	**2**	**3**
1	Weight of empty density bottle, W_1 (g)	36.0770	30.6400	36.1960
2	Weight of bottle plus dry soil, W_2 (g)	45.6770	42.6870	47.0610
3	Weight of bottle plus soil plus water, W_3 (g)	89.6520	87.2361	90.9936
4	Weight of bottle plus distilled water, W_4 (g)	83.5781	79.5944	84.1173

Mass of water having the same volume as that of solids =
$(W_4 - W_1) - (W_3 - W_2) = (W_2 - W_1) - (W_3 - W_4)$

5	$G_s = \left[\frac{(W_2 - W_1)}{(W_2 - W_1) - (W_3 - W_4)} \right]$	2.722	2.734	2.723

Difference between any two G-values is less than 0.03, hence acceptable

6	Average value of G_{sT} at Lab. Temp (T°C)	2.73		
7	Density of water. of water @ 24°C (g/cm³)	0.997296		
8	Density of water. of water @ 20°C (g/cm³)	0.9982		
9	Value of G_s at 20°C	2.73*(0.997296/0.9982) = 2.72		

TABLE 3.2
Density of Water (g/cm³) at T temperatures from 0°C (Liquid State) to 30.9°C by 0.1°C Increment

Example: The density of water at 4.4°C is 0.999972 g/cc

T°C	0	0.1	0.2	0.3	0.4	0.5	0.6	0.7	0.8	0.9
0	0.999841	0.99985	0.99985	0.99986	0.999866	0.99987	0.99988	0.99988	0.99989	0.9999
1	0.9999	0.99991	0.99991	0.99991	0.999918	0.99992	0.99993	0.99993	0.99993	0.99994
2	0.999941	0.99994	0.99995	0.99995	0.999953	0.99996	0.99996	0.99996	0.99996	0.99996
3	0.999965	0.99997	0.99997	0.99997	0.99997	0.99997	0.99997	0.99997	0.99997	0.99997
4	**0.999973**	**0.99997**	**0.99997**	**0.99997**	**0.999972**	**0.99997**	**0.99997**	**0.99997**	**0.99997**	**0.99997**
5	0.999965	0.99996	0.99996	0.99996	0.999957	0.99996	0.99995	0.99995	0.99995	0.99994
6	0.999941	0.99994	0.99994	0.99993	0.999927	0.99992	0.99992	0.99992	0.99991	0.99991
7	0.999902	0.9999	0.99989	0.99989	0.999883	0.99988	0.99987	0.99987	0.99986	0.99986
8	0.999849	0.99984	0.99984	0.99983	0.999824	0.99982	0.99981	0.9998	0.9998	0.99979
9	0.999781	0.99977	0.99977	0.99976	0.999751	0.99974	0.99973	0.99973	0.99972	0.99971
10	0.9997	0.99969	0.99968	0.99967	0.999664	0.99965	0.99965	0.99964	0.99963	0.99962
11	0.999605	0.9996	0.99959	0.99957	0.999564	0.99955	0.99954	0.99953	0.99952	0.99951
12	0.999498	0.99949	0.99948	0.99946	0.999451	0.99944	0.99943	0.99942	0.9994	0.99939
13	0.999377	0.99936	0.99935	0.99934	0.999326	0.99931	0.9993	0.99929	0.99927	0.99926
14	0.999244	0.99923	0.99922	0.9992	0.999188	0.99917	0.99916	0.99914	0.99913	0.99911
15	0.999099	0.99908	0.99907	0.99905	0.999038	0.99902	0.99901	0.99899	0.99898	0.99896
16	0.998943	0.99893	0.99891	0.99889	0.998877	0.99886	0.99884	0.99883	0.99881	0.99879
17	0.998774	0.99876	0.99874	0.99872	0.998704	0.99869	0.99867	0.99865	0.99863	0.99861
18	0.998595	0.99858	0.99856	0.99854	0.99852	0.9985	0.99848	0.99846	0.99844	0.99842
19	0.998405	0.99839	0.99837	0.99835	0.998325	0.99831	0.99829	0.99827	0.99824	0.99822

(Continued)

TABLE 3.2 (Continued)

Example: The density of water at 4.4°C is 0.999972 g/cc

	0.998203	**0.99818**	**0.99816**	**0.99814**	0.99812	**0.9981**	**0.99808**	**0.99806**	**0.99804**	**0.99801**
20	0.998203	0.99818	0.99816	0.99814	0.99812	0.9981	0.99808	0.99806	0.99804	0.99801
21	0.997992	0.99797	0.99795	0.99793	0.997904	0.99788	0.99786	0.99784	0.99782	0.99779
22	0.99777	0.99775	0.99772	0.9977	0.997678	0.99766	0.99763	0.99761	0.99759	0.99756
23	0.997538	0.99751	0.99749	0.99747	0.997442	0.99742	0.99739	0.99737	0.99735	0.99732
24	0.997296	0.99727	0.99725	0.99722	0.997196	0.99717	0.99715	0.99712	0.9971	0.99707
25	0.997044	0.99702	0.99699	0.99697	0.996941	0.99691	0.99689	0.99686	0.99684	0.99681
26	0.996783	0.99676	0.99673	0.9967	0.996676	0.99665	0.99662	0.99659	0.99657	0.99654
27	0.996512	0.99649	0.99646	0.99643	0.996401	0.99637	0.99635	0.99632	0.99629	0.99626
28	0.996232	0.9962	0.99618	0.99615	0.996118	0.99609	0.99606	0.99603	0.996	0.99597
29	0.995944	0.99591	0.99589	0.99586	0.995826	0.9958	0.99577	0.99574	0.99571	0.99568
30	0.995646	0.99562	0.99559	0.99556	0.995525	0.99549	0.99546	0.99543	0.9954	0.99537

3.4 SPECIFIC GRAVITY OF SOIL SOLIDS BY PYCNOMETER METHOD

References: IS: 2720-part 3(1); IS: 2720-part 3(2); ASTM D854-14; ASTM D5550-14; BS 1377: part 2, 1990

3.4.1 DEFINITIONS AND THEORY

The pycnometer method is widely used to determine the specific gravity of coarse grained soils (sandy soils), which is computed by the following equation (Figure 3.3):

$$G_s = \left[\frac{(W_2 - W_1)}{(W_2 - W_1) - (W_3 - W_4)} \right]$$

Where:
$W_1 = W_b$ = Weight of empty pycnometer bottle or volumetric flask.
W_2 = Weight of the pycnometer bottle with dry soil
W_3 = Weight of the pycnometer bottle and soil and water
$W_4 = W_{bw}$ = Weight of pycnometer bottle filled with water only, and
$G = G_s$ = Specific gravity of solids.

3.4.2 METHOD OF TESTING

Specific gravity of coarse grained soil solids passing through a 4.75 mm IS sieve is determined by using the pycnometer method.

3.4.3 SOIL TESTING MATERIAL

The soil specimen is collected from the field at selected locations in plastic bags and transported to a geotechnical testing laboratory. The soil sample is air-dried, pulverized, and sieved through desired sieves required for testing. About 250 g of an oven-dried soil sample passing through a 4.75 mm IS sieve is taken for the specific gravity test by the pycnometer method.

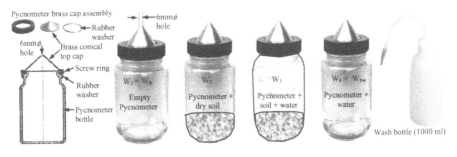

FIGURE 3.3 Determination of G-value by pycnometer method.

3.4.4 TESTING EQUIPMENT AND ACCESSORIES

The following equipment and accessories are required for the specific gravity test by the pycnometer method:

1. Pycnometer bottle (vol. flask) of about 1 L capacity
2. 4.75 mm IS sieves
3. Weighing balance, with an accuracy of 0.01 g
4. Distilled water
5. A glass plate (at least 150 mm long and 3 mm diameter) for mixing soil.
6. Porcelain dish
7. Wash bottle
8. Thermometer
9. Spatula
10. Desiccator
11. Hot water bath or sand bath
12. Oven
13. Vacuum pump
14. Moisture content cans
15. Glass rod
16. Duster and grease

3.4.5 TESTING PROGRAM

1. Clean and dry the pycnometer bottle. Screw its cap and take its weight W_1 to the nearest 0.01 g (Table.3.3).
2. Unscrew the cap and add a sufficient amount of de-aired water in the pycnometer up to one-third of its height.
3. Pour about 250 g of oven-dried soil (passing 4.75 mm sieve) in the pycnometer bottle. Add more water to cover the soil fully (up to one-half of the pycnometer height) and mix the soil water suspension. Screw the cap and determine the weight W_2 to the nearest of 0.01 g (Table 3.3).
4. Shake the pycnometer bottle and connect it to a vacuum pump to remove the entrapped air by vacuum or by boiling a water bath or sand bath for about 20 minutes for fine grained soils (silt dominated) and about 10 minutes for coarse grained soils (sandy soils).
5. When all the entrapped air is removed, disconnect the vacuum pump.
6. Remove the cap and fill the pycnometer bottle with de-aired water about three-fourths full. Reapply the vacuum for about 5 minutes until air bubbles stop appearing on the surface of the water.
7. Disconnect the vacuum pump and fill the pycnometer bottle with water completely and screw the cap on again. Dry it from the outside and take its weight W_3 to the nearest 0.01 g (Table 3.3).
8. Empty the pycnometer bottle and clean it thoroughly and dry it. Cross-check the weight of the cleaned, dry pycnometer bottle as in Step 1.

TABLE 3.3
Observation Data Sheet and Analysis for Specific Gravity by Pycnometer Method

Project/Site Name:	Date:
Test Site: Job. No.	Sample No.
Sample Sample Recovery Method:	
Recovery	
Depth:	

Sample Description:

Tested By:

Lab. Temp.: $\rho_w = 1.00$ g/cm^3	Density of water (27°C):
24°C	1 g/cm^3
Density of water at test lab. Temperature	0.997296 g/cm^3

Sl. No. Observations and Calculations	Trial Ns.		
	1	2	3

Observation

1   Pycnometer No.

2   Room Temperature

3   Weight of empty pycnometer bottle, W_1 (g)

4   Weight of pycnometer bottle + dry soil, W_2 (g)

5   Weight of pycnometer bottle + soil + water, W_3 (g)

6   Weight of pycnometer bottle + water, W_4 (g)

Calculations

7   Weight of dry soil, $W_d = W_2 - W_1$ (g)

8   Wt. of water, $W_w = W_3 - W_2$ (g)

     Wt. of water, $W_w = W_4 - W_1$ (g)

9   Weight of water having the same volume as that of solids $[(W_4-W_1)-(W_3-W_2)] = [W_2-W_1)-(W_3-W_4)]$

10   Calculate G_s using formula: $G_s = \dfrac{W_2 - W_1}{(W_2 - W_1) - (W_3 - W_4)}$

11   Average value of $G_{slab\ temp}$

Difference between any two G-values should be less than 0.03, else repeat the test

12 Density of water. of water @ 24°C (g/cm^3)	0.997296
13 Density of water. of water @ 20°C (g/cm^3)	0.9982
14 Value of G_s at standard temp. (20°C)	

9. Fill the pycnometer bottle with de-aired water only. Screw on the cap tightly. Wipe it dry from outside and take its weight $W_4 = W_{bw}$ to the nearest of 0.01 g (Table 3.3).

10. Record the temperature during the test and plot a calibration curve (Figure 3.4) between temperatures versus corresponding weights of the pycnometer bottle with water.

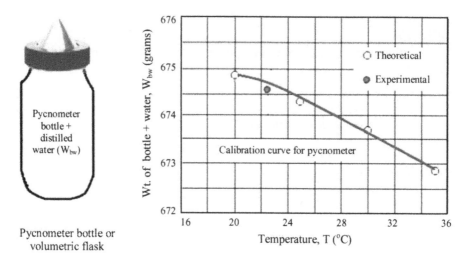

Pycnometer bottle or
volumetric flask

FIGURE 3.4   Typical calibration curve of pycnometer bottle.

Note: Calibration curves help in determining values of W_{bw} for any desired water temperatures (T°C) during the test. The calibration curve can be plotted by using the following equation for various temperatures against the weight of the pycnometer bottle and water:

$$W_{bw}(T_x°C) = \left[\frac{\rho_w * T_x}{\rho_w * T_{std}} * [(W_{bwTstd}) - W_b] \right] + W_b \qquad [3.4]$$

Where:

ρ_w (T_x°C) = Density of water identified at temperature (T_x°C see Table 6.2)
ρ_w (T_{std}°C) = Density of water identified at standard temperature (20°C)
W_{bw} = Weight of pycnometer bottle and water (g)
$W_b = W_1$ = Weight of pycnometer bottle (g)
T_{std}°C = Standard temperature of water (20°C), and
T_x°C = Any other desired or room temperature (°C).

3.4.6   OBSERVATION DATA SHEET AND ANALYSIS

Observations, test data analysis, and calculations for determination of specific gravity of soil solids by the pycnometer method are given in Table 3.3.

If room temperature is different than standard ($T_{std.}$= 20°C), then the corrected specific gravity is obtained as:

$$G_{s(T_{std})} = \left[G_{s(Lab.\ Temp)} * \frac{\text{Density of water at lab. Temperature (T°C)}}{\text{Density of water at standard Temperature (20°C)}} \right]$$

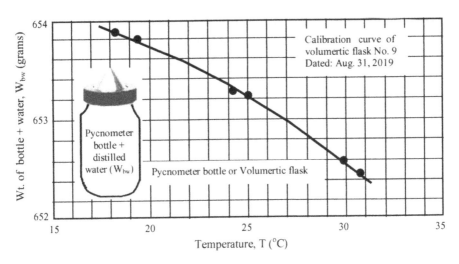

FIGURE 3.5   Calibration curve for pycnometer No. 9 for determination of specific gravity.

Generally, a calibration curve is plotted for a particular pycnometer bottle as the empty weight of bottles differs from one another. Figure 3.4 is a typical calibration curve and not valid for all bottles. Figure 3.5 shows a calibration curve for pycnometer No. 9, which will be used when determining the G-value using this pycnometer in the future.

3.4.7   RESULTS AND DISCUSSIONS

Based on test results, specific gravity of solids at ___°C = _____.

Soil type: ....

3.5   GENERAL COMMENTS

1. Specific gravity of soil solids is dependent upon on soil particles packing.
2. Generally, the specific gravity of granular soils varies in the narrow range between 2.61 to 2.67, with standard value of 2.65, which is the specific gravity of quartz mineral. However, granular soils exhibit specific gravity on the lower side compared to cohesive soils, which varies in the range of 2.69 to 2.84.
3. It may be noted that entrapped air in highly plastic and organic soils cannot be removed by vacuum. Therefore, the entrapped air may be removed by boiling the suspension in the density bottle for about 30 minutes gently. However, in such a case, distilled water may be added from time to time so that suspension may not dry during prolonged boiling. Also, vigorous boiling may be avoid loss of soil material out of the density bottle.
4. Presence of entrapped air can be detected by the movement of the surface of suspension upon the application and release of the vacuum.

5. The boiling to remove the air bubbles should not be too vigorous since soil may be carried out of the neck of the density bottle.
6. The heating of a pycnometer bottle to a high temperature is unwise because of the possibility of changing the shape of the bottle.
7. The presence of foreign matter such as soluble salts from the soil raises the boiling point slightly.

3.6 APPLICATIONS OF SPECIFIC GRAVITY

1. Specific gravity is often used for the computation of various parameters such as degree of saturation, theoretical dry density, void ratio, and concentration of substances in aqueous solutions indirectly. The specific gravity is also used in most of the laboratory tests such as to compute the grain-size distribution in a hydrometer analysis and the soil's void ratio in the compaction tests:
   • For hydrometer analysis, the G-value is used for analysis of particle size as:

$$D = \sqrt{\frac{30\mu}{(\rho_s - \rho_w)}} * \sqrt{\frac{l}{T}} = \sqrt{\frac{30\mu}{(G_S - 1)\rho_w}} * \sqrt{\frac{L}{t}} = K\sqrt{\frac{L}{t}} \qquad [3.5]$$

Where: D = Diameter of sphere/particle (mm).
   • Void ratio in the compaction tests:

$$e = \left[\frac{G_s(1 + w)}{\rho_b} * \rho_w - 1 \right] = \left[\frac{G_s * \rho_w}{\rho_d} - 1 \right] \qquad [3.6]$$

Where: ρ_b = Bulk/moist density of soil mass, and
   • ρ_d = Dry density of soil mass.
   • Its value also helps in identification and classification of soils.
   • It gives an idea about the suitability of the soil as a construction material; e.g. the higher the value of "G_s" the more strength of the soil material.
2. Specific gravity is also required for design applications in hydraulic structures.
3. Specific gravity is often used by geologists and mineralogists for the determination of mineral content of a rock samples, generally known as relative density.

3.7 SOURCES OF ERROR

• Two sources of important experimental errors are non-uniform temperatures and incomplete removal of air entrapped in the soil. Non-uniform temperature difficulties can easily be prevented by allowing a warmed bottle to stand overnight to adjust itself to room temperature.
• The boiling procedure is normally sufficient to remove air entrapped within the soil. However, the boiling process by about 10 minutes should not be too vigorous since the soil may be out the neck of the bottle.

- Incomplete de-airing of suspension may lead to an underestimate of the value of the specific gravity, which is a serious error and should be avoided.
- Since the soil material is complex and heterogeneous and the specific gravity values for all the individual soil grains are not the same, it is therefore recommended that the specific gravity should be determined in the laboratory as per standard codal procedures rather than assuming according to given typical values.

3.8 PRECAUTIONS

- Make sure that the soil sample is dry.
- The density bottle should be carefully cleaned and dried before the test, along with its stopper.
- The density bottle with a dry soil sample should be carefully filled with water just to cover the soil sample.
- The air bubbles should be removed completely from the density bottle and the pycnometer bottle.
- Once the air bubbles are removed, then carefully clean the density bottle and the pycnometer bottle completely for filling distilled water.
- Record temperature fluctuations during the test carefully.

REFERENCES

ASTM D854-14. 2014. "Standard Test Methods for Specific Gravity of Soil Solids by Water Pycnometer." ASTM International, West Conshohocken, PA, www.astm.org.

ASTM D5550-14. 2014. "Standard Test Method for Specific Gravity of Soil Solids by Gas Pycnometer." ASTM International, West Conshohocken, PA, www.astm.org.

BS 1377-2. 1990. "Methods of Test for Soils for Civil Engineering Purposes-Part 2: Classification Tests for Determination of Water Content." British Standards, UK.

IS: 2720 (Part 3(1)). 1980. "Method of Test for Soils: Determination of Specific Gravity of Fine Grained Soils." Bureau of Indian Standards, New Delhi.

IS: 2720 (Part 3(2)). 1980. "Method of Test for Soils: Determination of Specific Gravity of Fine, Medium and Coarse Grained Soils." Bureau of Indian Standards, New Delhi.

Reddy, Krishna R. 2000. "*Engineering Properties of Soils Based on Laboratory Testing.*" Department of Civil and Materials Engineering University of Illinois at Chicago, USA.

4 Particle or Grain Size Distribution of Soils by Sieve Analysis

References: IS: 1498; IS: 2720-Part 1; IS: 2720-Part 4; ASTM D421; ASTM D 422;
ASTM D 2217; ASTM D 2487; ASTM D2488; BS 1377: Part 2, 1990

4.1 OBJECTIVES

The main objective of performing the sieve or soil grading is to analyze, measure, and classify the soil particles in a given soil specimen. The soil grading or PSD is also indicative of engineering properties wherein it can be assessed whether a soil mass is stable or not. Thus, PSD plays a vital role in the identification and classification of soils for sustainable development.

4.2 INTRODUCTION

Sieve analysis or particle size analysis or soil grading of a soil sample is helpful to analyze, measure, and classify the soil particles in a given soil specimen in soil engineering practice. In a given soil mass, there will be soil particles of different sizes, and the size of these soil particles is measured and quantified by means of PSD. The PSD plays a vital role in the identification and classification of soils to ascertain their performance for the sustainable development in civil engineering practice (Reddy, 2000). In the process of soil grading, any given soil sample is sieved through a desired set of sieves for the identification and classification of different soil particles such as gravels (4.75–80 mm), sands (0.075–4.75 mm), silts (0.002–0.075 mm), and clays (<0.002 mm). A prominent sieve size of 75 µm or 0.075 mm sieve (No. 200) is used to classify a given soil sample either as fine grained or as coarse grained on the basis of the percentage passing through this sieve in the process of particle size distribution analysis. Therefore, a given soil sample can be classified as a fine grained (silt + clay) if the percentage passing through 75 µm sieve is more than 50%. However, if the percentage passing through the 75 µm sieve is less than 50% or if the percentage retained in a 75 µm sieve is more than 50%, the soil sample is considered coarse grained (sand + gravels). Likewise, another prominent sieve size of 4.75 mm is used to classify a coarse grained soil sample either as sand dominated or gravel dominated on the basis of the percentage passing through this sieve in PSD. Therefore, a given coarse grained soil sample can be classified as sand dominated if the percentage passing through a

4.75 mm sieve is more than 50%. However, if the percentage passing through 4.75 mm sieve is less than 50% or if the percentage retained in a 4.75 mm sieve is more than 50%, the soil sample is considered gravel-dominated coarse grained soil. It may be noted that the above-mentioned sieve analysis can be conducted either as a dry sieve analysis if it is presumed that there is negligible fine content present in the soil sample. If it seems that a large amount of fine content is present in the soil sample and larger particles are covered with appreciable amount of fine contents which cannot be separated by dry sieving, then it is recommended that wet sieve analysis be carried out. Thus, in the case of wet sieve analysis, the sample is first washed over a 75 μm sieve to remove silt and clay particles sticking with sand/gravel particles. A washed sample retained in a 75 μm sieve is dried in an oven (@ 105°C–110°C temp.) and then further subdivided by the set of sieves. Dry sieve analysis is essential to obtain a correct particle size distribution of washed and dried sand since during wet sieving, material just smaller than the sieve opening is often retained on a wet sieve due to surface tension.

However, it may also be noted that fine grained soils (silt + clay) cannot be separated by means of sieve analysis as of now. These can be separated by hydrometer or sedimentation analysis (Test No.5) to ascertain whether a given fine grained soil sample is silt dominated or clay dominated on the basis of percentage proportion (>50%).

4.3   DEFINITIONS AND THEORY

In soil engineering practice, a given soil sample is sieved through a specified set of sieves by means of soil grading, generally referred to as particle size distribution (PSD) to ascertain the distribution of soil particles over a wide range of sizes present in the soil mass. The sieving process is generally carried out in three methods:

a. **Dry sieve analysis or mechanical analysis:** This method is usually adopted for coarse grained soils (sands and gravels) retained in a 75 μm sieve. This is also referred to as dry sieve analysis when it is presumed that the soil particles present in a given soil sample are clean and no appreciable fine content is stuck on their surface (Figure 4.1).

b. **Wet sieve analysis:** This method is also adopted for coarse grained soils (sands and gravels) retained in a 75 μm sieve when it is presumed that there is appreciable fine content present in a given soil sample, which cannot be separated by mechanical sieve analysis and soil particles need to be washed in a 75 μm sieve until clean water flows through the sieve. The wet soil particles are then dried in the oven as per standard procedure before further sieving process.

c. **Hydrometer or sedimentation analysis:** This method is usually adopted for fine grained soils (silts and clays) passing through a 75 μm sieve. Unlike coarse grained soils (sands and gravels), fine grained soils (silts and clays) cannot be separated by sieve analysis as of now; therefore, hydrometer analysis is carried out to ascertain whether a fine grained soil sample is silt or clay dominated. The hydrometer analysis of fine grained soils is given in Test No. 5.

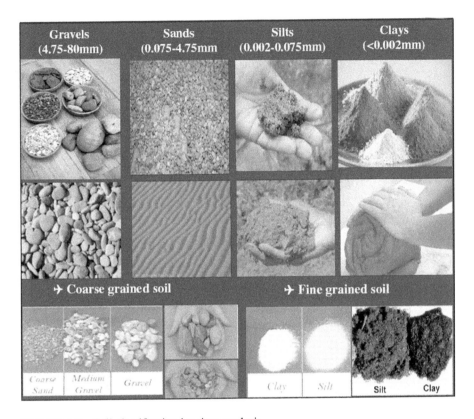

FIGURE 4.1   Soil classification by sieve analysis.

In soil engineering practice, a coarse grained soil sample comprises different soil constituents such as gravels and sands of different shapes such as angular, sub-rounded, rounded, bulky, etc.. However, coarse grained soil comprised of hard rock particles such as boulders (>300 mm) and cobbles (<300 >80 mm) are not considered in the sieve analysis process. The sizes of coarse grained soils based on soil grading are given as below (Figure 4.2):

- **Coarse gravel:** less than 80 mm and greater than 60 mm (<80 mm >60 mm sieve size)
- **Medium gravel:** less than 60 mm and greater than 20 mm (<60 mm >20 mm sieve size)
- **Fine gravel:** less than 20 mm and greater than 4.75 mm (<20 mm >4.75 mm sieve size)
- **Coarse sand:** less than 4.75 mm and greater than 2.00 mm (<4.75 mm >2.00 mm sieve size)
- **Medium sand:** less than 2.00 mm and greater than 0.0425 mm (<2.00 mm >0.425 mm sieve size)
- **Fine sand:** less than 0.425 mm and greater than 0.075 mm (<0.425 mm >0.075 mm sieve size)

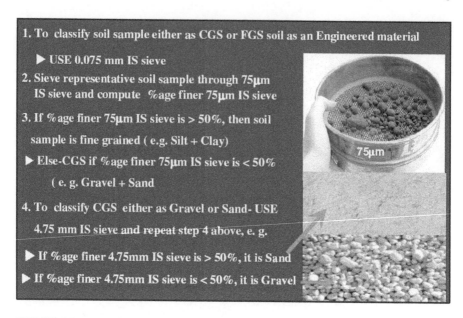

1. To classify soil sample either as CGS or FGS soil as an Engineered material

   ▶ USE 0.075 mm IS sieve
2. Sieve representative soil sample through 75μm
   IS sieve and compute %age finer 75μm IS sieve
3. If %age finer 75μm IS sieve is > 50%, then soil

   sample is fine grained (e.g. Silt + Clay)
   ▶ Else-CGS if %age finer 75μm IS sieve is < 50%

   (e. g. Gravel + Sand
4. To classify CGS either as Gravel or Sand- USE

   4.75 mm IS sieve and repeat step 4 above, e. g.

   ▶ If %age finer 4.75mm IS sieve is > 50%, it is Sand
   ▶ If %age finer 4.75mm IS sieve is < 50%, it is Gravel

FIGURE 4.2 Classification of coarse grained soils by sieve analysis.

Further, based on sieve analysis, soils are classified either as well graded or poorly graded and assigned distinct symbols (according to geotechnical language e.g. assigning a distinct symbol to a soil type according to their presence and behavior) as per standard codal procedures (ASTM D 2487, BS 1377-2 (1990), IS: 1498, IS: 2720-part 4) as below.

  i. Less than 5% particles of size smaller than 0.075 mm, the soil are classified as:
- GW— well-graded gravels if $C_u > 4$ and $C_c = 1 - 3$ (If $C_u = 1$: uniform soil)
- SW— well-graded sands if $C_u > 6$ and $C_c = 1 - 3$
- GP— poorly graded gravels if $C_u > 4$ and $C_c \neq 1 - 3$
- SP— poorly graded sands if $C_u > 6$ and $C_c \neq 1 - 3$

Where:

$$C_u = \left[\frac{D_{60}}{D_{10}} \right] \ \& \ C_c = \left[\frac{D_{30}^2}{D_{10} * D_{60}} \right] \qquad [4.1]$$

D_{10} = Diameter at 10% finer, also known as effective size
D_{30} = Diameter at 30% finer, and D_{60} = Diameter at 60% finer, respectively.
  ii. *"C_u" expresses the qualitative uniformity of a soil, whereas "C_c" expresses general shape of particle size distribution curve.*
 iii. If more than 12% of particles are smaller than 0.075 mm, the soil is classified as:

- GM: silty gravel if plasticity index falls below A-line of plasticity chart
- GC: clayey gravels if plasticity index falls above A-line of plasticity chart
- SM: silty sand if plasticity index falls below A-line of plasticity chart
- SC: clayey sand if plasticity index falls above A-line of plasticity chart

iv. Soils passing a 75 μm sieve between 5% and 12% are designated by dual symbols, depending upon the amount of maximum component of soil type such as:
- GW-SW: well-graded gravels with well-graded sand binder
- GW-GC: well-graded gravels with clay binder
- Other dual symbols used could be: GM-GC, GW-GP, SW-SP, SW-SC, and so on.
  Thus, we see that both gravels and sands are divided into three categories depending on the percentage of fines (percentage passing 0.075 mm sieve size):

- If <5% fines: Use two-letter group symbol to describe gradation (well or poorly graded), for example, GW: well-graded gravels with little or no fines
- If 5–12% fines: Use four-letter group symbol to describe both gradation and type of fines (e.g. Silt-M and clay-C, example GW-GC: well-graded gravels with clay binder)
- If >12% fines: Use two-letter group symbol to describe type of fines (silty or clay), example CH: inorganic clay of high compressibility

Fine grained soils may be designated by symbols: ML, CL, OL, HM, CH, MI, CI, respectively. These symbols are selected on the basis of liquid limit, plasticity index, and plasticity chart. Fine grained soils are also designated by dual symbols like CL-CI, OL-OI, CI-OI, CL-OL, CL-ML, CI-MI, MI-OI, respectively, when the plasticity index (PI) is between 4 and 7 and limits plotting above (C) or below (M) A-line as per the plasticity chart.

4.4 METHOD OF TESTING

Standard sieves of different sizes (from 100 mm to 0.075 mm) specified in different groups (e.g. set of sieves for coarse gained soils comprises of: 80, 60, 20, 10, 6.3, 4.75, and bottom pan; set of sieves for fine grained soils comprises of: 4.75, 2.0, 1.0, 0.6, 0.425, 0.075 mm, and bottom pan) are used for soil grading as per standard codal procedures.

4.5 SOIL TESTING MATERIAL

A disturbed or remolded soil sample of about 20 kg in the particle size range of 80 mm to 75 μm is required for sieve analysis. Soil particles above 80 mm size are replaced or rejected.

4.5.1 QUANTITY OF SOIL SPECIMEN

A sufficient amount of material is required to obtain a representative gradation, and depending on the maximum particle size present in substantial quantities, shall not be less than the amount shown in tabular form as below:

Diameter of Largest Particle Present in Substantial Quantities (mm)	Approximate Minimum Weight of Portion (Kg)	Diameter of Largest Particle Present in Substantial Quantities (mm)	Approximate Minimum Weight of Portion (Kg)
80	60.00	10.00	1.50
40	25.00	6.50	0.75
25	13.00	4.75	0.50
20	6.50	2.00	0.10
12.50	3.50	0.075	0.05

If substantial portion of material just passes the 4.75 mm IS sieve, then 200 g of the test sample shall be taken. This quantity may further be reduced if the largest size is smaller.

4.6 TESTING EQUIPMENT AND MATERIALS

1. Soil sample passing through a 80 mm IS sieve and retaining 0.075 mm IS sieve
2. 1st set of sieves of sizes 100 mm, 80 mm, 60 mm, 40 mm, 20 mm, 10 mm, 6.3 mm, and 4.75 mm
3. 2nd set of sieves of sizes 4.75 mm, 2 mm, 1 mm, 0.60 mm 0.425 mm, 0.30 mm, 0.15 mm, and 0.075 mm (set of sieves of standard sizes are given in Table 4.1)
4. Laboratory reagents: sodium hexametaphosphate, hydrogen peroxide, and hydrochloric acid (for cohesive soils)
5. Mechanical sieve shaker (optional)
6. Rubble pestle for pulverizing soil samples
7. Balances – Balance with a sensitivity and readability to 0.5 g and another with a sensitivity and readability to 0.01 g
8. Weights and weight box
9. Oven, desiccators, drying crucibles, trays/buckets, stirring apparatus, brush
10. Water (distilled)
11. Stopwatch

4.7 TESTING PROGRAM

IS 2720 (Part-4)—1985 lays down the step-by-step procedure for sieving soils with a particle size greater than a 0.075 mm sieve in two ways:

TABLE 4.1
Standard Sieve Equivalents

I.S. Sieves (460–1962)		ASTM Sieves (1961)		B.S. Sieves (1962)		International (I.S.O.)
Designation	Aperture (mm)	Designation	Aperture (mm)	Designation	Aperture (mm)	Aperture (mm)
50 mm	50	2 in.	50.80	2 in.	50.80	–
40 mm	40	1.5 in.	38.10	1.5 in.	38.10	–
20 mm	20	¾ in.	19.00	¾ in.	19.05	–
10 mm	10	3/8 in.	9.51	3/8 in.	9.52	10.00
5.6 mm	5.60	3.5 No	5.60	–	–	5.60
4.75 mm	4.75	4	4.76	3/16 in.	4.76	–
4.00 mm	4.00	–	–	–	–	4.00
2.80 mm	2.80	7	2.80	6	2.80	2.80
2.36 mm	2.36	8	2.38	7	2.40	–
2.00 mm	2.00	10	2.00	8	2.00	2.00
1.40 mm	1.40	14	1.41	12	1.40	1.40
1.00 mm	1.00	18	1.00	16	1.00	1.00
850 μ	0.85	20	0.85	18	0.85	0.85
710 μ	0.71	25	0.707	22	0.710	0.710
600 μ	0.600	30	0.595	25	0.600	–
500 μ	0.500	35	0.500	30	0.500	0.500
425 μ	0.425	40	0.420	36	0.420	–
355 μ	0.355	45	0.354	44	0.355	0.355
300 μ	0.300	50	0.297	52	0.300	–
250 μ	0.250	60	0.250	60	0.250	–

(Continued)

TABLE 4.1 (Continued)

I.S. Sieves (460–1962)		ASTM Sieves (1961)		B.S. Sieves (1962)		International (I.S.O.)
Designation	Aperture (mm)	Designation	Aperture (mm)	Designation	Aperture (mm)	Aperture (mm)
212 μ	0.212	70	0.210	72	0.210	–
180 μ	0.180	80	0.177	85	0.180	0.180
150 μ	0.150	100	0.149	100	0.150	–
125 μ	0.125	120	0.125	120	0.125	0.125
90 μ	0.090	170	0.088	170	0.090	0.090
75 μ	0.075	200	0.074	200	0.075	–
63 μ	0.063	230	0.063	240	0.063	–
45 μ	0.045	325	0.044	350	0.045	–
37 μ	0.037	400	0.037	400	0.037	–
25 μ	0.025	500	0.025	500	0.025	–

1 μ (micron)
= 10^{-3} mm
= 10^{-4} cm
= 10^{-6} m

1 Å = 10^{-10} m = 3.28×10^{-10} ft

a. Dry Sieve Analysis: to be adopted for those soils with less than 5% fine content or generally soils with no appreciable fine content stuck to its particles

b. Wet Sieve Analysis: applicable to all soils

IS 2720 (Part-4)—1985 lays down the step-by-step procedure for sieving soils with a particle size smaller than a 0.075 mm sieve in three ways:

1. Hydrometer or sedimentation analysis
2. Pipette analysis
3. Plummet balance method

For determination of soil grading, the following steps are followed:

1. Take suitable quantity of oven-dried soil sample (of about W = 3,000 g)
2. If it is seen that the soil sample taken contains more than 5% of fines (silt-clay), then spread out the soil sample in a large tray and soak in distilled water.
3. Add about 2 g of sodium hexametaphosphate to the soil for every liter of water used in soaking and stir it thoroughly and leave for further soaking.

   Note: One gram of sodium hydroxide and 2 g of sodium carbonate per liter of water, or any other suitable dispersing agent can also be used. The amount of dispersing agent may vary depending on the type of soil. A dispersing agent may not be required in the case of all soils; in such cases, the wet sieve analysis may be carried out without the addition of a dispersing agent.
4. Wash the soaked soil sample in a 0.075 mm (75 μm) sieve until the water passing through the sieve is clean.
5. The soil particles retained in the 75 μm sieve are carefully collected in a metal tray and dried in the oven at 105°C–110°C for 24 hours and weighed.
6. The loss in soil weight will give the percentage passing through a 75 μm sieve, which is kept for hydrometer analysis.
7. Part-A. Soil grading for coarse grained soils (>4.75 mm size) (e.g. coarse grained soils: CGS)
   a. Clean the 1st set of sieves and the bottom pan properly and stack these in descending order with the pan at the bottom of all sieves.
   b. Sieve the soil sample through the 1st set of sieves (e.g. from 80 mm to 4.75 mm as shown in Figure 4.3a) manually or using a sieve shaker (Figure 4.3) for about 10 minutes and ensure that the top sieve is tightly covered and there is no loss of soil from any of these sieves.
   c. After sieving, take the weight of the soil sample retained in each sieve carefully (nearest to 1 g accuracy).
   d. Also, check the weight of the soil sample passed through the sieves and retained in the bottom pan.

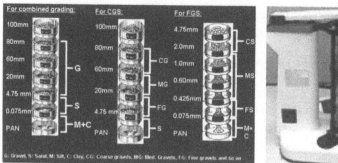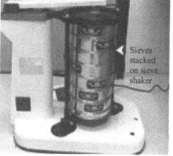

FIGURE 4.3   Set of IS sieves and a sieve shaker for analysis of particle size distribution of soils.

> **Note:** In the case of the dry sieving process, check that the weight of the soil sample retained and passed is checked against the original weight of sample taken for sieving to ascertain any loss of soil material while sieving.
>
> e. Now compute the percentage of soil mass or weight retained on each sieve, as shown in Table 4.2a (col. 3), followed by the cumulative percentage retained on each sieve, as shown in Table 4.2a (col. 4).
>
> f. Now compute the percentage finer or percentage passing through each sieve by deducting the cumulative percentage retained in each sieve from 100, as shown in Table 4.2a (col.5).
>
> g. Plot the particle size distribution curve between particle size (along x-axis) and percentage passing (along y-axis), as shown in Figure 4.4. This completes the sieve analysis for coarse grained soils. From this curve, the percentage of different soil constituents and different diameters (D_{85}, D_{60}, D_{50}, D_{45}, D_{30}, D_{20}, D_{10}, etc..) obtained through various design criterion parameters (C_u, C_c, S_n, filter criteria, etc..) can be calculated.
>
> 8. Part-B. Sieving for soil passing through a 4.75 mm sieve and retained in a 75 µm sieve
>
>   a. Clean the 2nd set of sieves (Figure 4.3b) and pan with brush and stack these in descending order with the pan at the bottom of all sieves.
>
>   b. Take a suitable quantity of oven-dried soil sample (of about $W_1 = 200$ g).
>
>   c. Sieve the soil sample through the 2nd set of sieves, as shown in Figure 4.3b (4.75 mm, 2 mm, 1 mm, 0.60 mm, 0.425 mm, 0.30 mm, 0.15 mm, and 0.075 mm for grading of fine grained soils), manually or using a sieve shaker for about 10 minutes and ensure that the top sieve is tightly covered and there is no loss of soil from any of these sieves.
>
>   d. After sieving, take the weight of the soil sample retained on each sieve carefully (nearest to 0.1 g accuracy).
>
>   e. Now compute the percentage retained on each sieve as shown in Table 4.2b (col. 3), followed by the cumulative percentage retained on each sieve, as shown in Table 4.2b (col. 4).

TABLE 4.2A

Observations and Data Sheet for Grain Size Analysis (for Sample OP3 at Depth 1.5 m: See Figure 4.4) for Soils Coarser than a 4.75 mm Sieve. Part-A: Soil Coarser than a 4.75 mm IS Sieve

Weight of soil sample taken, W (g)3000					Sample No. OP3	Depth: 1.5 m

Sieve size (mm)	Weight retained (g)	%age retained	Cum. %age retained	%age finer	Remarks
50	15	52	1.73	98	1. %age finer = 100-cum.%age retained
40	21	73	2.43	98	2. Particle size distribution (PSD)
31.5	52	125	4.17	96	curve is plotted between col. 1 and 5.
25	110	235	7.83	92	3. From test data, it is seen that soil is
20	135	370	12.33	88	fine grained with 71% passing
16	115	485	16.17	84	through a 4.75 mm IS sieve.
12.5	137	622	20.73	79	4. Sieve analysis for soils passing
10	109	731	24.37	76	through a 4.75 mm IS sieve is given
6.3	119	850	28.33	72	in Table 4.4b.
4.75	31	881	29.37	71	

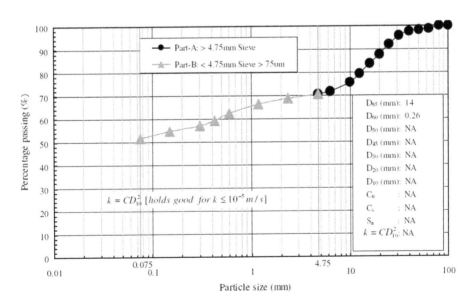

FIGURE 4.4 Particle size distribution curve by sieve analysis.

TABLE 4.2B

Part-B: Observations and Data Sheet for Grain Size Analysis (Figure 4.4) for Soils Passing 4.75 mm Sieve and Retained in 75 μm Sieve

Weight of soil sample taken: $W_1 = 200$ g

Sieve size (mm)	Weight retained (g)	%age retained	Cum. %age retained	%age finer w.r.t. (W_1)	%age finer w.r.t. (W_M)
Col. 1	2	3	4	5	6
4.75	0	0	0	100	71
2.36	5	5	2.5	98	69
1.18	7	12	6	94	66
0.6	11	23	11.5	89	63
0.425	9	32	16	84	59
0.3	6	38	19	81	57
0.15	7	45	22.5	78	55
0.075	9	54	27	73	52

Sample No. OP3   Depth: 1.5 m

Remarks 1.%age passing w.r.t. (W_M) = %age passing w. r. t. W_1 (Col. 5)*%age passing 4.75 IS sieve from Table 4.4a, e.g.= Col.5*0.712. %age passing 75 μm sieve is 52%. Therefore, hydrometer analysis needs to be done to specify the size of silt and clay on the material passing 75 μm (Refer practical No. 5 on hydrometer analysis)

%age passing 4.75 mm sieve (from Table 4.2a) = 71% = 0.71
Total weight of soil sample taken (Table 4.2a) = 3,000 g
Weight of soil sample taken for sieve analysis passing 4.75 mm sieve, W_1 = 200 g
Data observations and calculations are given in Table 4.2b as below.

f. Now compute the percentage finer or percentage passing through each sieve by deducting the cumulative percentage retained on each sieve from 100, as shown in Table 4.2b (col. 5).

g. Now, calculate the %age passing w.r.t. total weight for plotting combined particle size distribution curve for soil sample coarser than and passing through 4.75 mm, as explained in Table 4.2b (col. 6), e.g. %age finer w.r.t. Total weight (W_M) is given as:
    %age passing w.r.t. (W_M) = %age passing w.r.t. W_1 (col. 5)* %age passing 4.75 mm IS sieve from Table 4.4a
    = Col.5 (Table 4.4b)*0.71 reported in Col.6 (Table 4.4b)

h. Thus, the data for plotting combined particle size distribution curve for soils coarser than 4.75 mm sieve and passing 4.75 mm sieve is given in Table 4.2c.

i. Plot the particle size distribution curve between particle size (along x-axis) and percentage passing (along y-axis), as shown in Figure 4.4. This completes the sieve analysis for coarse grained soils. From this curve, the percentage of different soil constituents and different diameters (D_{85}, D_{60}, D_{50}, D_{45}, D_{30}, D_{20}, D_{10}, etc.) can be obtained through various design criterion parameters (C_u, C_c, S_n, filter criteria, etc.) and can be calculated.

TABLE 4.2C
Observations and Data Sheet for Combined Grain Size Analysis (Figure 4.5)

Total weight of soil sample taken, W (g): 3,000; Sample No. OP3; Depth: 1.5 m

Sieve size (mm)	Weight retained (g)	%age retained	Cum. %age retained	%age finer	Remarks
50	15	52	1.73	98	From PSD curve,
40	21	73	2.43	98	find:
31.5	52	125	4.17	96	
25	110	235	7.83	92	D_{85} (mm):
20	135	370	12.33	88	D_{60} (mm):
16	115	485	16.17	84	D_{50} (mm):
12.5	137	622	20.73	79	D_{45} (mm):
10	109	731	24.37	76	D_{30} (mm):
6.3	119	850	28.33	72	D_{20} (mm):
4.75	31	881	29.37	71	D_{10} (mm):
2.36	5	5	2.5	69	C_u :
1.18	7	12	6	66	C_c :
0.6	11	23	11.5	63	C_c :
0.425	9	32	16	59	S_n :
0.3	6	38	19	57	k :
0.15	7	45	22.5	55	Filter design
0.075	9	54	27	52	parameters & so on

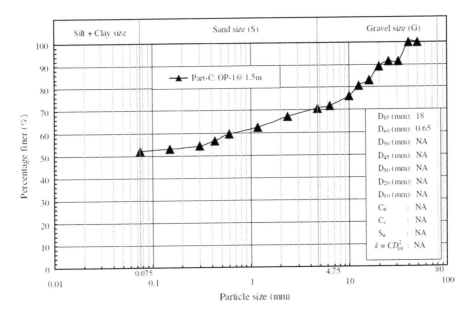

FIGURE 4.5 Particle size distribution curves for different types of fine grained soils (since % passing 75 μm is > 50%).

TABLE 4.3

Observations and Data Sheet for Grain Size Analysis for Combined Set of Sieves

Total weight of soil sample taken, W (g): 3,000; Sample No.: OP1; Depth: 1.5 m

Sieve size (mm)	Weight retained (g)	%age retained	Cum. %age retained	%age finer	Remarks
50	0	0	0	100	From particle size distribution
40	0	0	0	100	curve, it is seen that %age
31.5	252	252	8.4	92	passing 75 μm (0.075 mm IS
25	0	252	8.4	92	sieve) is 52%, which is more
20	73	325	10.8	89	than 50%. Therefore, based on
16	179	504	16.8	83	soil grading, the soil sample is
12.5	75	579	19.3	81	classified as poorly graded fine
10	140	719	24.0	76	grained soil.
6.3	125	844	28.1	72	
4.75	33	877	29.2	71	
2.36	111	988	32.9	67	
1.18	137	1,125	37.5	63	
0.6	93	1,218	40.6	59	
0.425	87	1,305	43.5	57	
0.3	67	1,372	45.7	54	
0.15	36	1,408	46.9	53	
0.075	29	1,437	47.9	52	

9. Part-C. Else, a combined set of sieves (Figure 4.3c) can be used for sieve analysis and a single curve (Figure 4.5) will be plotted directly. The data for single particle size distribution sieve analysis is given in Table 4.3.

10. The material passing 75 μm is used for hydrometer analysis.

4.8 OBSERVATION DATA SHEET AND ANALYSIS

The test results of sieve analysis are presented in a graphical plot between particle size (mm) along the x-axis versus percent finer (%) along the y-axis. It may be noted that in this graphical plot, the particle size (mm) is plotted on a logarithmic scale. This is because a soil may have particles larger than 200 mm to less than 0.01 μm. Plotting of these sizes on a natural scale would require a very large sheet of paper. Hence, it is customary to plot particle sizes on the logarithmic scale on the abscissa (x-axis) and percentage finer on the natural scale as the ordinate (y-axis). Such a plot is known as the particle size distribution curve. A semi-log graph of four or five cycles is usually adequate for this purpose. Therefore, to find the percentage of soil particles passing through each sieve, first find the percent retained on each sieve by using the following equation:

TABLE 4.4

Observations and Data Sheet for Grain Size Analysis (Practice Sheet)

Total weight of soil sample taken, W (g):   ; Sample No.   ; Depth:

Sieve size (mm)	Weight retained (g)	%age retained	Cum. %age retained	%age finer	Remarks

$$\% \text{ Retained} = \left[\frac{W_{\text{retained on sieve}}}{W_{\text{total weight of sample}}} * 100 \right] \qquad [4.2]$$

Where: $W_{\text{retained on Sieve}}$ is the weight of soil sample in the sieve retained after sieving, and $W_{\text{Total weight of sample}}$ is the total weight of the soil sample taken for sieve analysis.

Once the percent retained on each sieve is computed, then the cumulative percentage of soil particles retained on each sieve is found by adding up the total amount of percentage retained on each sieve (Table 4.2a & b). The cumulative percent passing of the aggregate is found by subtracting the percent retained from 100%:

$$\% \text{cumulative passing} = 100 - \% \text{cumulativer retained} \qquad [4.3]$$

4.9 RESULTS AND DISCUSSIONS

Based on test results, the following soil grading parameters from the PSD curve are:

1. Gravel size ranges between (percentage passing): 100–71 = 29%
2. Sand size ranges between: 71–52 = 19%
   a. Coarse sand ranges between: 71–69 = 02%
   b. Medium sand ranges between: 69–60 = 09%
   c. Fine sand ranges between: 60–57 = 03%
3. %age passing 0.075 mm sieve (silt + clay content): 52% (>50%: fine grained soils)

⇒ Silt-clay content 52% >12%: Hydrometer analysis is required

Soil classification, probably: GC/GM (insufficient data to correctly classify the sample)

The data for single particle size distribution sieve analysis is given in Table 4.3 (refer section 4.7, step 9). Similarly, a practice sheet is given for sieve analysis in Table 4.4 to be completed by the students.

4.9.1 CLASSIFICATION PARAMETERS

Some important parameters are obtained from the PSD curve, which are useful in specifying certain engineering criteria in terms of quantitative measures such as:

1. **Effective size (D_{10}):** The size usually defined for sands corresponding to 10% finer is known as Hazen's effective size. Higher effective size indicates coarser soil.
2. **Coefficient of uniformity (C_u):** The coeff. of uniformity expresses the qualitative uniformity/ spread of the range of the particle sizes of a soil, which is defined as the ratio of D_{60} to D_{10}, given as:

$$C_u = [D_{60}/D_{10}] \qquad [4.4]$$

A larger uniformity coefficient means well-graded soil and its value close to one means uniform or poorly graded soil.

3. **Coefficient of curvature or gradation (C_c):** The coeff. of curvature "C_c" expresses general shape of particle size distribution curve, which is defined as the ratio of:

$$C_c = [D_{30}^2/(D_{60} * D_{10})] \qquad [4.5]$$

Where: D_{10} = Diameter at 10% finer, also known as effective size
D_{30} = Diameter at 30% finer, and
D_{60} = Diameter at 60% finer, respectively.

Thus, the data for plotting a combined PSD curve for soils coarser than 4.75 mm sieve and passing 4.75 mm sieve is given in Table 4.2c.

Based on test results, the following soil grading parameters from the PSD curve are:

1. Gravel size ranges between: 100–71 = 29%
2. Sand size ranges between: 71–52 = 19%
   a. Coarse sand ranges between: 71–68 = 3%
   b. Medium sand ranges between: 68–56 = 12%
   c. Fine sand ranges between: 56–52 = 4%

3. %age passing 0.075 mm sieve (silt + clay content): 52%

⇒ Silt-clay content >50%: Hydrometer analysis is required
Soil classification-probably: Fine grained soil with appreciable gravel content
From data observations, it is seen that the soil is poorly graded with appreciable fines (silt + clay content >50%). Therefore, a hydrometer analysis is required for complete gradation analysis and classification of the soil sample.

Generally, based on sieve analysis, typical particle size distribution curves are described as shown in Figure 4.6.

4.9.2 Practice Sheet

An observations and data sheet (Table 4.4) and graph sheet (Figure 4.7) is also given so that students can plot a PSD curve and obtain various parameters for

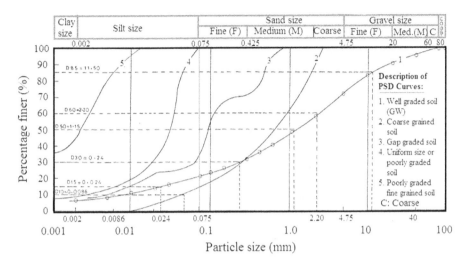

FIGURE 4.6 Typical particle size distribution curves for different types of soils.

classification of soil. After finding all the gradation parameters (e.g. soil components and diameters: D_{85}, D_{60}, D_{50}, D_{45}, D_{30}, D_{20}, D_{10}, S_n, k, etc.), soil can be classified and recommended as per requirement.

From PSD curve:

1. Gravel size ranges between: %
2. Sand size ranges between: %
   a. Coarse sand ranges between: %
   b. Medium sand ranges between: %
   c. Fine sand ranges between: %

3. %age passing 0.075 mm sieve (silt + clay content):

⇒ Silt-clay content %: Hydrometer analysis is required

Soil classification, probably: ....

4.10 GENERAL COMMENTS

a. Soil grading is a very tedious process of separating of soil particles of different sizes and classifying these soils either as coarse grained or fine grained soils. Therefore, it is suggested that based on visual inspection of soil samples and the amount of fine content present, a set of sieve sizes of 4.75 mm, 2 mm, 0.425 mm, and 0.075 mm may be chosen for a soil sample with little or no gravel contents to avoid a bulky set of sieves from 100 mm to 0.075 mm. However, if it is felt that the soil sample includes appreciable amount of gravel content, then the standard set of sieves (1st set of sieves of sizes

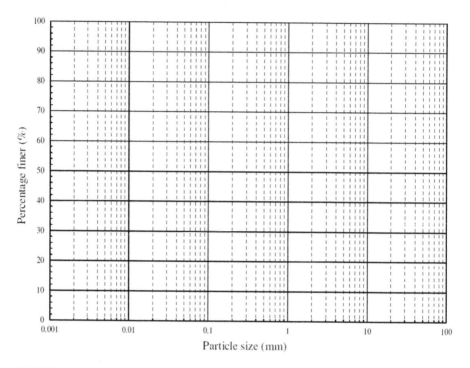

FIGURE 4.7 Semi-log graph for PSD by sieve analysis (practice sheet).

100 mm, 80 mm, 60 mm, 40 mm, 20 mm, 10 mm, 6.3 mm, 4.75 mm, 2 mm, 0.425 mm, and 0.075 mm) may be used.

b. Further, the PSD curve plotted from the results of dry or wet sieve analysis cannot quantify soil particle sizes of fine grained soils, especially clays. Therefore, hydrometer or sedimentation analysis is recommended for soils with appreciable particles of smaller sizes. This is due to the reasons that firstly, sieves smaller than the 0.075 mm size are not available, and secondly the most important reason for conducting hydrometer analysis is that smaller size soil particles such as clays are charged particles and form flocs when imbibed with water, cannot be sieved through smaller sieve sizes. Therefore, a dispersing agent is necessarily used in the preparation of soil suspension for hydrometer analysis to avoid formation of clay flocs.

c. It may be noted that the minimum weight of the representative soil sample increases both for dry and wet sieve analysis, as illustrated in Figure 4.8.

d. The coefficient of uniformity (C_u) signifies the range of particle size distribution in a given soil sample.

e. The coefficient of curvature, C_c, is a parameter which expresses the general shape of the PSD curve.

f. The D_{10}, D_{20}, and D_{50} sizes are used for ascertaining the suitability soil as backfill material.

4.11   APPLICATIONS/ROLE OF SIEVE ANALYSIS IN SOIL ENGINEERING

The main role of soil grading is to analyze, measure, and classify the soil particles in a given soil specimen. The soil grading or PSD is also indicative of engineering properties wherein it can be assessed whether a soil mass is stable or not. Various parameters such as effective size, coefficient of uniformity, coefficient of curvature, suitability number, etc. are derived from the PSD curve in terms of different particle sizes (e.g. D_{10}, D_{15}, D_{20}, D_{30}, D_{45}, D_{50}, D_{60}, D_{85}). These parameters are used for assessing the suitability of in situ soil in various geotechnical applications as briefly outlined below:

- **Soil gradation:** Soil grading helps in the identification of soil particles of different sizes and their representation in terms of coefficient of uniformity and coefficient of curvature.

- **Coefficient of permeability of a soil sample:** The coefficient of permeability of a soil has direct bearing on particle size (D_{10}). One of the earliest and simplest empirical formulae was proposed by Hazen (1911). For clean sands (with less than 5% fines) with D_{10} size between 0.1 and 3 mm, the coefficient of permeability k (cm/s) is given by:

$$k = CD_{10}^2 \quad [\textit{holds good for } k \leq 10^{-5} \; m/s] \qquad [4.6]$$

Where: C = 100 = Constant (1/cm^{-s}) valid for clean sands (with less than 5% fines)

C = Between 100 (1/cm-s) and 150 (1/cm-s) when D_{10} and k are taken in cm and cm/sec, respectively

C = Between 10 (1/mm-s) and 15 (1/mm-s) when D_{10} and k are taken in mm and mm/sec, respectively

D_{10} = Effective side obtained from PSD curve against 10% finer (cm), and

k = Coeff. of permeability (cm/s), respectively.

However, in case of other soils, the value of "C" is taken as:

C = 4–8 (1/mm-s) for very fine well graded or with appreciable fines less than 75 μm sieve

C = 8–12 (1/mm-s) for medium, coarse, poorly graded, clean, coarse but well-graded soils

C = 12–15 (1/mm-s) for very coarse, very poorly graded, gravely, clean soils

D_{10} = Effective side obtained from PSD curve against 10% finer (mm), and

k = Coeff. of permeability (mm/s), respectively.

- **Suitability of backfill material:** A soil sample with good draining characteristics plays a vital role in the design of earth retaining structures to avoid development of pore pressure. The suitability of backfill material depends on the gradation. Brown (1977) has developed a rating system to assess the suitability of the backfill material. The rating system is based on a suitability number (S_N) defined as:

$$S_N = 1.7\sqrt{\frac{3}{D_{50}^2} + \frac{1}{D_{20}^2} + \frac{1}{D_{10}^2}} \qquad [4.7]$$

Where: D_{10}, D_{20}, and D_{50} are the particle sizes corresponding to 10%, 20%, and 50% finer of the backfill material. Table 4.5 presents the rating description (Brown 1977).

- **Design of filters:** For design of filters, filter requirements for filter material depend on gradation of soil material are as illustrated in Table 4.6:
  The filter specifications based primarily on grain size distributions of the filter material and the base (protected) material are given as below:

$$\left[\frac{D_{15}\ of\ Filter\ Material}{D_{85}\ of\ Base\ material} < 5\right] \qquad [4.8]$$

$$\left[4 < \frac{D_{15}\ of\ Filter\ Material}{D_{15}\ of\ Base\ Material} < 20\right] \qquad [4.9]$$

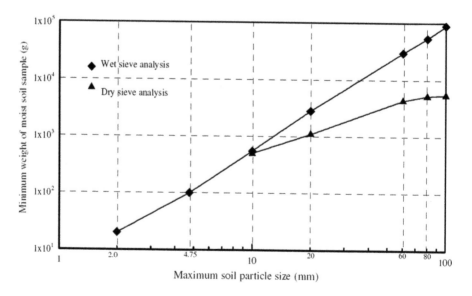

FIGURE 4.8 Minimum weight of soil sample to be taken for dry and wet sieving.

TABLE 4.5
Evaluation Criterion for Backfill Materials (Brown 1977)

Sr. No.	Suitability Number, S_N	Description of Rating for Backfill Material
1	0 – 10	Excellent backfill material
2	10 – 20	Good backfill material
3	20 – 30	Fair backfill material
4	30 – 50	Poor backfill material
5	> 50	Unsuitable

TABLE 4.6
Evaluation Criterion for Filter Materials

Sr. No.	Character of Filter Materials	Ratio R_{50}	Ratio R_{15}	Remarks
1	Uniform grain size distribution (C_u = 3–4)	5–10	–	$R_{50} = \dfrac{D_{50}\ of\ Filter\ Material}{D_{50}\ of\ Base\ Matereial}$
2	Well graded to poorly graded (non-uniform); sub rounded grains	12–58	12–40	$R_{15} = \dfrac{D_{15}\ of\ Filter\ Material}{D_{15}\ of\ Base\ Material}$ Where D_{15} and D_{50} are the particle sizes corresponding to 15 and 50% finer of the corresponding materials.
3	Well graded to poorly graded (non-uniform); angular particles	9–30	6–18	

$$\left[\frac{D_{50}\ of\ Filter\ Material}{D_{50}\ of\ Base\ Material} < 25 \right] \qquad [4.10]$$

Where D_{15}, D_{50}, and D_{85} are the particle sizes corresponding to 15%, 50%, and 85% finer of the corresponding filter/base materials.

- **Pavement mix design:** As per the Ministry of Road Transport and Highways (MoRTH-2000) specifications, the gradation requirements of gravel/soil aggregate for use in base and surface courses of a gravel road are given in Tables 4.7 and 4.8, respectively. It may be noted that any of the three gradings given in Table 4.7 for Base Course can be adopted depending upon the availability of materials. These gradings are recommended in case the gravel is sealed by chip sealing or surface dressing using bituminous material.

The percentages of gravel, sand, and fines (silt and clay) in the gradings A, B, and C of Table 4.7 are as under:

Soil Type	Grading A (%)	Grading B (%)	Grading C (%)
Gravel	53–67	47–61	41–53
Sand	25–43	31–49	39–55
Silt and Clay	4–8	4–8	4–8

TABLE 4.7

Grading Requirements for Base Course

Sieve Size (mm)	Percent by Mass Passing IS Sieve Grading Designation		
	A	B	C
53	100	–	
37.5	97–100	100	–
26.5	–	97–100	100
19	67–81	–	97–100
9.5	–	56–70	67–79
4.75	33–47	39–53	47–59
0.425	10–19	12–21	12–21
0.075	4–8	4–8	4–8

The material for granular sub-base (non-bituminous) should generally conform to the gradings indicated in Tables 4.9 and 4.10 or a combination thereof.

It may be noted that according to IS:2720 (Part V), the coarse graded granular sub-base materials shall have a liquid limit and plasticity index not more than 25% and 6%, respectively, in the three gradings above.

- **Soil aggregate mixtures:** Soil aggregate mixtures may be in the form of naturally occurring materials like soil-gravel, or soil purposely blended with suitable aggregate fractions. The primary criteria for acceptability of such materials are plasticity characteristics and gradation. The material should be smoothly graded for achieving the maximum possible dry density. Fuller's grading rule could be used as a guide to work out the optimum grading in different cases. A few typical gradings are given in Table 4.11. The first three gradings indicated in Table 4.11 are especially suited for base courses, whereas the remaining two are suitable both for base course and for surfacing.

TABLE 4.8

Grading Requirements for Surface Course

Sieve Size (mm)	Percent by Mass Passing Designated Sieve
26.5	100
19	97–100
4.75	41–71
0.425	12–28
0.075	9–16

TABLE 4.9

Grading for Coarse-Graded Granular Sub-Base Materials

Sr. No.	Sieve Size (mm)	Percent by Weight Passing the IS Sieve		
		Grading I	Grading II	Grading III
1	75.0	100	–	–
2	53.0	80–100	100	–
3	26.5	55–90	70–100	100
4	9.50	35–65	50–80	65–95
5	4.75	25–55	40–65	50–80
6	2.36	20–40	30–50	40–65
7	0.425	10–25	15–25	20–35
8	0.075	3–10	3–10	3–10
9	CBR Value (Minimum)	30	25	20

4.11.1 WEIGHTED MEAN DIAMETER (D_M) FOR DETERMINATION OF SILT FACTOR (F)

The weighted mean diameter (d_m) is obtained from soil grading (e.g. from particle size analysis) for determination of silt factor given by:

$$f = 1.76 * \sqrt{d_m} \qquad [4.11]$$

The weighted mean diameter (d_m) is obtained from soil grading by the following expression:

$$d_m = \left[\frac{\sum_{i=1}^{n} p_i d_i}{\sum_{i=1}^{n} p_i} \right] \left[\begin{array}{l} Where: d = particle\ size\ (mm) \\ p = percentage\ finer\ against\ particle\ size\ (d) \end{array} \right] \qquad [4.12]$$

The computation of weighted mean diameter (d_m) is shown in Table 4.12.

TABLE 4.10

Grading for Coarse-Graded Granular Sub-Base Materials

Sr. No.	Sieve Size (mm)	Percent by Weight Passing the IS Sieve		
		Grading I	Grading II	Grading III
1	75.0	100	–	–
2	53.0	–	100	–
3	26.5	55–75	50–80	100
4	9.50	–	–	–
5	4.75	10–30	15–35	25–45
6	2.36	–	–	–
7	0.425	–	–	–
8	0.075	< 10	<10	<10
9	CBR Value (Minimum)	30	25	20

Once the silt factor (f) is computed, the mean scour depth can be computed as:

$$D_{sm} = \left[1.34 \frac{Q_d^2}{f} \right]^{1/3} \qquad [4.13]$$

Where: D_{sm} = Mean scour depth (m) below design flood level

Q_d = Design flood discharge intensity in m³/s/m allowing for concentration of flow, and

f = Silt factor based on particle size analysis of the soil.

The value of the silt factor (f) for various grades of sandy bed is given below for ready reference and adoption (IRC: 78-2014):

Sr. No.	Type of bed material	d_m (mm)	Silt factor (f)
1	Coarse silt	0.04	0.35
2	Silt/fine sand	0.081–0.158	0.5–0.7
3	Medium sand	0.223–0.505	0.85–1.25
4	Coarse sand	0.725	1.5
5	Fine bajri and sand	0.988	1.75
6	Heavy sand	1.29–2.00	2.0–2.42

TABLE 4.11

Typical grading limits for soil aggregates mixture

Sieve Size (mm)	Percent by weight passing the sieve (IRC: 63 – 1976)				
	Nominal Maximum Size				
	80 mm	40 mm	20 mm	10 mm	5 mm
80	100	–	–	–	–
40	80–100	100	–	–	–
20	60–80	80–100	100	–	–
10	45–65	55–80	80–100	100	–
4.75	30–50	40–60	50–75	80–100	–
2.36	–	30–50	35–60	50–80	80–100
1.18	–	–	–	40–65	50–80
0.60	10–30	15–30	15–35	–	30–60
0.30	–	–	–	20–40	20–45
0.075	5–15	5–15	5–15	10–25	10–25

The silt factor (f) is also used for the computation of the following parameters:

$$\text{Velocity of canal flow: } V = \left[\frac{Q_d * f^2}{140} \right]^{1/6} \quad\quad [4.14]$$

$$\text{The area of the canal: } A = \frac{Q_d}{V} \quad\quad [4.15]$$

$$\text{Hydraulic mean depth: } R = \frac{5V^2}{2f} \quad\quad [4.16]$$

$$\text{The bed slope (S) value: } S = \frac{f^{5/3}}{3340Q_d^{1/6}} \quad\quad [4.17]$$

4.11.2 SILT FACTOR FOR CLAY BED MATERIAL

In the case of clayey bed material, the following guidelines should be followed for determining the silt factor:

1. When ϕ <15° and c >20 kPa, the silt factor (f) is computed as:

$$f = F[1 + \sqrt{c}] \quad\quad [4.18]$$

TABLE 4.12
Weighted mean diameter (d_m) from particle size analysis

Sr. No.	Particle size, d (mm)	Percentage finer (p)	$d_i = \sqrt{d_1 \cdot d_2}$ (mm)	$p_i = p_1 - p_2 = (p_{n-1} - p_n)$	$p_i * d_i$	Remarks
1	4.75	41	$\sqrt{4.75 * 3} = 3.77$	$41 - 32 = 9$	$3.77 * 9$ $= 33.93$	$d_1 = 4.75$; $d_2 = 3.00$ $p_1 = 41$; $p_2 = 32$
2	3.00	32	$\sqrt{3 * 2} = 2.45$	$32 - 25 = 7$	17.15	$d_1 = 3.00$; $d_2 = 2.00$ $p_1 = 32$; $p_2 = 25$
3	2.00	25	$\sqrt{2 * 1} = 1.41$	$25 - 16 = 9$	12.69	So on....
4	1.00	16	$\sqrt{1 * 0.5} = 0.71$	$16 - 9 = 7$	4.97	—
5	0.50	9	$\sqrt{0.5 * 0.3} = 0.39$	$9 - 6 = 3$	1.17	—
6	0.30	6	$\sqrt{0.3 * 0.15} = 0.21$	$6 - 4 = 2$	0.42	—
7	0.15	4	$\sqrt{0.15 * 0.075} = 0.11$	$4 - 2.5 = 1.5$	0.165	$d_1 = 0.15$; $d_2 = 0.075$ $p_1 = 4$; $p_2 = 2.5$
8	0.075	2.5	From gradation results, we get:	$\Sigma_{p_i} = 38.5$	$\Sigma_{p_i d_i} = 70.495$	$d_m \frac{\Sigma p_i d_i}{\Sigma p_i} = 1.83$ (mm)

Silt factor, $f = 1.76\sqrt{1.83} = 2.38$

Where: F = 1.50 for ϕ >10° and <15°
    = 1.75 for ϕ >5° and <10
    = 2.00 for ϕ <5°
c = Cohesion of soil (kPa) and
ϕ = Angle of shearing friction/internal friction.
2. When ϕ >15° and c >2 N/mm², the silt factor (f) is computed as:

$$f = [1.76 * \sqrt{d_m}]$$ [4.19]

4.12   SOURCES OF ERROR

It may be noted that the following sources of error may affect the PSD results if proper precautionary measures are not taken:

1. If damaged or defective sieves are used for soil grading
2. There may be loss of soil material during sieving process if top cover and bottom pan are not properly fixed with sieves
3. Make sure that soil particles are clean and no appreciable fines are sticking in case of dry sieving
4. Make sure that dispersing agent is properly used in the case of sieving clayey soils
5. Make sure that soil retained after washing during wet sieving is properly dried and weighed for further sieving process

4.13   PRECAUTIONS

The following precautions may be followed while sieving:

1. Adequate amount of dispersing agent added if required
2. Rigorous shaking should be avoided for loss of soil material during sieving
3. For better representation of soil fraction with large variation in its size, particle size should be plotted on semi-log paper versus percentage passing on natural scale
4. It should be clearly visualized whether appreciable fines are coated with coarser soil particles or not while mechanical sieving. If yes, then wet sieving should be done by washing the soil sample until clean water comes down the sieve.

4.14   LIMITATIONS OF SIEVE ANALYSIS

The main limitation of either dry or wet sieve analysis is that these cannot quantify the particle size distribution of fine grained soils adequately. Therefore, particle qsize distribution analysis has no significance for fine grained soils. Therefore, hydrometer analysis is recommended for these soils.

REFERENCES

ASTM D421. 1985. "Standard Practice for Dry Preparation of Soil Samples for Particle-Size Analysis and Determination of Soil Constants." *Annual Book of ASTM Standards*, Philadelphia, sec. 4, Vol. 04–08, pp. 10–16, 1963, www.astm.org.

ASTM D 422–63(07e2). 1963. "Standard Test Method for Particle Size Analysis of Soils." *Annual Book of ASTM Standards*, Philadelphia, sec. 4, Vol. 04–08, pp. 10–16, www.astm.org.

ASTM D 2487. 2006. "Standard Practice for Classification of Soils for Engineering Purposes (USCS)." *Annual Book of ASTM Standards*, Vol. 04–08. American Society for Testing and Materials, Philadelphia, United States, www.astm.org.

ASTM D2488-00. 2000. "Standard Practice for Description and Identification of Soils (Visual-Manual Procedure)." *Annual Book of ASTM Standards*, Vol. 04–08. American Society for Testing and Materials, Philadelphia, United States, www.astm.org.

ASTM E11-2016. 1961. Standard Specification for Woven Wire Test Sieve Cloth and Test Sieves. ASTM International, West Conshohocken. DOI: 10.1520/E0011-6. www.astm.org

Brown, R. E. 1977. "Vibroflotation Compaction of Cohesionless Soils." *J. Geotech Eng. Div.* 103(GT12): 1437–1451.

BS 1377-2. 1990. "Methods of Test for Soils for Civil Engineering Purposes-Part 2: Classification Tests for Determination of Water Content." British Standards, UK.

BS 410. 1962. Specification for Test Sieves. British Standards, UK.

Hazen, A. 1911. "Discussion: Dams on Sand Foundations."*Transactions of the American Society of Civil Engineers* 73: 199-203.

IRC: 78. 2014. "Standard Specifications and Code of Practice for Road Bridges, Section: VII – Foundations and Substructure." Indian Roads Congress Kama Koti Marg, R.K. Puram, New Delhi.

IRC: 63. 1976/2011. "Tentative Guidelines for the Use of Low Grade Aggregates and Soil Aggregate Mixtures in Road Pavement Construction." The Indian Roads Congress, Jamnagar House, Shahjahan Road, New Delhi.

IS: 149.- 1970. "Classification and Identification of Soils for General Engineering Purposes." Bureau of Indian Standards, New Delhi.

IS: 2720 (Part 1). 1980. "Indian Standard Code for Preparation of Soil Samples." Bureau of Indian Standards, New Delhi.

IS: 2720 (Part 4). 1985. "Method of Test for Soils: Determination of Grain Size Distribution." Bureau of Indian standards, New Delhi.

MoRTH. 2000. "Ministry of Road Transport and Highways – Specifications for Road and Bridge Works." The Indian Roads Congress, Jamnagar House, Shahjahan Road, New Delhi.

Reddy, Krishna R. 2000. "Engineering Properties of Soils Based on Laboratory Testing." Department of Civil and Materials Engineering University of Illinois at Chicago, USA.

5 Particle Size Distribution Analysis by the Hydrometer Method

References: IS: 1498; IS: 2720-Part 1; IS: 2720-Part 4; ASTM D421; ASTM D 422; ASTM D 2487; ASTM D2488; ASTM E100–05; ASTM E126-05a; ASTM D1298; BS 1377: Part 2, 1990

5.1 OBJECTIVES

The objectives of this experiment are to quantify soil particle sizes of fine grained soils finer than 0.075 mm (especially clays < 0.002 mm) and to avoid formation of clay flocs by adding the desired quantity of dispersing agent during sedimentation analysis. The hydrometer analysis also determines the density (or specific gravity) of the soil suspension at the center of its bulb. Thus, sedimentation analysis is the soil grading process in which "silts and clays" are quantified, e.g. to determine the percentage of silt and clay content contained within fine grained soil less than 75 μm.

5.2 INTRODUCTION

It has been observed that the PSD curve plotted from the results of dry or wet sieve analysis cannot quantify soil particle sizes of fine grained soils, especially clays. Therefore, hydrometer or sedimentation analysis is recommended for soils with appreciable particles of smaller sizes. This is due to the reasons that firstly, sieves smaller than 0.075 mm size are not available, and secondly the most important reason for conducting hydrometer analysis is that smaller size soil particles such as clays are charged particles and form flocs when imbibed with water, and cannot be sieved through smaller sieve sizes. A dispersing agent is used in the preparation of soil suspension for hydrometer analysis to avoid formation of clay flocs. Therefore, hydrometer analysis or sedimentation analysis is a process of quantification of percentage proportion of the constituents of the fine grained soils (e.g. silts and clays). The sedimentation process measures the specific gravity or the density of water suspension in which the soil particles are initially suspended but settle down with time. A hydrometer with a heavily weighted bulb at its bottom and a graduated stem is used to measure the density at the center of its bulb with time (Figure 5.1).

It may be noted that the hydrometer is a very fragile device with two unbalanced components (e.g. it has a heavy bulb and a lightweight graduated stem). Therefore, it should be handled with utmost care during sedimentation analysis. Figure 5.2 illustrates how to handle the hydrometer device during the tests.

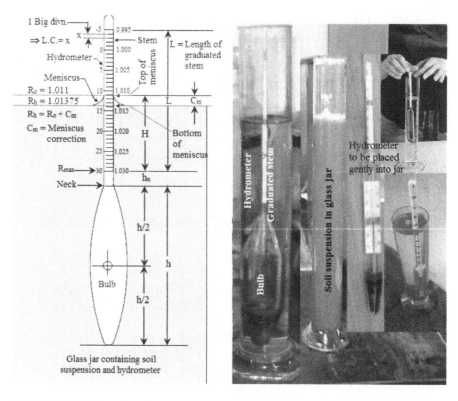

FIGURE 5.1   Schematic of a hydrometer device.

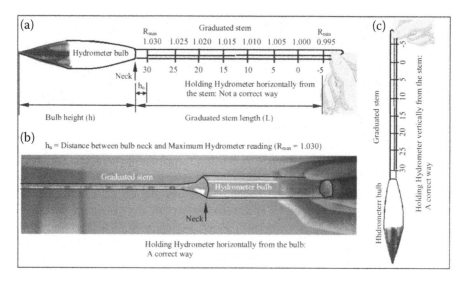

FIGURE 5.2   Handling of a hydrometer: (a). Holding hydrometer horizontally from the stem: Not a correct way; (b). Holding hydrometer horizontally from the bulb: A correct way; (c). Holding hydrometer vertically from the stem: A correct way.

5.3   DEFINITIONS AND THEORY

Dry or wet sieve analysis is performed to quantify the distribution of the coarse, larger-sized soil particles, and the sedimentation analysis is used to quantify the distribution of the finer particles (silt and clays). The complete grain size distribution curves for coarse and fine grained soils is required for engineering classification of soils (Reddy, 2000). The sedimentation analysis by hydrometer works on the principle of Stokes's Law (1856), which defines the rate of free fall of a sphere through a liquid given by the relation:

$$v = \left[\frac{g\,(\gamma_s - \gamma_w)}{18\eta} D^2 \right] = \left[\frac{\rho_s - \rho_w}{18\mu} D^2 \right] \qquad [5.1]$$

Where: v = Velocity of sphere also known as terminal velocity (cm/s)
$\rho_s = G_s * \rho_w$ = Mass of sphere (g/cm^3)
ρ_w = Mass of water/liquid (g/cm^3)
$\mu = \eta/g$ = Viscosity of liquid in absolute units (poise or dyne/sec-cm^2 or gm-sec/cm^2)
η = Viscosity or dynamic viscosity of liquid in (kN-s/m^2)
g = Acceleration due to gravity (cm/sec^2)
t = Time in seconds
D = Diameter of sphere (mm), which is given by the relation:

$$D = \left[\left(\sqrt{\frac{18\mu}{\rho_s - \rho_w}} \right) \sqrt{v} \right] = \left[\left(\sqrt{\frac{18\mu}{G_s\rho_w - \rho_w}} \right) \sqrt{\frac{H_e}{t}} \right] = \left[\left(\sqrt{\frac{18\mu}{(G_s - 1)\rho_w}} \right) \sqrt{\frac{H_e}{t}} \right]$$
$$[5.2]$$

H_e = Effective depth from center of hydrometer bulb to the hydrometer reading (cm)
   If H_e is in cm, t is in min, ρ_w in g/cc, μ in (g-sec)/cm^2, and D in mm, then Eq. (5.2) may be written as:

$$\frac{D\,(mm)}{10} = \left[\sqrt{\frac{18\mu}{(G_s - 1)\rho_w}} \sqrt{\frac{H_e}{t * 60}} \right] \ OR \ D = \left[\sqrt{\frac{30\mu}{(G_s - 1)\rho_w}} \sqrt{\frac{H_e}{t}} \right]$$

$$= \left[M \sqrt{\frac{H_e}{t}} \right] \qquad [5.3]$$

$$\text{Where Factor } M = \sqrt{\frac{30\mu}{(G_s - 1)}} \ \text{ and } \rho_w = 1 g/cm^3 \qquad [5.4]$$

The values of factor-M [Eq. 5.4] for several specific gravity values of soil solids and temperature are given in Table 5.1 and the values of viscosity of water for different

TABLE 5.1

Values of M for Several Specific Gravities of Solid and Temperature Combinations

Sp. Gravity, G_s of Soil Solids (Assuming $\rho_w = 1$ g/cm^3)

T°C	2.50	2.55	2.60	2.65	2.70	2.75	2.80	2.85
16	0.0151	0.0148	0.0146	0.0144	0.0141	0.0139	0.0139	0.0136
17	0.0149	0.0146	0.0144	0.0142	0.0140	0.0138	0.0136	0.0134
18	0.0148	0.0144	0.0142	0.0140	0.0138	0.0136	0.0134	0.0132
19	0.0145	0.0143	0.0140	0.0138	0.0136	0.0134	0.0132	0.0131
20	0.0143	0.0141	0.0139	0.0137	0.0134	0.0133	0.0131	0.0129
21	0.0141	0.0139	0.0137	0.0135	0.0133	0.0131	0.0129	0.0127
22	0.0140	0.0137	0.0135	0.0133	0.0131	0.0129	0.0128	0.0126
23	0.0138	0.0136	0.0134	0.0132	0.0130	0.0128	0.0126	0.0124
24	0.0137	0.0134	0.0132	0.0130	0.0128	0.0126	0.0125	0.0123
25	0.0135	0.0133	0.0131	0.0129	0.0127	0.0125	0.0123	0.0122
26	0.0133	0.0131	0.0129	0.0127	0.0125	0.0124	0.0122	0.0120
27	0.0132	0.0130	0.0128	0.0126	0.0124	0.0122	0.0120	0.0119
28	0.0130	0.0128	0.0126	0.0124	0.0123	0.0121	0.0119	0.0117
29	0.0129	0.0127	0.0125	0.0123	0.0121	0.0120	0.0118	0.0116
30	0.0128	0.0126	0.0124	0.0122	0.0	0.0118	0.0117	0.0115

temperatures are given in Table 5.2. Further, the density of water (g/cm^3) at temperatures from 0°C (liquid state) to 30.9°C by 0.1°C increment is given in Table 5.3.

It may be noted that soils are heterogeneous and the soil particles comprises of different shapes such as angular, sub-rounded, rounded, etc.. However, Stokes's law assumes that the soil particles are spherical with an equivalent diameter and the same density falling in a liquid of infinite extent. The spherical particles reach a constant terminal velocity within a few seconds after they are allowed to fall. It may also be noted that a soil particle is said to have an equivalent diameter, D_e, if a sphere of diameter D with the same density as the soil particle has the same velocity of fall as the soil particle. Since the sedimentation analysis is carried out in a relatively small glass jar of 1,000 cm^3 capacity, the effect of influence of one particle over the other is minimized by limiting the mass of soil for sedimentation analysis to 50 g in this jar.

5.4   METHOD OF TESTING

Particle size analysis of soils passing 75 µm is carried out using hydrometer analysis as per standard codal procedures.

TABLE 5.2
Coefficient of Viscosity of Water (μ-Absolute Viscosity)

T°C	Unit Weight of Water (g/cm^3)	Viscosity of Water (poise)	Viscosity of Water (gm-sec/cm^2)
4	1.00000	0.01567	$1.567 * 10^{-05}$
16	0.99897	0.01111	$1.111 * 10^{-05}$
17	0.99880	0.0108	$1.080 * 10^{-05}$
18	0.99862	0.0105	$1.050 * 10^{-05}$
19	0.99844	0.01030	$1.030 * 10^{-05}$
20	**0.99823**	**0.01005**	$\mathbf{1.005 * 10^{-05}}$
21	0.99802	0.00981	$9.810 * 10^{-06}$
22	0.99780	0.00958	$9.580 * 10^{-06}$
23	0.99757	0.00936	$9.360 * 10^{-06}$
24	0.99733	0.00914	$9.140 * 10^{-06}$
25	0.99708	0.00894	$8.940 * 10^{-06}$
26	0.99682	0.00874	$8.740 * 10^{-06}$
27	0.99655	0.00855	$8.550 * 10^{-06}$
28	0.99627	0.00836	$8.360 * 10^{-06}$
29	0.99598	0.00818	$8.180 * 10^{-06}$
30	0.99568	0.00801	$8.010 * 10^{-06}$

1 millipoise = 0.1 mN-s/m^2; 1 millipoise = 10^{-3} poise; 1 poise = 10^{-4} kN-sec/m^2;
1 gm-sec/cm^2 = 981 poises (1,000 poises); 1 pound-sec/sq. ft = 478.69 poises; 1 MN = 10^6 N;
1 dyne-s/sq. cm = 1 poise; 1 poise = 1,000 millipoises; 1 g/cm^3 = 1 t/m^3 = 10 kN/m^3;
1 μ = 10^{-3} mm = 10^{-6} m; 1 N = 10^6 μN, 1 kg/cm^2 = 10 N/cm^2 = 0.1 N/mm^2 = 10 t/m^2 = 10^5N/m^2 =
0.1 MN/m^2 = 100 kN/m^2

5.5   SOIL TESTING MATERIAL

About 50 g of oven-dried soil sample passing 0.075 mm IS sieve is taken for hydrometer analysis for identification of constituents of silt and clay present in the soil sample.

5.6   TESTING EQUIPMENTS AND ACCESSORIES

The following equipments and accessories are needed for hydrometer analysis (Figure 5.3):

1. Oven-dried fine-grained soil passing 0.075 mm sieve of known specific gravity
2. Hydrometer calibrated at standard temperature
3. Rubber bung for the cylinder (jar)
4. Glass measuring cylinder (jar), 1,000 ml
5. Weighing balance, accuracy 0.01 g

TABLE 5.3
Density of Water (g/cm³) at Temperatures from 0°C (Liquid State) to 30.9°C by 0.1°C Increment

T°C	0	0.1	0.2	0.3	0.4	0.5	0.6	0.7	0.8	0.9
0	0.999841	0.99985	0.99985	0.99986	0.999866	0.99987	0.99988	0.99988	0.99989	0.9999
1	0.9999	0.99991	0.99991	0.99991	0.999918	0.99992	0.99993	0.99993	0.99993	0.99994
2	0.999941	0.99994	0.99995	0.99995	0.999953	0.99996	0.99996	0.99996	0.99996	0.99996
3	0.999965	0.99997	0.99997	0.99997	0.99997	0.99997	0.99997	0.99997	0.99997	0.99997
4	0.999973	0.99997	0.99997	0.99997	**0.999972**	0.99997	0.99997	0.99997	0.99997	0.99997
5	0.999965	0.99996	0.99996	0.99996	0.999957	0.99996	0.99995	0.99995	0.99995	0.99994
6	0.999941	0.99994	0.99994	0.99993	0.999927	0.99992	0.99992	0.99992	0.99991	0.99991
7	0.999902	0.9999	0.99989	0.99989	0.999883	0.99988	0.99987	0.99987	0.99986	0.99986
8	0.999849	0.99984	0.99984	0.99983	0.999824	0.99982	0.99981	0.9998	0.9998	0.99979
9	0.999781	0.99977	0.99977	0.99976	0.999751	0.99974	0.99973	0.99973	0.99972	0.99971
10	0.9997	0.99969	0.99968	0.99967	0.999664	0.99965	0.99965	0.99964	0.99963	0.99962
11	0.999605	0.9996	0.99959	0.99957	0.999564	0.99955	0.99954	0.99953	0.99952	0.99951
12	0.999498	0.99949	0.99948	0.99946	0.999451	0.99944	0.99943	0.99942	0.9994	0.99939
13	0.999377	0.99936	0.99935	0.99934	0.999326	0.99931	0.9993	0.99929	0.99927	0.99926
14	0.999244	0.99923	0.99922	0.9992	0.999188	0.99917	0.99916	0.99914	0.99913	0.99911
15	0.999099	0.99908	0.99907	0.99905	0.999038	0.99902	0.99901	0.99899	0.99898	0.99896
16	0.998943	0.99893	0.99891	0.99889	0.998877	0.99886	0.99884	0.99883	0.99881	0.99879
17	0.998774	0.99876	0.99874	0.99872	0.998704	0.99869	0.99867	0.99865	0.99863	0.99861
18	0.998595	0.99858	0.99856	0.99854	0.99852	0.9985	0.99848	0.99846	0.99844	0.99842
19	0.998405	0.99839	0.99837	0.99835	0.998325	0.99831	0.99829	0.99827	0.99824	0.99822
20	0.998203	0.99818	0.99816	0.99814	0.99812	0.9981	0.99808	0.99806	0.99804	0.99801

21	0.997992	0.99797	0.99795	0.99793	0.997904	0.99788	0.99786	0.99784	0.99782	0.99779
22	0.99777	0.99775	0.99772	0.9977	0.997678	0.99766	0.99763	0.99761	0.99759	0.99756
23	0.997538	0.99751	0.99749	0.99747	0.997442	0.99742	0.99739	0.99737	0.99735	0.99732
24	0.997296	0.99727	0.99725	0.99722	0.997196	0.99717	0.99715	0.99712	0.9971	0.99707
25	0.997044	0.99702	0.99699	0.99697	0.996941	0.99691	0.99689	0.99686	0.99684	0.99681
26	0.996783	0.99676	0.99673	0.9967	0.996676	0.99665	0.99662	0.99659	0.99657	0.99654
27	0.996512	0.99649	0.99646	0.99643	0.996401	0.99637	0.99635	0.99632	0.99629	0.99626
28	0.996232	0.9962	0.99618	0.99615	0.996118	0.99609	0.99606	0.99603	0.996	0.99597
29	0.995944	0.99591	0.99589	0.99586	0.995826	0.9958	0.99577	0.99574	0.99571	0.99568
30	0.995646	0.99562	0.99559	0.99556	0.995525	0.99549	0.99546	0.99543	0.9954	0.99537

Example: The density of water at 4.4°C is 0.999972 g/cc.

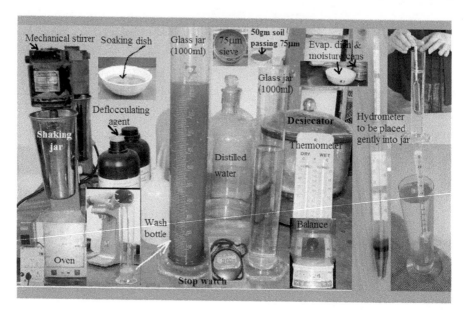

FIGURE 5.3   Test equipments/accessories for hydrometer sieve analysis.

6. Mechanical stirrer
7. Suspension shaking jar
8. Oven
9. Desiccator
10. Conical flask or beaker, 1,000 ml
11. Glass rod
12. Stopwatch
13. Wash bottle
14. Deflocculating agent
15. Thermometer
16. 75 μm sieve
17. Distilled water
18. Measuring scale
19. Soaking dish
20. Evaporating dish
21. Moisture cans

5.7 TESTING PROGRAM

5.7.1 CALIBRATION OF HYDROMETER AND SEDIMENTATION JAR

The following steps are followed to calibrate the hydrometer and sedimentation jar to find out their volume and cross-sectional area, respectively.

1. Take about 700 ml clean distilled water in a measuring jar of 1,000 ml capacity and place on the table.
2. Record the initial reading in the measuring jar R_1 (e.g. 700 ml).
3. Immerse the hydrometer in the jar slowly so that it cannot strike either side of the measuring jar or the bottom surface.
4. Record the final reading of water rise in the jar when the hydrometer is quite stable in the jar, R_2 (in this case, it is 770 ml).
5. Determine the increase in volume, which represents the volume of hydrometer as:

$$V_h = [R_2 - R_1] = [770 - 700] = 70 \ cc$$

Alternatively, the volume of the hydrometer (ml) is approximately equal to its mass in g (nearest to 0.1 g).

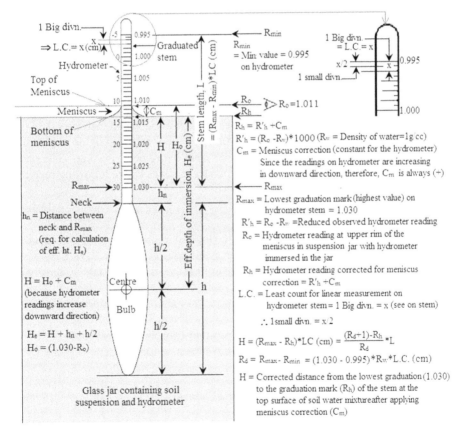

FIGURE 5.4 Calibration of hydrometer and sedimentation jar.

6. Measure the water depth (e.g. linear measurement) displaced by the hydrometer in the jar, h_j (cm) between initial reading (700 ml) and final reading graduations (770 ml) on the jar. Then the internal cross-sectional area (cm^2) of this measuring jar is equal to the volume between two graduations divided by the linear distance:

$$A_J = \left[\frac{V_h}{h_J} \right] (cm^2)$$

7. Measure the distance from the neck of the hydrometer bulb to the lowest graduation (1.030) designated by h_n as shown in Figure 5.4. It is optional and for a few hydrometers it may be zero. In the present case, $h_n = 0.8$ cm.
8. Measure the distance between the neck and the bottom of the bulb. Record it as the height of the bulb, h, as shown in Figure 5.4.
9. Measure the distance (H_o) between the neck to each mark on the hydrometer (R'_h), as shown in Figure 5.4, where $R'_h = (R_o - R_w)$ is reduced observed hydrometer reading after deduction of density of water in the suspension jar.
10. Correct the observed hydrometer reading (R'_h) for meniscus correction and record it as R_h.
11. Measure the distance between any two graduations and find the least count (L.C.) for every big division very carefully for finding linear measurements (e. g. length of stem, L and depth of immersion of hydrometer into jar for each reading, etc.). In the present case, each big division measures 0.38 cm and each small division measures 0.19 cm. This will help in the calculation of effective depth of the hydrometer from hydrometer readings. The parameter "H" is calculated from hydrometer readings by the following relation:

$$H = [(R_{max} - R_h) * L.\ C.] \tag{5.5(a)}$$

Where: R_{max} = Lowest graduation mark (highest value) on hydrometer stem
R_h = Hydrometer reading corrected for meniscus correction = $R'_h + C_m$
$R'_h = R_o - R_w$ = Reduced observed hydrometer reading after taking density of water
R_o = Hydrometer reading at upper rim of the meniscus in suspension jar with hydrometer immersed in the jar, and
$L.C.$ = Least count for linear measurement on hydrometer stem (cm).
Then, from Figure 5.4, effective depth of immersion hydrometer is given by:

$$H_e = \left[(R_{max} - R_h) * L.\ C. + h_n + \frac{1}{2}\left(h - \frac{V_h}{A_J} \right) \right] \tag{5.5(b)}$$

Where: h_n, h are constant measurable parameters for hydrometer as explained above.
V_h/A_J = Immersion correction ($\because$ Suspension level rises when the hydrometer is placed into suspension jar, explained in next section)

Alternatively, parameter "H" is can also be calculated as (Figure 5.4):

$$H = \left[\frac{(R_d + 1) - R_h}{R_d} * L \right]$$ [5.6(a)]

Where: $R_d = R_{max} - R_{min}$

R_{max} = Maximum value = 1.030 on hydrometer

R_{min} = Minimum value = 0.995 on hydrometer

L = Length of hydrometer stem (Linear distance between 1.030 and 0.995 values on hydrometer stem), and

$R_h = R'_h + C_m$ = Hydrometer reading corrected for meniscus correction.

However, it should be made sure that linear measurements of L and R_h are recorded very carefully. A simple method for determining value of "L" is to find the number of divisions between R_{max} (1.030) and R_{min} (0.995) and multiply by L.C.

In the present case, the number of divisions between R_{max} (1.030) and R_{min} (0.995) are 7 × 5 = 35 (∵stem has 7 parts and each part has five big divisions of L.C. of 0.38 cm). Therefore, the length of the stem is given by L = 35 × 0.38 = 13.3 cm.

However, this value of "L" should be cross-checked by taking the actual measurement of the stem on the hydrometer.

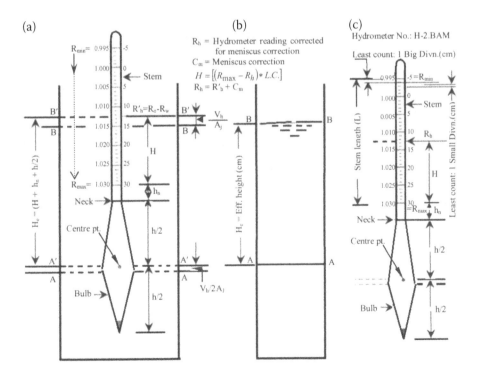

FIGURE 5.5 Effective depth of hydrometer.

Further, the value of "L" can also be calculated theoretically as:
$L = [(R_{max} - R_{min}) \times R_w] \times L.C. = [(1.030 - 0.995) \times 1000] \times 0.38 = 35 \times 0.38 =$
13.3 cm
Then, effective depth is given by:

$$H_e = \left[\frac{(R_d + 1) - R_h}{R_d} * L + h_n + \frac{h}{2} \right] \qquad [5.6(b)]$$

However, immersion correction has to be incorporated for computing the effective depth for each hydrometer reading as explained below.

5.7.2 EFFECTIVE HEIGHT (H_E) AND IMMERSION CORRECTION

The effective height from the center of the hydrometer bulb to the hydrometer reading (cm) is given by the relation (Figure 5.5):

$$H_e = \left[H + h_n + \frac{1}{2} \left(h - \frac{V_h}{A_J} \right) \right] \qquad [5.7]$$

Where: H = Distance from the lowest graduation (1.030) to the graduation mark (R_h) of the stem at the top surface of soil water mixture
  h = Height of hydrometer bulb
  V_h = Volume of hydrometer (cm^3)
  A_J = Internal cross- sec. area of jar containing hydrometer, and
  h_n = Distance between the neck of hydrometer to the lowest graduation mark (1.030). It is optional and for some hydrometers it may be zero.
    The mark on the hydrometer stem gives the specific gravity of the suspension at the center of the bulb. However, when the hydrometer is inserted into the suspension, the surface of the suspension rises and therefore, the immersion correction must be applied. As seen from Figure 5.5a, as the hydrometer is immersed into the suspension, the level B-B rises to B'-B' and the level A-A rises to A'-A'. If V_h is the volume of the hydrometer and A_J is the cross sectional area of sedimentation jar in which the hydrometer is immersed, then the rise of B-B to B'-B' level is V_h/A_J, because below the level of B'-B', only half of the hydrometer is in water.
    Similarly, the level the rise of A-A to A'-A' level is equal to $V_h/2A_J$ since below the level of A'-A', the hydrometer is fully immersed in water. It may be noted that A-A is the level situated at a depth "H_e" below the top surface level B-B at which the density measurements of the soil suspension are being taken (Figure 5.5b). The level A'-A' is now corresponding to the center of the hydrometer bulb, but the soil particles at A'-A' are of the same concentrations as they were at the A-A level. Therefore, the *Immersion Correction* is $V_h/2A_J$. Thus from Figure 5.5 a & b, effective height is given by the relation
$H_e = H + h_n + \frac{h}{2} + \frac{V_h}{2A_J} - \frac{V_h}{2A_J}$. However, in this relationship, it has been assumed that the rise in the suspension level from A-A to A'-A' at the center of the

hydrometer bulb is equal to half the total rise due to the volume of the hydrometer. Thus, the effective height is given as:

$$H_e = \left[H + h_n + \frac{h}{2} - \frac{V_h}{A_J} + \frac{V_h}{2A_J} \right] = \left[H + h_n + \frac{1}{2}\left(h - \frac{V_h}{A_J} \right) \right]$$

Thus, the depth of any level A-A from the free surface level B-B is the effective depth at which the specific gravity is measured by the hydrometer (Figure 5.5b). In Equation (5.7), there are two variables, the effective depth (H_e) and the depth of immersion of hydrometer (H), which depend upon the hydrometer reading R_h. Therefore, by selecting various hydrometer readings (R_h), the depth "H" can be measured with the help of an accurate scale, and corresponding depth "H_e" can be found. Therefore, hydrometer calibration curves can be developed to find the effective depth for each hydrometer reading.

5.7.3 HYDROMETER CALIBRATION CURVES

In order to obtain effective depth of immersion "H_e," the factor V_h/A_J is not applied when the hydrometer is not taken out from the jar containing soil mixture (suspension) when taking readings after the start of the sedimentation at 0.5, 1, 2, and 4 minutes. For subsequent depths after the 4 min reading, the factor V_h/A_J is applied for obtaining the effective depth of immersion.

Thus, the effective height, $H_e = H + h_n + h/2$ (without immersion correction)[5.8]

Thus, the effective height, $H_e = H + h_n + 1/2(h - V_h/A_J)$ (with immersion correction)[5.9]

The effective height is obtained from calibration curves (Figure 5.6) drawn between hydrometer readings corrected for meniscus correction and corresponding

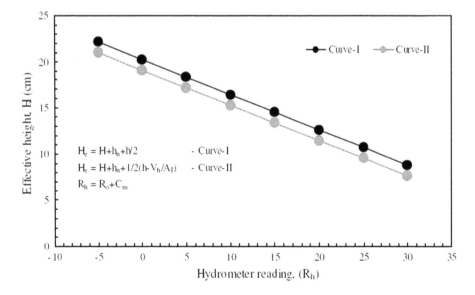

FIGURE 5.6 Hydrometer calibration curves.

effective height as given by Equations [5.8] and [5.9], respectively, as illustrated below:

- Hydrometer No.: 1_BAM (Figure 5.6c)
- Sedimentation Jar No.: SM3
- Height of hydrometer bulb, h (cm): 16
- Volume of hydrometer, V_h (cm^3): 70
- Depth of water displaced by hydrometer in jar, h_j (cm): 2.3
- X-sec area of jar, A_J (cm^2): 30.43
- Dist. between bulb neck and Max^m reading (1.030), h_n (cm): 0.8
- Least count for linear measurements on stem-1(big) div. = 0.38 cm, small div. = 0.19 cm
- Minimum reading on hydrometer stem = 0.995: (0.995 − 1) $*$ 1,000 = (−5) or = 995 − 1,000 = −5
- Maximum reading on hydrometer stem = 1.030: (1.030 − 1) $*$ 1,000 = (30) or 1030 − 1,000 = 30
- Calibration data for hydrometer are given in Table 5.4.

The effective depth "H_e" for various hydrometer readings corrected for meniscus correction at different time intervals is obtained graphically from calibration curves.

TABLE 5.4
Calibration Data for Hydrometer

Curve-I: $H_e = H + h_n + h/2$ (valid for readings taken @ 0.5, 1, 2, and 4 minutes)			Curve-II: $H_e = H + h_n + 1/2(h\text{-}V_h/A_J)$ (valid for readings taken after 4 minutes)		
R_h	H (cm)	H_e (cm)	R_h	H (cm)	H_e (cm)
30	0	8.8	30	0	7.65
25	1.9	10.7	25	1.9	9.55
20	3.8	12.6	20	3.8	11.45
15	5.7	14.5	15	5.7	13.35
10	7.6	16.4	10	7.6	15.25
5	9.5	18.3	5	9.5	17.15
0	11.4	20.2	0	11.4	19.05
-5	13.3	22.1	-5	13.3	20.95

h = 16 cm h_n = 0.8 cm
Volume of hydrometer, V_h = 70 cm^3
X-sec area of jar, A_J = 30.43 cm^2

5.7.4 HYDROMETER READING CORRECTIONS

5.7.4.1 Meniscus Correction (Cm)

Insert the hydrometer in the measuring jar containing about 900 ml of water.

1. Take the readings of the hydrometer at the top and at the bottom of the meniscus, as shown in Figure 5.7.
2. Determine the meniscus correction, which is equal to the difference between the two readings.
3. Since the readings on the hydrometer are increasing in a downward direction, therefore, the meniscus correction C_m is always positive and is constant for the hydrometer.
4. The reduced observed hydrometer reading R'$_h$ is corrected to obtain the corrected hydrometer reading R$_h$ as:

$$R_h = [R'_h + C_m] \tag{5.10}$$

Where: $R'_h = R_o - R_w$

R_o = Hydrometer reading at upper rim of the meniscus in suspension jar with hydrometer immersed in the jar

R_w = Density of water at test temperature

R_h = Corrected hydrometer reading for meniscus correction (C_m)

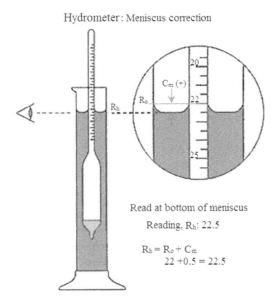

FIGURE 5.7 Meniscus correction for hydrometer.

Notes:
- For convenience, the hydrometer readings ($R_o = R'_h$) are recorded after subtracting 1 and multiplying the remaining by 1,000. This reduced observed reading is designated as R'_h.
- For example, if the hydrometer reading is $R_o = 1.027$, the density of water is assumed as $R_w = 1$ g/cc, then: $R'_h = (R_o - R_w) * 1000 = (1.027 - 1) * 1000 = 27$
- However, if the hydrometer reading is recorded as 1,027 instead of 1.027, then R_w is assumed as 1,000 and $R'_h = (1,027 - 1,000) = 27$ and so on.

It may be noted that Equation [5.8] is useful for finding the effective height (H_e) from the calibration chart corresponding to the reduced observed hydrometer reading (R'_h).

5.7.4.2 Dispersing Agent Correction (C_d)
Since the addition of dispersing agent in water increases its density, it is always negative.

1. The dispersing agent correction is determined by recording the hydrometer reading in clean distilled water in one jar and in the same water after adding the dispersing agent in another jar. Care must be taken that same quantity of dispersing agent be added as taken for the suspension.
2. To obtain the C_d correction, 100 ml of the dispersing agent solution (Sodium-hexa-meta-phosphate solution) is taken in a standard measuring cylinder and the volume is made equal to 1,000 ml by adding distilled water. The hydrometer is inserted and the reading is taken.
3. Similarly the hydrometer is inserted in a standard measuring jar filled with clean distilled water and a reading is taken. The difference between the two readings gives the dispersing agent correction (C_d) and it is always subtracted from the hydrometer reading after it is corrected for meniscus, e.g. C_d is applied after correcting observed hydrometer reading (R_o) for meniscus correction. Then the corrected hydrometer reading after applying C_d correction is:

$$R_h'' = [R_h - C_d] = [R'_h + C_m - C_d] = [(R_o - R_w) + C_m - C_d] \qquad [5.11]$$

5.7.4.3 Temperature Correction (C_t)
If the hydrometer is calibrated at a standard temperature ($20°C$), then the test temperature should also be the same as the standard, else, temperature correction is applied, which is either positive (if test temp > standard temp) or negative (test temp < standard temp). It is found as follows:

1. Record the test temperature carefully.
2. If the standard temperature at which the hydrometer is calibrated is less than the test temperature at which the test is conducted, then the density of the

suspension becomes lighter and lesser readings are recorded as the readings increase in a downward direction on the hydrometer stem. Therefore, the temperature correction (C_t) is positive.

3. If the standard temperature is greater than test temperature, then the density increases and the solution becomes denser, which gives higher hydrometer readings and hence the temperature correction (C_t) is negative.
4. A manufacturer's chart is available and the correction can be applied accordingly.
5. In the present case, temperature correction (C_t) was zero as the temperature was almost the same.

Thus, the corrected hydrometer reading will be:

$$R = [R''_h \pm C_t] = [R'_h + C_m - C_d \pm C_t] \qquad [5.12]$$

5.7.4.4   Composite Correction (C)

Instead of finding individual corrections, it is also convenient to find the composite correction (C) and then find the corrected hydrometer reading (R). This correction may be either positive or negative. This correction is determined as follows:

1. Dispersing agent solution is prepared as in usual manner in a measuring jar (1,000 ml) used for th hydrometer test.
2. Clean distilled water is taken in another comparison jar (1,000 ml) and the same quantity of dispersing agent is added as used in the test (for preparation of soil suspension).
3. The temperature of both measuring jars should be same.
4. The hydrometer is immersed in the comparison jar containing distilled water and dispersing solution and the hydrometer reading is taken at the top of the meniscus.
5. The negative of the hydrometer reading obtained gives the positive composite correction or vice versa for positive reading.

   For example, if the hydrometer reading in the comparison jar is +1.5, then the composite correction is negative, e.g. C = −1.5.

   Similarly, if the hydrometer reading in the comparison jars is −1.0. Then the composite correction is positive, e.g. C = + 1.0.
6. The composite correction is found before the start of test, and also at every time interval exceeding 30 minutes.

Thus, the three corrections can be combined into one composite correction (±C) and the corrected hydrometer reading after applying all corrections is computed as:

$$R = [R''_h \pm C_t] = [R'_h + C_m - C_d \pm C_t] \Rightarrow [R = R'_h \pm C] \qquad [5.13]$$

It may be noted as R'_h is useful for computation of %age finer (N) with respect to dry mass of 50 g after applying composite correction, e.g. R = R'_h ± C.

Whereas $R_h = [R'_h + C_m]$, Equation 5.8 is useful for finding the effective height (H_e) from the calibration chart corresponding to the observed hydrometer reading (R'_h) as illustrated in data observation sheet.

Then the percentage finer than the size D is given by:

$$N = \left[\left(\frac{G_s}{G_s - 1} \right) \times \frac{R}{M_s} \times 100 \right] (\%) \qquad [5.14]$$

Where: R = Corrected hydrometer reading $(R'_h \pm C)$

$R'_h = R_o$- R_w (as explained in section 5.8.4.1.) = reduced observed hydrometer reading

R_o = Observed hydrometer reading when hydrometer is immersed in suspension jar

R_w = Density of water (1 g/cm^3 of at test temperature as the case may be), and

M_s = Mass of dry soil in 1,000 ml suspension (50 g).

5.7.5 PRE-TREATMENT OF SOIL

The percentage of soluble salts shall be determined (IS: 2720 (Part-21)-1977). In this case, it is more than 1%, so the soil is washed with water before further treatment, taking care that soil particles are not lost. For details, refer to IS: 2720(Part-4)-1985, Clause No. 5.2.4.1.

5.7.6 PREPARATION OF SUSPENSION WITH DISPERSING AGENT

1. Take about 50 g or pre-treated dry fine soil passing through a 75 μm sieve.
2. Place the soil in an evaporating dish and cover it with 100 cm^3 of dispersing agent solution (3.3 g of sod. hexamataphosphate + 0.7 g of sodium carbonate) and warm gently for about 10 minutes.
3. Prepare sodium hexametaphosphate solution (dispersing agent) using distilled water and sodium hexametaphosphate. Add sodium hexametaphosphate to distilled water at a rate of 40 g sodium hexametaphosphate per liter of total solution.
4. Transfer the sample to the shaking jar with a mechanical stirrer using distilled water until the jar is three-fourths full and operate for about 5 minutes.
5. Meanwhile, immerse the hydrometer in the 1,000 cm^3 jar filled with distilled water and 100 cm^3 dispersing agent solution.
6. The soil suspension in the jar is then well stirred for about 15 minutes and then transferred into 1,000 cm^3 jar, making it 1,000 cm^3 using distilled water.

5.7.7 SEDIMENTATION TEST

1. Place the rubber stopper on the open end of the measuring cylinder containing the soil suspension. Shake the jar end-over-end by keeping your palm of your hand on rubber stopper for a period of about 1 to 2 minutes to

complete the agitation of the soil-water solution.

Note: The number of turns during this agitation time should be approximately 60, counting the turn upside down and back as two turns. Any soil remaining in the bottom of the jar during the first few turns should be loosened by vigorous shaking of the jar while it is in the inverted position.

2. Remove the rubber stopper after the shaking is complete. Place the measuring cylinder on the table and start the stop watch.
3. Immerse the hydrometer slowly into the liquid 20 to 25 seconds before each reading, and then allow it to float freely.

Note: It should take about 10 seconds to insert or remove the hydrometer to minimize any disturbance, and the release of the hydrometer should be made as close to the reading depth as possible to avoid excessive bobbing.

4. The reading is taken by observing the top of the meniscus formed by the suspension and the hydrometer stem. The hydrometer is removed slowly and placed back into another jar filled with distilled water at the same temperature as that of the test jar to remove any particles that may have adhered to it.
5. Insert the hydrometer into the suspension jar carefully and take hydrometer readings after elapsed time of 0.5, 1, 2, and 4 minutes and do not remove the hydrometer from the suspension jar up to 4 minutes.
6. Remove hydrometer from suspension jar only after 4th minute reading and after other subsequent readings very carefully and then float it in another jar as mentioned above.
7. Take hydrometer readings after elapsed time of 8, 15, 30, 60 minutes; 2, 4, 8, and 24 hours; and take out the hydrometer after each reading and keep in the clean water jar, respectively.
8. Record the temperature of the suspension once during the first 15 minutes and thereafter at the time of every subsequent reading.
9. Record the value of viscosity of water against recorded temperature.
10. After the final reading, pour the suspension in an evaporating dish, dry it in an oven and find its dry mass.
11. Determine the composite correction before the start of the test and also at 30 minutes, 1, 2, and 4 hours as explained in Section 5.7.4.4. Thereafter, just after each reading, composite correction is determined.
12. Complete the observation sheet as shown in Table 5.5.
13. Compute the required test parameters and the percentage passing through the 75 μm sieve with respect to dry weight of 50 g of soil as shown in Table 5.6.
14. Compute percentage passing with respect to total mass of soil as shown in Table 5.6.
15. Plot the combined particle size distribution curve as shown in Figure 5.8.

5.7.8   OBSERVATION DATA SHEET AND ANALYSIS

Percentage passing 75 μm sieve from sieve analysis data (see Table 5.5):

- Mass of dry soil (M_s) = 50 g

TABLE 5.5
Data sheet for hydrometer test: OBSERVATIONS

Practical Name: **Date:**
Project/Site Name:
Tested By:

Soil Data	Hydrometer analysis of soil sample				
%age finer 0.075 mm from wet sieve analysis	8.43	Specific gravity, G	2.65	ρ_w (g/cm³)	1.00
Mass of dry soil taken passing 0.075 mm sieve for hydrometer analysis, W_d (gm)	?			g (cm/sec²)	981
Particle size (dia), D = ? $M[SQRT\{H_e/t\}]$	BH No.	?	Depth (m)	?	Sample No ?
Hydrometer calibration parameters					
Hydrometer No.	1_BAM	Ht of hydrometer bulb, h (cm)	17.00	Dist. between bulb neck & Max^m reading (1.030), h_n (cm)	0.80
Vol of hydrometer, V_h (ml)	70.00	Depth of water displaced by hydrometer in jar, h_j (cm)	2.30	Area of jar, $A_j = V_h/h_j$ (cm²)	30.43
L.C., 1Divn Big (cm)	0.38	Least Count, 1Divn small (cm)	0.19	Stem length, L (cm)	?

Other parameters

H_e = effective ht. (in cm) = $H + 1/2\{h-V_h/A_j\}$, where $H = h_n + L.C.*$No. of Divns; and t = time (in min.)

H = distance from neck of hydrometer bulb to Hydrometer reading on stem

$M = SQRT[(30* \mu)/(G-1)\rho_w]$: units as: μ in (gm-sec/cm²), and ρ_w in g/cm³ respectively [1 poise = 10^{-3} gm-sec/cm²]

or $M = SQRT[(0.3* \eta/g)/(G-1)\rho_w]$; units as: η in poise, g = 981 cm/sec², and ρ_w in g/cm³ respectively

Else $M = SQRT[(3000*\eta)/(G-1)\gamma_w]$; units as: η in kN-s/m², γ_w = 9.81 kN/m³, and G = Sp. Gravity respectively

Neta (η) = viscosity of liquid in (kN-s/m²), μ = viscosity in absolute units of poise or dyne-s/cm²

1 poise = 10^{-3} gm-sec/cm²   1 poise = 1 dyne-s/cm²   1 millipoise = 10^{-3} poise   1 poise = 10^{-4} kN-s/m²
= 0.1 mN-s/m²

μ (poise) = $[(4.089 * 10^{-9} * (T°C)^2 -(4.1793 * 10^{-7} * T°C) + 1.7016 * 10^{-5}))/g]$ for any value of temperature (T°C) and g = 981 cm/s²

Corrections

Dispersing agent	Sodium Hexametaphosphate (Dispersing agent not included in sample dry mass)		C_d (-)	?
Lab. Temp.°C ?	Meniscus correction, C_m (+)	0.5	Temp. correction, C_t ?	Composite correction, C (±) −0.25

$C = [C_m + C_t - C_d]$; For C_t use + sign if room/lab. Temp. is more than standard temp., C_d is always -ve & C_m + ve

Corrected hydrometer readings

R_h = Corrected hydrometer reading for meniscus to be used in the calibration graph to obtain the effective depth, $H_e = R_h' + C_m$

R = Corrected hydrometer for composite correction (C) to be used for determination of %age finer: (N) = $R_h' - C$

TABLE 5.6
Data Sheet for Hydrometer Test: Calculations

Date and time for taking hydrometer readings	t (min)	Temp (°C)	Viscosity, μ (poise)	Observed hydrometer reading in jar, (Rw)	Density of water (ρw)	Observed hydrometer reading, Rh' = (Ro – Rw) * 1000	Rh = Rh' + Cm	Eff. Height, He (cm) = H + hn + h/2	Eff. Height, He (cm) = H + hn + 1/2(h-Vh/Aj)	SQRT (He/t)	M = √{(30 * μ)/(G-1)ρw}	D = (Col. 12 * Col. 11) (mm)	R = Rh' – C	%age finer w.r.t. Md, N (%) = [(100 * G)/((G-1) Md)] * R	Combined %age finer w.r.t. total soil sample, N' (%) = N * %age finer 0.075 mm sieve
1	2	3	4	5	6	7	8	9	10	11	12	13	14	15	16
	0.5	23.6	0.00925	1.0265	0.99739	29.11	29.61	8.95	NA	4.23	0.012703	0.054	29.36	92.85	48.28
	1	23.6	0.00925	1.0235	0.99739	26.11	26.61	10.09	NA	3.18	0.012703	0.040	26.36	83.36	43.35
	2	23.4	0.00930	1.022	0.99744	24.56	25.06	10.68	NA	2.31	0.012733	0.029	24.81	78.46	40.80
	4	23.7	0.00923	1.02	0.99737	22.63	23.13	11.41	NA	1.69	0.012687	0.021	22.88	72.37	37.63
	8	23.9	0.00919	1.0175	0.99732	20.18	20.68	NA	11.19	1.18	0.012657	0.015	20.43	64.62	33.60
	15	23.5	0.00927	1.015	0.99742	17.58	18.08	NA	12.18	0.90	0.012718	0.011	17.83	56.40	29.33
	30	23.3	0.00932	1.0125	0.99747	15.03	15.53	NA	13.15	0.66	0.012748	0.008	15.28	48.34	25.14
	60	26.5	0.00864	1.0095	0.99665	12.85	13.35	NA	13.98	0.48	0.012279	0.006	13.10	41.44	21.55
	120	26.7	0.00861	1.007	0.99659	10.41	10.91	NA	14.91	0.35	0.012251	0.004	10.66	33.70	17.53
	240	26.5	0.00864	1.0045	0.99665	7.85	8.35	NA	15.88	0.26	0.012279	0.003	8.10	25.62	13.32
	1440	26.5	0.00864	1.0015	0.99665	4.85	5.35	NA	17.02	0.11	0.012279	0.001	5.10	16.13	8.39

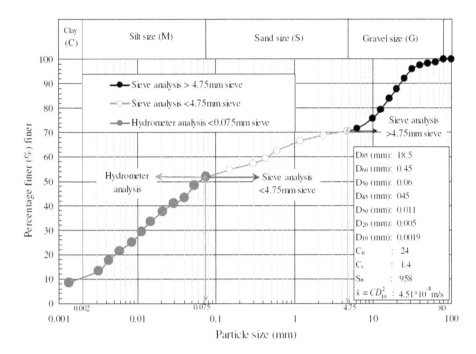

FIGURE 5.8   Combined particle size distribution curve for the soil sample (sieve + hydrometer analysis).

- Meniscus correction (C_m) = 0.5 (Refer Section 5.7.4.1)
- Specific gravity of solids (G) = 2.65 (to be obtained from sp. gravity test for the same soil taken for hydrometer analysis)
- Composite correction (C) = −0.25 (refer to Section 5.7.4.4)
- Record observed $(R_o$ and $R_{h'})$ and corrected hydrometer readings $(R_h$ & R) are given in Table 9.6
- $R_h = R_{h'} + C_m$ to be used for calculation of effective height [See Equation (5.8)]
- $R = R_{h'} \pm C$ to be used for calculation of %age Passing, N (See [5.11])
- Calculate Eff. Height, H_e (cm) using Equation (9.7) or Figure 5.6 as desired
- Calculate Factor – M using[5.4]
- Calculate particle size (dia), D using Equation (5.3)
- Calculate percentage passing (N) with respect to dry mass of soil using [5.14]
- Calculate Combined %age finer with respect to total soil sample, N' (%) = N * %age finer 0.075 mm sieve
- For calculation details, refer to Table 5.6
- Combined data for combined particle size distribution curve is given in Table 5.7
- Plot particle size distribution curve as shown in Figure 5.8
- Calculate various diameters such as D_{10}, D_{20}, D_{30}, D_{50}, D_{60}, D_{85}, etc.
- Calculate C_u, C_c, S_n, k, and filter design parameters
- Classify the soil and comment on the suitability of this soil sample either as a construction material or as a foundation medium for construction of various infrastructures.

TABLE 5.7

Data for combined particle size distribution curve (sieve + hydrometer analysis)

Sieve size (mm)	Mass retained (g)	%age retained	Cum. %age retained	%age finer	Remarks
50	0	0	0	100	1. %age finer = 100-cum.%age
40	0	0	0	100	retained
31.5	252	252	8.4	92	2. Particle size distribution
25	0	252	8.4	92	(PSD) curve is plotted
20	73	325	10.8	89	between col. 1 and 5.
16	179	504	16.8	83	3. From test data, it is seen that
12.5	75	579	19.3	81	soil is fine grained with 52%
10	140	719	24	76	passing 0.075 mm IS sieve.
6.3	125	844	28.14	72	4. Hydrometer analysis is
4.75	33	877	29.24	71	carried out on the soil sample
2.36	96	973	32.44	69	passing 75 μm.
1.18	122	1,095	36.5	66	
0.6	83	1,178	39.3	63	
0.425	9	32	16	59	
0.3	77	1,255	41.83	57	
0.15	20	1,275	42.5	55	
0.075	10	1,285	42.833	52	
0.054	For data analysis refer to			48	Data from hydrometer analysis test.
0.040	Tables 5.5 and 5.6			43	Combined %age finer with respect
0.029				41	to total soil sample, N' (%) = N*%
0.021				38	age finer 0.075 mm sieve (refer to
0.015				34	Table 5.6). Based on soil grading
0.011				29	and analysis of test data, soil is
0.008				25	classified as poorly graded fine
0.006				22	grained with appreciable gravel
0.004				18	content.
0.003				13	
0.001				8	

5.7.9 RESULTS AND DISCUSSIONS

Based on test results and analysis, the PSD curve can be plotted using col. 13 and 16. From the combined PSD curve find:

1. D_{10} = _____ (mm)
2. D_{30} = _____ (mm)
3. D_{60} = _____ (mm)
4. $C_u = D_{60}/D_{10}$

5. $C_c = (D_{30})^2/D_{10} \times D_{60}$ based on C_u and C_c, classify the soil
6. $D_{20} = \rule{1cm}{0.4pt}$ (mm)
7. $D_{50} = \rule{1cm}{0.4pt}$ (mm)
8. $S_N = 1.7 \sqrt{\dfrac{3}{D_{50}^2} + \dfrac{1}{D_{20}^2} + \dfrac{1}{D_{10}^2}}$ required for fill material (see Expt 4)
9. D_{15}, D_{50} & D_{85} for filter design criteria (see Expt 4)
10. $k = CD_{10}^2 = 100D_{10}^2$

5.8   GENERAL COMMENTS

1. The material that passed through the 75 μm sieve during the mechanical sieve analysis is used to perform the hydrometer analysis.
2. This method shall not be applicable if less than 10% of the material passes the 75 μm IS sieve.
3. Prepare sodium hexametaphosphate solution (dispersing agent) using distilled water and sodium hexametaphosphate. Add sodium hexametaphosphate to distilled water at the rate of 40 g sodium hexametaphosphate per liter of total solution.

5.9   APPLICATIONS OF HYDROMETER ANALYSIS

In a given soil, the percentage of different soil particles up to 75 μm is determined by sieve analysis, but the percentage of various soil particles finer than 75 μm is determined by hydrometer analysis. Hence, the hydrometer analysis is useful in knowing the percentage of silts and clays.

Furthermore, activity of clays may be estimated if clay fraction finer than 2 μm is known, which is determined using hydrometer analysis. Therefore, activity of clay is given by the relation:

$$A = \frac{Plasticity\ Index\,(PI)}{Clay\ fraction\ finer\ 2\ micron\ sieve} \qquad [5.15]$$

Activity is a soil parameter, which describes higher compressibility and higher swelling and shrinkage characteristics of clays. Clays can be grouped into three qualitative categories depending upon their activity such as 0.75 for inactive clays, 0.75–1.25 for normal clays, and greater than 1.25 for active clays, respectively.

5.10   SOURCES OF ERROR

1. One of the biggest sources of error shall arise if an adequate quantity of dispersing agent is not used in wet sieve analysis or hydrometer analysis.
2. Any change in temperature will result in a big error. Therefore, temperature may be recorded during the test and compared with the standard temperature.
3. Any deviation in hydrometer readings in opaque soil suspension may result in error. Therefore, the hydrometer readings may be correctly taken on the top surface of the meniscus formed by the soil suspension.
4. Hydrometer readings may be taken when the hydrometer settles in the suspension and there is no movement. Otherwise, incorrect readings will be recorded, which will result in error.

5.11   PRECAUTIONS

1. The insertion of the hydrometer should be done carefully.
2. The hydrometer should float at the center of the jar and should not touch the wall of the sedimentation cylinder.
3. The stem of the hydrometer should be dry and clean.
4. There must be no vibrations in the vicinity of setup.
5. Minimize the temperature variations by keeping both jars away from any local course of heat and direct sunlight.

REFERENCES

ASTM E100-05. "Standard Specification for ASTM Hydrometers." *Annual Book of ASTM Standards*, © ASTM International, 100 Barr Harbor Drive, PO Box C700, West Conshohocken, PA 19428-2959, United States, www.astm.org.

ASTM E126-05a. "Standard Test Method for Inspection, Calibration, and Verification of ASTM Hydrometers." *Annual Book of ASTM Standards*, © ASTM International, 100 Barr Harbor Drive, PO Box C700, West Conshohocken, PA 19428-2959, United States, www.astm.org.

ASTM D 1298. "Test Method for Density, Relative Density (Specific Gravity), or API Gravity of Crude Petroleum and Liquid Petroleum Products by Hydrometer Method." *Annual Book of ASTM Standards*, © ASTM International, 100 Barr Harbor Drive, PO Box C700, West Conshohocken, PA 19428-2959, United States, www.astm.org.

ASTM D421. 1985. "Standard Practice for Dry Preparation of Soil Samples for Particle-Size Analysis and Determination of Soil Constants." *Annual Book of ASTM Standards*, Philadelphia, sec. 4, Vol. 04–08, pp. 10–16, 1963, www.astm.org.

ASTM D 422–63(07e2). 1963. "Standard Test Method for Particle Size Analysis of Soils." *Annual Book of ASTM Standards, Philadelphia*, sec. 4, Vol. 04–08, pp. 10–16, www.astm.org.

ASTM D 2487. 2006. "Standard Practice for Classification of Soils for Engineering Purposes (USCS)." *Annual Book of ASTM standards*, Vol. 04–08.American Society for Testing and Materials, Philadelphia, United States, www.astm.org.

ASTM D2488-00. 2000b. "Standard Practice for Description and Identification of Soils (Visual-Manual Procedure)." *Annual Book of ASTM Standards*, Vol. 04–08, American Society for Testing and Materials, Philadelphia, United States, www.astm.org.

ASTM D 2937-(00). "Standard Test for Density of Soil in Place by the Drive-Cylinder Method." *Annual Book of ASTM Standards*, Vol. 04–08, American Society for Testing and Materials, Philadelphia, United States, www.astm.org.

IS: 1498. 1970. "Classification and Identification of Soils for General Engineering Purposes." *Bureau of Indian Standards*, New Delhi.

IS: 2720 (Part 1). 1980. "Indian Standard Code for Preparation of Soil Samples." *Bureau of Indian Standards*, New Delhi.

IS: 2720 (Part 4). 1985. "Method of Test for Soils: Determination of Grain Size Distribution." *Bureau of Indian Standards*, New Delhi.

Reddy, Krishna R. 2000. "Engineering Properties of Soils Based on Laboratory Testing." Department of Civil and Materials Engineering University of Illinois at Chicago, USA.

Weatherly, W. C. 1929. "The Hydrometer Method for Determining the Grain Size Distribution Curve of Soils." Master of Science Thesis, Department of Civil Engineering, Massachusetts Institute of Technology (MIT), USA.

6 Atterberg Limits of a Fine-Grained Soil Sample

References: IS: 2720-Part 5 (1985); IS: 2720-Part 6 (1972); IS: 10077; ASTM D 4318-2010; ASTM D427-04; BS 1377: Part 2 (1990)

6.1 OBJECTIVES

To determine the range of water content that exhibits consistency of given fine grained soils at different consistency limits. The Atterberg limits or index properties are indicative of engineering properties of soils.

6.2 INTRODUCTION

Swedish soil scientist Albert Atterberg (1911) defined the five consistency limits in terms of water content used to evaluate the behavior of fine grained soils. Out of these five limits, the liquid limit (LL), plastic limit (PL), and shrinkage limit (SL), are determined on a soil specimen passing through a 425 μm (0.425-mm) sieve in the laboratory. LL and PL generally indicate the consistency states of fine grained soils in a qualitative manner with change in water content. SL helps in the identification of expansive soils, which swell and shrink with changing water content. LL represents the state of consistency at which the soil changes from the liquid state to the plastic state and behaves as plastic material with small shear strength of about 2 kPa. LL indicates water content against 25 blows to close about 12 mm soil groove in Casagrande's limit device for clayey soils or clay-dominated fine grained soils. It may be noted that the groove cut in the soil pat (with about 2 mm wide and 12 mm in height) by a standard cutting tool in the Casagrande's cup closes (or fails by closing the groove) by flowing towards the center rather than sliding under given blows. The number of blows (N) are recorded by dropping the Casagrande's cup from a 10 mm height onto a hard rubber base repeatedly untl the cut groove closes by 12–13 mm in length at a rate of about 2 drops per second. When the groove closes about 12–13 mm, the number of blows are recorded and the test is stopped. The soil specimen along the closed groove is collected for the determination of water content (w %) against the recorded number of blows (N). Since it may be very difficult to determine the LL against 25 blows exactly in a single test, it is recommended to perform at least four to five tests to determine the water content against specified recorded number of blows, a flow curve is plotted between number of blows (N) along the x-axis on a semi-log graph paper and water content (w %) along the y-axis on natural graph paper. The water content against 25 blows from the flow curve represents LL of the soil specimen. The flow curve plays a vital role in the characterization of fine grained soils. The slope of the flow curve (generally known as flow index) indicates the rate of loss of shear strength of a soil specimen in the field under changing water content.

However, LL for silt-dominated fine grained soils is determined using a fall cone penetrometer device as these soils fail by sliding in the Casagrande cup rather than flowing towards the center under given blows. In this test, a soil pat is prepared and filled in a cup which is then placed under a falling cone, which is allowed to penetrate into soil pat for a given time interval and the penetration is measured for each such test. Likewise a flow curve is plotted between penetration (mm) along the x-axis and water content (%) along the y-axis on a natural graph paper. The water content against 20 mm penetration from the flow curve represents the LL for the soil specimen tested. Various indices derived from consistency limits play a vital role in identification and characterization of fine grained soils.

6.3 DEFINITIONS AND THEORY

6.3.1 ATTERBERG LIMITS

Atterberg limits, generally referred to as consistency limits, represent the range of water contents at which the fine grained soils (e.g. clays) change their consistency from one state to another with an increase or decrease of water content, such as from liquid state to plastic state, plastic state to semi-plastic state etc. (Reddy 2000). Therefore, the behavior of the fine-grained soils is dependent upon the water-holding capacity of the soil system. Swedish soil scientist Albert Atterberg (1911) defined the five consistency limits in terms of water content for fine grained soils:

1. **Liquid limit (LL):** Defined as the upper limit of water content at which soil posses minimum shear strength and behaves as a plastic material. LL is also known as a boundary between plastic and liquid states.
2. **Plastic limit (PL):** Defined as the water content below which the soil is non-plastic. PL is also known as boundary between plastic and semi-solid states.
3. **Shrinkage limit (SL):** Defined as the water content at which soil becomes saturated and below which no volume change occurs. SL is also known boundary between the semi-solid state and the solid states.
4. **Cohesion limit:** Defined as the moisture content at which soil lumps just stick together.
5. **Sticky limit:** Defined as the moisture content at which soil just sticks to a metal surface such as a spatula blade.

A schematic of the relative locations of the consistency limits on the water content scale are shown in Figure 6.1.

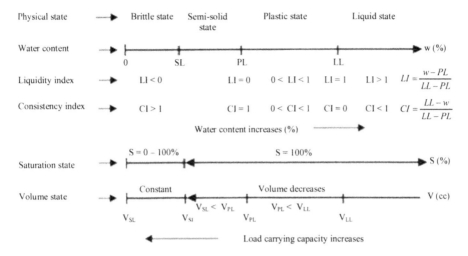

FIGURE 6.1 A schematic of the relative locations of the consistency limits.

Further, out of these five consistency limits, LL and PL generally indicate the consistency states of fine grained soils in qualitative manner with change in water content. These depend upon the amount and type of clay present in the soil, e.g., a soil with high clay content usually has high LL and PL.

6.3.2 SOIL CONSISTENCY

Soil consistency is generally a term associated with clayey soil which describes the degree of firmness of the soil specimen qualitatively and the ease with which it can be deformed under varying water content. The schematic of soil consistency and the variation of volume change with water content is shown in Figure 6.2.

Typical values of LL and PL fine soils are given in Table 6.1.

Shrinkage limit and plasticity index can be used to determine the swell potential of the fine grained soil, which can be detrimental to the lightweight structures. Typical values of "degree of expansion" of expansive soils are given in Table 6.2.

6.4 METHOD OF TESTING

The consistency limits (LL) for fine grained soils passing through a 425 μm IS sieve are determined using Casagrande's apparatus and falling cone penetrometer method.

6.5 DETERMINATION OF LIQUID LIMIT OF A REMOLDED FINE-GRAINED SOIL SAMPLE BY CASAGRANDE'S METHOD (IS: 2720-PART 5; ASTM D4318; BS 1377: PART 2)

6.5.1 SOIL TESTING MATERIAL

Disturbed or remolded fine grained soil sample of about 300 passing Sieve No. 40 (0.425-mm IS sieve) is taken for determination of consistency limits.

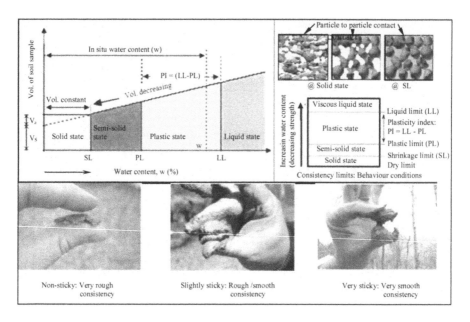

FIGURE 6.2   Consistency limits and variation of volume change with water content.

TABLE 6.1
Typical Values of LL and PL for Fine Grained Soils

Sr. No.	Soil type	Liquid limit, LL (%)	Plastic limit, PL (%)
1	Silts	24–27	16–20
2	Clays	80–100	45–54
3	Colloidal clays	300–400	45–50
4	Kaolinite	35–100	15–60
5	Illite	55–120	20–70
6	Montmorillonite	100–800	50–700

6.5.2   TESTING EQUIPMENT AND ACCESSORIES

The following equipment and accessories are required for the LL test (Figure 6.3):

1. Casagrande's liquid limit device
2. 0.425 mm IS sieve
3. Grooving tolls of Casagrande and ASTM types
4. Disuntiled water
5. Glass Perspex sheet for mixing soil
6. Porcelain dish
7. Weighing balance accuracy 0.01 g

TABLE 6.2

Typical Values for Degree of Expansion Based on Consistency Limits (Holtz and Kovacs 1981)

Sr. No.	Plasticity Index, PI (%)	Shrinkage Limit, SL (%)	Volume Change	Degree of Expansion
1	> 35	< 11	> 30	Very High
2	25–41	7–12	20–30	High
3	15–28	10–16	10–20	Medium
4	< 18	> 15	< 10	Low

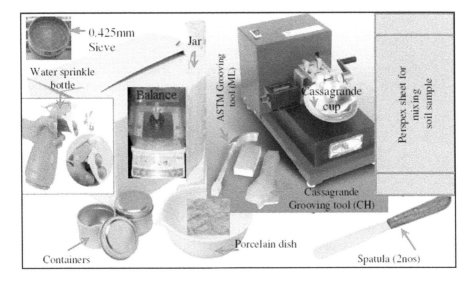

FIGURE 6.3 Casagrande apparatus and allied accessories for LL test.

8. Wash bottle
9. Duster and grease
10. Oven, moisture content cans9. A plastic bag
11. Spatula
12. Desiccator

6.5.3 TESTING PROGRAM

1. Inspect the LL device and ensure that cups drops exactly by 10 mm onto a rubber base (Figure 6.4).
2. Take about 300 g of the air-dried soil sample passing through the 0.425 mm sieve. It is worth mentioning here that the same soil paste should be used for all the three consistency limit tests (LL, PL, and SL).

3. Spread the soil sample on a glass Perspex sheet uniformly, sprinkle water on it, and mix it thoroughly. Add more water and keep mixing until a uniform paste is prepared such that materials in Casagrande's cup under repeated drops in the range of 10 to 15 blows.

   Note: It should be ensured that the consistency limit tests are done from "wet to dry method." For the liquid limit test, each trial is to done from wet to dry side by drying the soil paste. Care should be taken that no dry soil is to be added into the paste to dry it in haste. This will not produce a homogeneous mix as the clay particles have a high surface area and need more time to saturate thoroughly.

4. Keep the soil paste in an airtight plastic bag for sufficient time until the water content is uniform in the whole soil paste.

5. Take out the soil paste from the plastic bag and remix it thoroughly before testing.

6. Take about 200 g of the soil paste for the LL test and preserve remaining soil paste for plastic and shrinkage limit tests.

7. Fill the paste in Casagrande's cup as required with the help of a spatula and level it carefully with a straight edge. Make sure that soil specimen has about 10 to 12 mm in the center of the cup (Figure 6.5) and excess soil paste is transferred back to the glass Perspex sheet.

8. Now take either Casagrande's tool or ASTM tool as per requirement and cut a groove along the centerline through the soil pat in Casagrande's cup. Make sure that the groove is cut straight with at least 2 mm wide and 10 mm in height.

   Note: It may be noted that ASTM tool is used for low plastic soils (silty soils) to cut a groove, whereas Casagrande tool is used for high plastic soils (clays or clayey soils). Soils can be classified either as clayey soils or silty soils by soil grading (sieve and hydrometer analysis).

9. Now give repeated drops by turning the handle of Casagrande's cup at 2 revolutions per second and count the number of blows until the two sides of the groove flow and join for a distance of about 12 mm at the bottom of the groove in the cup.

10. Record the number of blows for the first trial as N_1 (Table 6.3) when the two parts of the groove join together for about a 12 mm length.

11. Take a representative sample of the soil paste (15–20 g) from the two parts of the groove, which joined together by repeated drops for determination of water content. Record this value of water content (as per codal procedure) as w_1 against N-value of N_1 for first trial of test (Table 6.3).

12. Transfer the remaining soil paste from the cup to the glass Perspex sheet and clean the cup thoroughly for the next trial of LL test.

13. Mix the remaining soil paste thoroughly for about 10–15 minutes (in some cases, it may take more time to dry the soil paste for next trial test) so as to change the water content.

14. Repeat steps 7 to 13 and record the number of blows (N) and determine the water content (w %) in each case for about 5 to 6 trials (data sheet in Table 6.3).

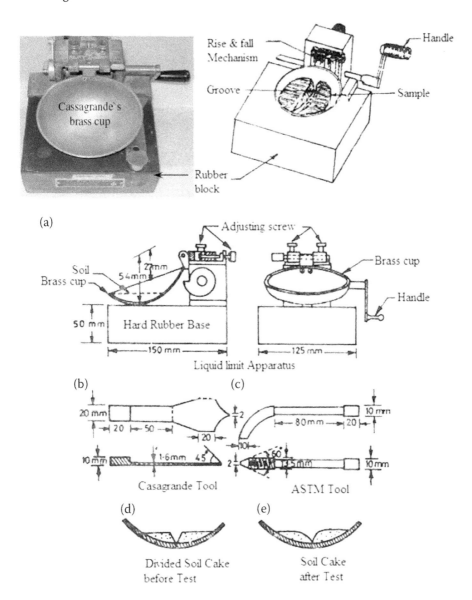

FIGURE 6.4 Liquid limit (LL) test apparatus: (a). Casagrande's cup (b). Casagrande grooving tool, (c). ASTM grooving tool, (d). Divided soil cake before LL test, and (e). Soil cake after LL test.

15. Make sure that N_1 is in the range of 12–15 for first trial test. If N_1 is less than 10, ignore it and dry the soil paste by mixing it thoroughly so as to get N_1 value within above the range.

For guidance, N_1 should be in the range of 12–15, N_2: 15–24, N_3: 24–30, N_4:30–36 and N_5: 35–45, respectively (in no case should N exceed 45). This is due to the fact

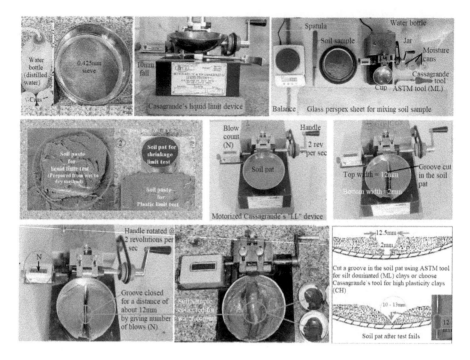

FIGURE 6.5   Schematic procedure for LL test.

that as the water content reduces by drying the soil paste, the soil paste becomes stiffer and resistance to soil paste flow increases (e.g. N value increases).

6.5.4   OBSERVATION DATA SHEET AND ANALYSIS

Based on test results and data analysis as given in Table 6.3, plot a flow curve between $\log_{10}N$ (along x-axis) and w (along y-axis) on a semi-log paper and determine water content as the liquid limit corresponding to N = 25 (Figure 6.6). The water content corresponding to 25 blows (i.e. *LL*) is derived by interpolation in case the N-value is not exactly 25. A sample of the flow curve is shown in Figure 6.6 and the outcome of the test results is discussed in detail. A practice data sheet is given in Table 6.4 and Figure 6.P-1. The students are advised to conduct the liquid limit test as per codal procedures given above and plot the flow curve and discuss the outcome of the test results.

6.5.5   RESULTS AND DISCUSSIONS

From flow curve between ($\log_{10}$ N) and (w %), the following parameters are obtained:

1. Water content against N = 25 blows = 35.5% ⇒ Liquid limit, LL = 35.5% (a)

The liquid limit helps in classifying the soils in terms of degree of plasticity as given in tabular form below.

TABLE 6.3
Data sheet for liquid limit test by Casagrande method

Project/Site Name: Date:

Client Name: Job. No. Sample No.

Sample Recovery Depth: Sample Recovery Method:

Sample Description:

Tested By:

Test Type: on gravimetric basis/on volumetric basis

Sl. No.	Observations and Calculations	Trial No.					
		1	2	3	4	5	6
Observations							
1	Number of blows recorded (N)	12	19	27	34	42	NA
2	Moisture Container No.	C101	C105	C107	C109	C111	NA
3	Weight of empty container, W_1 (g)	9.31	9.89	9.5	9.84	9.62	NA
4	Weight of container + wet soil, W_2 (g)	46.72	49.19	47.18	48.63	45.58	NA
5	Weight of container + dry soil, W_3 (g)	36.58	38.81	37.49	38.66	36.41	NA

Calculations (For water content determination as per Codal Procedures, refer Practical No. 1)

5	Weight of water, $W_w = W_2 - W_3$ (g)	10.14	10.38	9.69	9.97	9.17	NA
6	Weight of solids, $W_s = W_3 - W_1$ (g)	27.27	28.92	27.99	28.82	26.79	NA
7	Water content, $w = (W_w)/(W_s) * 100$ (%)	37.18	35.89	34.62	34.59	34.23	NA
8	Plot flow curve between Row No. 1 and Row No. 7						

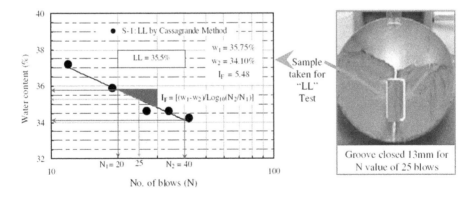

FIGURE 6.6 Flow curve for fine grained soil by LL test.

TABLE 6.4
Practice data sheet for liquid limit test by Casagrande method

Project/Site Name: Date:

Client Name: Job. No.   Sample No.

Sample Recovery Depth: Sample Recovery Method:

Sample Description:

Tested By:

Test Type: on gravimetric basis/on volumetric basis

Sl. No.	Observations and Calculations	Trial No.					
		1	2	3	4	5	6

Observations

1   Number of blows recorded (N)

2   Moisture Container No.

3   Weight of empty container, W1 (g)

4   Weight of container + wet soil, W_2 (g)

5   Weight of container + dry soil, W_3 (g)

Calculations (For water content determination as per Codal Procedures, refer Practical No. 1)

5   Weight of water, $W_w = W_2 - W_3$ (g)

6   Weight of solids, $W_s = W_3 - W_1$ (g)

7   Water content, $w = (W_w)/(W_s) * 100$ (%)

8   Plot flow curve between Row No. 1 and Row No. 7 (Figure 6.P-1)

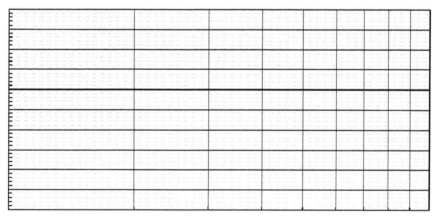

10   100

FIGURE 6P-1   Flow curve (based on liquid limit Test).

Note: Higher values of liquid limit also indicate higher compressibility of soils under loads.

As per Indian standard soil classification system (ISSCS)		As per unified soil classification system (USCS)	
Liquid limit, LL (%)	**Description of soil**	**Liquid limit, LL (%)**	**Description of soil**
< 35	Clayey soil of low plasticity (CL)	< 35	Clayey soil of low plasticity (CL)
	Silty soil of low plasticity (ML)		Silty soil of low plasticity (ML)
35–50	Clayey soil of medium compressibility (CI)	> 50	Clayey soil of high plasticity (CH)
	Silty soil of medium plasticity (MI)		Silty soil of high plasticity (MH)
> 50	Clayey soil of high plasticity (CH)	C: Clayey soil M: Silty soil L: Low; I:	
	Silty soil of high plasticity (MH)	Medium; H: High	

The *flow index* or slope of the flow curve may be calculated from the relation:

$$I_F = \frac{w_2 - w_2}{Log_{10}\frac{N_2}{N_1}} = 5.48 \tag{b}$$

Where w_1 = water content (%) at N_1 blows = 35.75%, and
w_2 = water content (%) at N_2 blows = 34.10%.

The slope of the flow curve indicates *the rate of loss of shear strength* of the soil sample. This is also used in association with the plasticity index (PI = LL-PL) to get an idea about the shear strength (as toughness index, I_T) of a soil at plastic limit (e.g. $I_T = PI/I_F \cong 0$–3).

6.5.6 PRECAUTIONS

1. Use distilled water for consistency tests
2. Use only air-dried (not oven-dried) soil for consistency tests. However, if organic content is present in the soil sample, then take oven-dried soil for the LL test.
3. Make sure that the groove closes by flowing towards center in the cup under repeated drops and not by slipping for a distance of about 10–12 mm.
4. Soil sample should be collected from the groove area for determination of water content.
5. Make sure that sufficient time is given to the soil paste to attain uniform water content distribution before LL test.

6. Make sure that the test is carried out from the wet to the dry side of soil paste.
7. At least 5 to 6 trials should be conducted for liquid limit test as given in Table 6.4.

6.5.7 GENERAL COMMENTS

- Air-drying of soils may decrease their liquid limit by 2 to 6% (Casagrande 1932, Bowels 1970). Soils which are likely to pass the 425 μm (0.425 mm) sieve may be tested in their natural conditions after removing a few coarse particles present by hand during mixing. If so tested, the results should state that the soil has been used in the natural condition.
- It is also observed that most of the air-dried soils regain their original limits if they are allowed to mature for 24 to 48 hours (Bowles 1970).
- At the liquid limit, a majority of soils posses a more or less constant shear strength of 2.0 to 2.5 kPa (20 to 25 g/cm^2), i.e., about 1 g/cm^2 per blow (Bowles 1970, Casagrande 1932). A variation from 0.8 to 3kPa has also been reported by others researchers (Norman 1958, Karlsson 1961, Skempton and Northey 1952).
- It has been found that the locus of points for any one soil tends to fall on a straight line by plotting blows (N) to a logarithmic scale (along x-axis) versus water content (w, %) to a natural scale (along y-axis), resulting in a linear relationship. However, it may also be noted that this linear relationship only holds across one cycle of the semi-log plot (10 to 100 blows) with about 4 to 6 points at different water contents together with the corresponding no. of blows (N-values).

6.5.8 SOURCES OF ERROR

One of the main sources of error in LL test is if the soil paste is not mixed properly and if the soil paste does not mature with uniform water content.

Another source of error in the LL test is if the test is not performed from the wet to dry side of water content. In this case, the soil paste will not have uniform water content and hence erratic values of liquid limit will be obtained.

Furthermore, if the silt-dominated soil specimen is tested in Casagrande's cup, it will result in erratic test data as the two halves of the groove fail by sliding towards the center rather than flowing. In such cases, the fall cone penetrometer device is recommended for LL test.

6.5.9 LIMITATIONS OF CASAGRANDE METHOD FOR DETERMINATION OF LIQUID LIMIT

The determination of liquid limit by the Casagrande method has the following limitations:

1. Tendency of low plastic soils to slide in the cup rather than to flow plastically

2. Sensitivity to small differences in the apparatus such as the form of the grooving tool, the hardness of base, wear of tool and the cup
3. Sensitivity to operator's technique

Thus, as an alternative, the cone penetrometer method is recommended to perform to determine the "LL" of low plastic soils.

6.6 DETERMINATION OF PLASTIC LIMIT OF A REMOLDED FINE GRAINED SOIL SAMPLE (IS: 2720-PART 5; ASTM D4318; BS 1377: PART 2)

6.6.1 OBJECTIVES

The main objective for determination of plastic limit is to ascertain the plasticity behavior of fine grained soils. It represents the water content below which the soil is non-plastic. PL is also known as boundary between plastic and semi-solid states.

6.6.2 DEFINITIONS AND THEORY

The plastic limit (PL) is the water content of the soil sample at it begins to crumble when rolled into a thread of 3 mm diameter. About 50 to 60 g of soil paste is required for the PL test, which is preserved from the soil paste prepared for the LL test. In no case is preparation of separate soil paste recommended for PL test. This is due to the fact that both "LL" and "PL" are used together for classification and identification of soils. In case of some unavoidable circumstances, if it is required to prepare soil paste for "PL" test separately, then it should be made sure that the soil sample taken is same as taken for "LL" test and with same other properties such as natural moisture content and sp. gravity values. Further, the soil paste should be prepared exactly in the same way as prepared for "LL" test as per codal procedures. Otherwise, this will not produce a homogeneous mix as the clay particles have a high surface area as shown in SEM images of fine grained soils (Figure 6.7a) and need more time to saturate thoroughly. The plasticity has direct bearing on the properties of fine grained soils and it is used to describe the ease with which soil can be deformed or reworked causing permanent deformation without rupturing. The degree of relative ease or classes of plasticity is described by forming a 40 mm long thread or wire of different diameters of soil pat at a water content where maximum plasticity is expressed (Figure 6.7b).

a. **Soil is non-plastic:** In this class of plasticity, the water content is the dry side of PL (e.g. plastic limit) and when the soil pat is rolled, it will not form a 40 mm long thread of 6 mm diameter.
b. **Soil is slightly plastic:** In this class of plasticity, the water content is the dry side of PL and when the soil pat is rolled, it will form a 40 mm long thread of 6 mm diameter. However, it will not form a 40 mm long thread of 4 mm diameter.

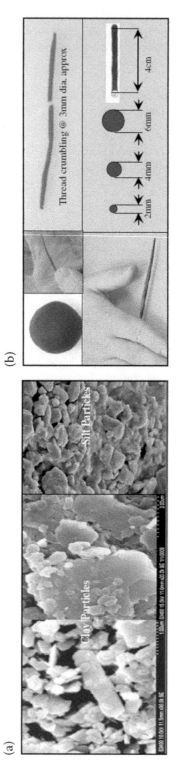

FIGURE 6.7 (a) SEM images of fine grained soils, and (b) classes of plasticity.

    c. **Soil is moderately plastic:** In this class of plasticity, the water content is the dry side of PL and when the soil pat is rolled, it will form a 40 mm long thread of 4 mm diameter. However, it will not form a 40 mm long thread of 2 mm diameter.

    d. **Soil is very plastic:** In this class of plasticity, the water content is the wet side of PL and when the soil pat is rolled, it will form a 40 mm long thread of 2 mm diameter.

6.6.3   SOIL TESTING MATERIAL

About 50 to 60 g of soil paste prepared from fine grained soil passing Sieve No. 40 (0.425-mm IS sieve) preserved from the "LL" test is taken for determination of plastic limit as per standard codal procedures.

6.6.4   TESTING EQUIPMENT AND ACCESSORIES

Perform the "PL" test in a laboratory environment using the following equipment and accessories.

   1. Glass Perspex sheet for mixing soil
   2. Spatula
   3. Metallic rod 3 mm diameter and 100 mm long
   4. Moisture content cans
   5. Weighing balance accuracy 0.01 g
   6. A plastic bag
   7. Wash bottle
   8. Oven
   9. Desiccator
  10. Duster and grease

6.6.5   TESTING PROGRAM

   1. Take about 50 g of soil paste from the soil specimen prepared for the LL test and mix it thoroughly on the glass Perspex sheet until it cannot stick to your fingers or palm.
   2. Take about 12 to 15 g of soil pat from the 50-g mass and roll it on a finely ground glass plate with your fingers or palm at the rolling rate of about 80 to 90 strokes per minute to form a uniform thread diameter of 3 mm within 2 minutes (Figure 6.8).

Note: One stroke may be taken as one complete motion of the hand palm from the starting position to forward motion and back to the starting position. However, the rolling rate should be decreased for very fragile soil.

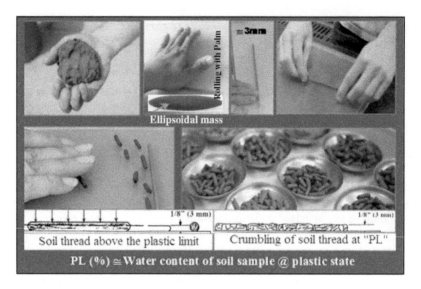

FIGURE 6.8 Schematic of plastic limit test procedure.

TABLE 6.5
Data Sheet for Plastic Limit Test

Project/Site Name: Date:
Client Name: Job. No.   Sample No.
Sample Recovery Depth: Sample Recovery Method:
Sample Description:
Tested By:
Test Type: on gravimetric basis/on volumetric basis

Sl. No.	Observations and Calculations	Trial No.		
		1	2	3
Observation				
1	Moisture Container No.			
2	Weight of empty container, W_1 (g)	10.25	10.35	10.15
3	Weight of container + wet soil, W_2 (g)	18.58	18.36	16.67
4	Weight of container + dry soil, W_3 (g)	16.97	16.81	15.41
Calculations				
5	Weight of water, $W_w = W_2 - W_3$ (g)	1.61	1.55	1.26
6	Weight of solids, $W_s = W_3 - W_1$ (g)	6.72	6.72	6.72
7	Water content, $w = (W_w)/(W_s) * 100$ (%)	23.96	23.99	23.95
8	Average water content, w (PL), %	23.97		

3. Note that if the thread cracks at about 3 mm diameter, stop the test and collect the cracked pieces of thread for water content determination, which represents the PL of the soil specimen (Table 6.5).
4. However, note that if the thread diameter becomes less than 3 mm diameter, then it shows that water content is more than PL and the soil pat needs to be dried by kneading further.

Note: It may be noted that the soil pat when rolled into thread, the whole thread may not crumble or crack at the same time. The thread may break into small pieces. Therefore it is suggested to keep rolling the piece until 3 mm diameter thread is formed. Further, the thread may not crumble exactly at 3 mm diameter. If the thread crumbles when the thread is nearly equal or slightly greater than 3 mm diameter, the test may be stopped and the water content may be determined for PL.

5. Repeat the above test procedure at least three times with fresh samples taken from the 50 g (step 1) and take crumbled soil pieces from each trial for water content determination and determine the average value of water content as the PL of the soil specimen (Table 6.5).

TABLE 6.6
Practice Data Sheet for Plastic Limit Test

Project/Site Name:			**Date:**		
Client Name:	**Job. No.**		**Sample No.**		
Sample Recovery Depth:	**Sample Recovery Method:**				
Sample Description:					
Tested By:					
Test Type: on gravimetric basis/on volumetric basis					
Sl. No.	**Observations and Calculations**			**Trial No.**	
			1	**2**	**3**
Observation					
1	Moisture Container No.				
2	Weight of empty container, W_1 (g)				
3	Weight of container + wet soil, W_2 (g)				
4	Weight of container + dry soil, W_3 (g)				
Calculations					
5	Weight of water, $W_w = W_2 - W_3$ (g)				
6	Weight of solids, $W_s = W_3 - W_1$ (g)				
7	Water content, $w = (W_w)/(W_s) * 100$ (%)				
8	Average water content, w (PL), %				

6.6.6   Observation Data Sheet and Analysis

Based on test data observations and analysis, calculations are given in Table 6.5.

6.6.7   Results and Discussions

From the test results, the average water content of the soil sample = 23.97%.
Therefore, the plastic limit of the soil sample = 23.97%.
From the "LL" test, LL of this soil sample = 35.5%.
Therefore, plasticity index, PI, of this soil sample = LL-PL = 11.53%.

A data sheet is given in Table 6.6 for student's practice.

6.6.8   General Comments

The plastic limit is a measure of shear strength of a soil. "LL" and "PL" are used together for classification and identification of soils.

6.6.9   Sources of Error

One of the biggest sources of error in the determination of PL is the rolling of soil pat into a thread of 3 mm diameter. If the thread is not formed as per standard codal procedures, then the PL will not be correctly determined.

6.6.10   Precautions

1. Use disuntiled water for consistency tests
2. Use only air-dried (not oven-dried) soil for consistency tests
3. Use the soil paste from the preserved sample of "LL" test
4. Dry the soil pat by mixing until a consistency with which the soil can be rolled without sticking to the hands is achieved
5. Roll the soil pat on a glass Perspex with hand palm to a uniform thread of 3 mm diameter
6. Metallic rod 3 mm diameter and 100 mm long to check the diameter of crumbled soil thread pieces
7. The soil thread should crumble at 3 mm diameter
8. Take crumbled soil pieces for determination of average water content as per codal procedures

6.7   DETERMINATION OF SHRINKAGE LIMIT OF A REMOLDED FINE GRAINED SOIL SAMPLE REFERENCES: IS: 2720 (PART 6); IS: 10077; ASTM D 427; BS 1377: PART 2 (1990)

6.7.1   Objectives

The main object of this experiment is to determine shrinkage characteristics (shrinkage limit, shrinkage index, shrinkage ratio, and volumetric shrinkage) of a remolded soil.

6.7.2 DEFINITIONS AND THEORY

Shrinkage limit (SL) is defined as the water content at which soil becomes saturated and below which no volume change occurs when the water content is reduced. SL is also known boundary between the semi-solid state and the solid states. It can be computed from the relation:

$$w_s = SL = \left[\frac{(M_1 - M_s) - (V_1 - V_2)\rho_w}{M_s} * 100 \right] \tag{6.1}$$

Where: M_1 = Initial wet mass of soil pat
 V_1 = Initial volume of soil pat
 M_s = Dry mass of soil pat, and
 V_2 = Volume after drying of soil pat.

The fundamental concept/relative location of the SL are presented in Figure 6.1. Figure 6.2 illustrates the mechanism of how a saturated soil specimen loses volume with decreasing water content at different consistency states. However, Figure 6.2 also illustrates that during the drying process a stage is reached when there is no volume change with decreasing water content. This water content beyond which there is not volume change despite decrease in water content, represents the "shrinkage limit" of the soil specimen. Therefore, the shrinkage limit is the upper limit (maximum limit) of water content at which there is no appreciable reduction in the volume of soil mass with further re-duction in water content. This is due to the fact that on further reduction in water content, there is particle to particle contact between the soil particles thereby creating void pores between the soil particles, which are filled by air and the volume at shrinkage limit remains constant (Figure 6.2). Thus, it is observed that if the water content is decreased below the shrinkage limit, there is no change in the volume; however, if the water content is increased above shrinkage, volume

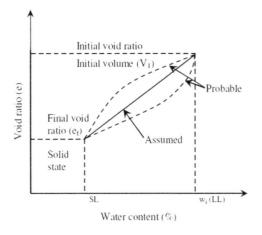

FIGURE 6.9 Qualitative plot of volume changes between water content versus void ratio.

changes will occur with an increase in water content. The phenomenon of this volume change is illustrated in terms of graphical plot between void ratio and water content, as shown in Figure 6.9. In Figure 6.9, dotted lines indicate possible shrinkage paths, which are non-linear as compared with the assumed linear path.

Therefore, it may be noted that the soils which exhibit volume change with variation in water content cannot be used as a foundation medium for various engineering structures such as embankments, highways, dams, etc. It may also not be that the volume change in such troublesome soils will not be uniform and there will chance of differential settlement. Therefore, foundations match on or in such soil deposits need to be designed properly. The consistency limits of LL and Pl may be used to predict the arbitrary compressibility characteristics in these soils due to change in water content. However, shrinkage limit can be used to forecast the quantitative indication of moisture change that can cause appreciable volume change.

6.7.3 Soil Testing Material

About 30 g of soil paste prepared from fine grained soil passing Sieve No.40 (0.425-mm IS sieve) preserved from the "LL" test is taken for determination of shrinkage limit as per standard codal procedures.

6.7.4 Testing Equipment and Allied Accessories

Perform the "SL" test in a laboratory environment using the following equipment and accessories.

1. Glass Perspex sheet for mixing soil
2. Two glass plates, one plain and other with prongs, 75 mm × 75 mm × 3 mm size
3. Shrinkage dish having flat bottom, 45 mm diameter and 15 mm height
4. Two large evaporating dishes about 120 mm diameter with a pour out and flat bottom
5. Glass cup 50 mm diameter and 25 mm height 425 μm sieve
6. Porcelain dish
7. One small mercury dish 60 mm diameter
8. Desiccator
9. Spatula
10. Straight edge
11. Mercury
12. Weighing balance accuracy 0.01 g
13. Duster and grease
14. Oven
15. Wash bottle

6.7.5 TESTING PROGRAM

The procedure for determination of shrinkage limit involves *two phases*.

In the first phase, initial water content and wet volume of the soil paste is determined.

In the second phase, the volume of dry soil pat is determined by the mercury displacement method and the loss of water content between "PL" and "SL." The difference between initial water content and the loss of water content between two phases gives the shrinkage limit. The procedure is explained below:

Phase I Determination of initial water content of soil paste for shrinkage limit test

1. Take about 30 g of soil paste from the soil specimen prepared for the LL test and mix it thoroughly on the glass Perspex sheet.

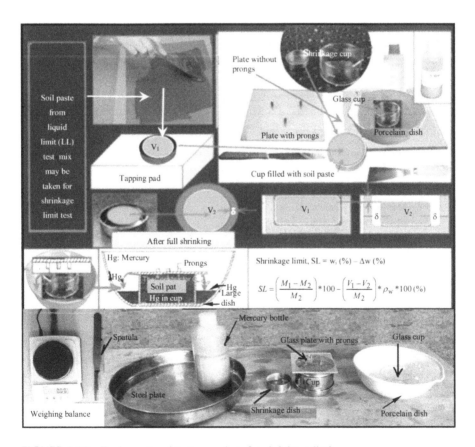

FIGURE 6.10 Equipment and test procedure for shrinkage limit.

Note: The amount of required water could be about the liquid limit in low plastic (friable) soils; otherwise it could be about 1.1 to 1.2 times the liquid limit in medium to high plastic soils.

2. Take the shrinkage cup. Clean it and apply a thin layer petroleum jelly/grease inside the cup and then determine its empty mass (M_1) and initial volume [$V_1 = \pi/4(d)^2*h$], which will be taken as the volume of wet soil, where: d is diameter and h is height of the shrinkage cup. Else, determine the volume of shrinkage cup by mercury displacement method.

3. Fill the shrinkage cup with the soil paste about one-third full and tap gently on a tapping pad to remove entrapped air (Figure 6.10).

4. Fill the shrinkage full by simultaneously tapping it so that cup is filled completely without entrapping any air bubbles (Figure 6.10).

5. Remove any excess soil paste in the cup and level the top surface of the cup with a straight edge and clean the cup from outside and weigh it with full soil paste inside (M_2).

6. Compute the mass of wet soil pat of volume (V_1), $M_3 = M_2 - M_1$ (g).

7. Allow the shrinkage cup to air-dry until the color of the soil pat changes from dark to light (it may take time from few hours to three to four days). Then put the shrinkage cup with the soil into the oven set at 105 to 110°C for 24 hours to dry until its weight becomes constant.

TABLE 6.7

Data sheet for shrinkage limit test

Sl. No.	Observations and Calculations	Trial No.		
		1	2	3
1	Mass of shrinkage cup, M_1 (g)			
2	Mass of shrinkage cup + wet soil, M_2 (g)			
3	Mass of wet soil pat, $M_3 = M_2 - M_1$ (g)			
4	Dry mass of the shrinkage cup with dry soil pat, M_4 (g)			
5	Mass of dry soil pat, $M_s = M_4 - M_1$ (g)			
6	Mass of water in Stage-I, $M_w = M_3 - M_s$ (g)			
7	Initial water content of wet soil pat, $w_i(\%) = (M_w/M_s)*100$			
8	Volume of shrinkage cup $V_1 (cm^3)$			
	a. Mass of shrinkage cup, M_1 (g)			
	b. Mass of shrinkage cup filled with mercury, M_{SCHg} (g)			
	c. Mass of mercury, $M_{Hg} = M_{SCHg} - M_1$ (g)			
	d. Sp. Gravity or mass density of mercury = 13.6 g/cm^3			
	e. Volume of shrinkage dish, $V_1 = M_{Hg}/13.6$ (cm^3)			
9	Volume of dry soil pat, $V_2 (cm^3)$			
	a. Mass of mercury weighing dish, M_{HgDish} (g)			

(Continued)

TABLE 6.7 (Continued)

Sl. No.	Observations and Calculations	Trial No.		
		1	2	3
	b. Mass of mercury weighing dish filled with mercury displaced by dry soil pat, M_{DHgpat} (g)			
	c. Mass of mercury displaced by dry soil pat, $M_{Hgpat} = M_{DHgpat} - M_{HgDish}$ (g)			
	d. Volume of dry soil pat, $V_2 = M_{Hgpat}/13/6$ (cm^3)			
10	Shrinkage limit $SL = w_{SL} = \frac{M_w}{M_s} = \frac{(M_4 - M_s) - (V_1 - V_2)\rho_w}{M_s} * 100(\%)$			
11	Shrinkage ratio, $SR = \frac{M_s}{V_2 \rho_w} = \frac{\rho_d}{\rho_w}$			
12	Volumetric shrinkage, $V_s = \frac{V_1 - V_2}{V_2} * 100(\%)$			
13	Linear shrinkage, $L_s = 100\left[1 - \sqrt[3]{\frac{100}{V_s - 100}}\right]$			
14	Shrinkage Index, SI. = PL − SL (%)			

8. After 24 hours or once the dry weight of shrinkage with dry soil pat is constant, take out the shrinkage cup from the oven and keep it in desiccators to cool it.

9. Remove the shrinkage cup with dry soil pat from the desiccator and take the dry weight of the shrinkage cup with dry soil pat (M_4).

10. Compute mass of dry soil pat, $M_s = M_4 - M_1$ (g) and preserve the dry soil pat for determination of its dry volume by mercury displacement method in phase II.

11. Determine the initial water content as per codal procedures (refer Practical No. 1 for reference). The data observations are given in Table 6.7.

PHASE II DETERMINATION OF DRY VOLUME OF DRIED SOIL PAT BY MERCURY DISPLACEMENT METHOD AND SHRINKAGE LIMIT (SL)

12. Volume of shrinkage cup (V_1).

a. Clean the shrinkage cup and keep it in porcelain dish and fill it full with mercury. Remove the excess mercury by pressing the plain glass plate (without prongs) over the top of the shrinkage cup. Care should be taken that the plate is flushed with the top of the shrinkage cup and no air should be entrapped (Figure 6.11).

b. Transfer the mercury filled in the shrinkage cup to a mercury weighing dish and determines the mass of the mercury to an

accuracy of 0.1 g (M_{Hg}). Or, determine the mass of shrinkage cup
filled with mercury to an accuracy of 0.1 g (M_{SC+Hg}). Then cal-
culate mass of mercury to an accuracy of 0.1 g ($M_{Hg} = (M_{SC+Hg} - M_1)$), whichever way is convenient.

c. The volume of the shrinkage cup is equal to the mass of mercury
    in grams divided by the specific gravity of the mercury of 13.6,
    e.g. $V_1 = M_{Hg}/13.6$ (g/cm^3).

13. Volume of dry soil pat (V_2)

a. Place a glass cup (50 mm diameter and 25 mm height) in a por-
    celain dish and fill it with mercury. Remove the excess mercury
    by pressing the glass plate with prongs firmly over the top of the
    cup and adhering to the outside of the cup. Transfer the glass cup
    full of mercury into another porcelain dish without spilling out
    any mercury from the cup.

b. Now immerse the dry soil pat in the glass cup full of mercury

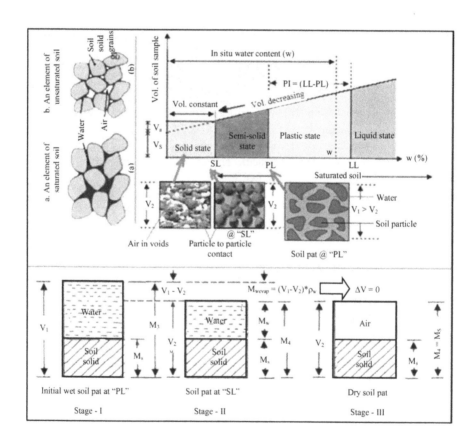

FIGURE 6.11 Equipment and test procedure for shrinkage limit.

slowly carefully so that air bubbles are not entrapped under the pat.

c. Place the glass plate with prongs on the top of the dry soil pat in cup firmly and press down slowly so that the mercury is displaced by the dry soil pat out of the glass cup into the porcelain dish.

d. Collect the mercury displaced by the dry soil pat in the porcelain dish and transfer it to the mercury weighing dish. Determine the mass of the mercury to an accuracy of 0.1 g (M_{Hgpat}). The volume of the dry soil pat (V_2) is equal to the mass of the mercury divided by the specific gravity of the mercury (13.6 g/cm^3), e. g.$V_2 = M_{Hgpat}/13/6$.

e. Repeat the test at least three times to get an average value. The data observations are given in Table 6.7.

6.7.6 OBSERVATION DATA SHEET AND ANALYSIS

Based on test observations and data analysis, shrinkage limit is determined as below:

Loss of water from "PL" and "SL" (or from Stage I to Stage II) due to evaporation by air-drying with oven-dried sample dry weight M_s and volume V_2.

a. The loss of water from "PL" and "SL" Phase II is illustrated in Figure 6.11.

b. Stage – I. Represents initial soil sample in "PL" state with volume V_1 and mass of wet soil pat M_3 (Step 7) in Figure 6.11.

c. Stage – II. Represents soil sample with volume V_2 and corresponding mass of soil pat M_4 (Step 10) in Figure 6.11.

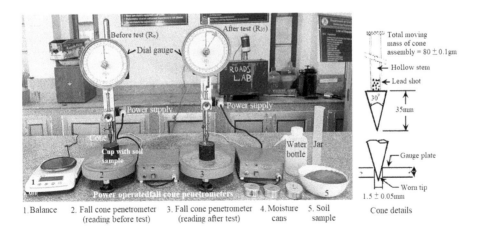

1. Balance   2. Fall cone penetrometer   3. Fall cone penetrometer   4. Moisture   5. Soil
(reading before test)   (reading after test)   cans   sample

Cone details

FIGURE 6.12 Schematic of liquid limit by cone penetrometer method.

d. Stage – III. Represents dried sample with volume V_2 and mass of dry soil pat = $M_s = M_4$ in Figure 6.11.

Note: $M_s = M_4$ The is due to the fact any moisture changes below the shrinkage limit do not cause soil volume changes and mass remains constant.

e. Now, according to definition of water content, shrinkage limit will be water content in Stage – II (in Figure 6.12), i.e., $SL = w_i = w = M_w/M_s$

f. But, the mass of water at "SL" in Stage-II $= M_w - M_{wevap} = (M_4 - M_s) - (V_1-V_2)\rho_w$ (6.3) Where: $M_{wevap} = (V_1-V_2)\rho_w =$ Mass of water evaporated between "PL" and "SL" in Figure 6.11.

Thus, loss of water from "PL" and "SL" or from Stage-I to Stage-II due to evaporation by air drying is: $\Delta M_W = (V_1-V_2)\rho_w$

g. Shrinkage limit of soil, SL – Substitute Equation (10.3) into Equation (6.2), we get:

$$SL = w_{SL} = \frac{M_w}{M_s} = \frac{(M_4 - M_s) - (V_1 - V_2)\rho_w}{M_s} * 100(\%)$$

$$\text{Or} \quad SL = \frac{M_w}{M_s} * 100 - \frac{(V_1 - V_2)\rho_w}{M_s} * 100 = w_{initial} - \frac{(V_1 - V_2)\rho_w}{M_s} * 100(\%)$$

$$(6.5)$$

Where: $w_{initial} =$ Initial water content of wet soil pat (Refer to Step 12 and Table 6.6).

Therefore, shrinkage limit, SL = water content in stage II expressed as percentage.

6.7.7 RESULTS AND DISCUSSIONS

From test results, shrinkage limit=_____%.

6.7.8 GENERAL COMMENTS

The ratio of the liquid limit to the shrinkage limit (LL/ SL) of a soil gives a good idea about the shrinkage properties of the soil. If the ratio of LL/SL is large, the soil in the field may undergo undesirable volume change due to change in moisture. New foundations constructed on these soils may show cracks due to shrinking and swelling of the soil that result from seasonal moisture change.

6.7.9 PRECAUTIONS

1. Apply thin layer of grease inside the shrinkage cup before filling it with wet soil pat to avoid soil pat sticking to the shrinkage cup.
2. Give sufficient tapping to shrinkage while filling with soil pat to avoid entrapping of air bubbles

3. Keep the shrinkage cup filled with wet soil pat for air drying for sufficient time so that dark color changes to light.
4. Weigh the dry soil pat after cooling in the desiccator immediately, otherwise it will pick up moisture from the air
5. Make sure that no air is entrapped under the dry soil pat while pressing it to the glass plate with prongs.
6. Test should be repeated three times so as to get an average value of shrinkage limit.

6.7.10 DETERMINATION OF SHRINKAGE LIMIT FROM KNOWN VALUE OF SPECIFIC GRAVITY OF A SOIL SAMPLE

When the specific gravity of a soil specimen is known, then the shrinkage limit of the said soil specimen can be computed from known properties (e.g. M, V, ρ) of this soil specimen, as illustrated in Figure 6.11.

From Figure 6.11 (Stage-II), shrinkage limit is given by:

$$SL = w_{SL} = \frac{M_w}{M_s} = \frac{(V_2 - V_s)\rho_w}{M_s} * 100(\%) \tag{6.6}$$

Where: V_s is the volume of soil specimen.

Now, the Equation (6.6) can also be written as: $w_{SL} = \left(\frac{V_2}{M_s} - \frac{V_s}{M_s}\right)\rho_w =$
$\left(\frac{V_2}{M_s} - \frac{V_s}{G\rho_w * V_s}\right)\rho_w$ Where: $M_s = \rho_s * V_s$ and $G = \rho_s/\rho_w \Rightarrow \rho_s = G * \rho_w$
$\therefore M_s = G\rho_w * V_s = \rho_s/\rho_w * \rho_w * V_s = \rho_s * V_s$

$$\text{Or} \quad w_{SL} = \left(\frac{V_2\rho_w}{M_s} - \frac{1}{G}\right) \tag{6.7}$$

Also, from the definition of dry mass density, $\rho_d = M_s/V_2, \Rightarrow Ms = \rho_d * V_2, \therefore$ substituting M_s in Equation (6.7), we get:

$$w_{SL} = \left(\frac{\rho_w}{\rho_d} - \frac{1}{G}\right)$$

Thus, Equations (6.7) and (6.8) can be used for determination of shrinkage limit for known values of M_s and V_2.

6.7.11 DETERMINATION OF SPECIFIC GRAVITY OF A SOIL SAMPLE FROM KNOWN VALUE OF SHRINKAGE LIMIT

Method I: The specific gravity of solids (G) can be computed from the known value of shrinkage limit by using Equation (6.8) as:

$$G = \frac{1}{(V_2\rho_w/M_s) - w_{SL}} \tag{6.9A}$$

OR Equation (6.8) can be used to determine the mass specific gravity (G_m) in a dried state by taking $G_m = \rho_d/\rho_w$ as:

$$G = \frac{1}{1/(G_m) - w_{SL}} \tag{6.9B}$$

Method II: The observations made in the shrinkage limit test (Section 6.7.5) can be used to compute the approximate value of G as summarized below:

1. By taking the volume of solids (V_s) in Stage-III (Figure 6.11) given by: $V_s = M_s/G\rho_w$ [a]
2. The volume of solids can also be determined from V_1 in Figure 6.11 (Stage-I) as:

$V_s = V_1 - V_w$ (V_w = volume of water in Stage-I)

$$\Rightarrow V_s = V_1 - \frac{(M_3 - M_s)}{\rho_w} \tag{b}$$

$\therefore$ from Equations [a] and [b], we get: $\dfrac{M_s}{G\rho_w} = V_1 - \dfrac{(M_3 - M_s)}{\rho_w}$

Or $\quad \dfrac{M_s}{G} = V_1\rho_w - (M_3 - M_s) \Rightarrow G = \dfrac{M_s}{V_1\rho_w - (M_3 - M_s)} \tag{6.10}$

The parameters V_1, M_3, and M_s can be determined as explained in Section 6.7.5 for the determination of shrinkage limit.

6.7.12 APPLICATIONS/ROLE OF SHRINKAGE LIMIT IN SOIL ENGINEERING

Problematic soils such as black cotton soils generally referred to as expansive soils worldwide exhibit high swell-shrink characteristic. When there is moisture change in these soil deposits, they cause tremendous damage to structures built on or with them. Therefore, shrinkage limit is a useful parameter used to predict the swell-shrink characteristics of expansive soils to avoid any loss of property or damage to various engineering structures.

6.8   TO DETERMINE LIQUID LIMIT OF A REMOLDED SOIL SAMPLE BY CONE PENETROMETER METHOD REFERENCES: IS: 11196-1985; ASTM D 427; BS 1377: PART 2 (1990)

6.8.1   OBJECTIVES

The main objective of using the cone penetrometer method is to avoid failure of low plastic soils (silty soils) by sliding under repeated blow counts in the Casagrande cup. The cone penetrometer method applicable to a wider range of low plastic soils (Sherwood and Ryley 1968).

6.8.2   INTRODUCTION

The change in natural moisture content has a direct bearing on the shear strength of soils. However, as the depth of ground water level recedes down from the ground level, the penetration depth indicates inverse reflection of the shear strength with changing moisture contents. There exists unique linear relationship between depth of penetration and water content, which is also independent of soil type with depth. This unique linear relationship helps to determine the LL corresponding to 20 mm cone penetration depth in low plasticity soils (SP 36-1987). The basic idea of observing the ground penetration depth at various initial water contents of a metal cone of specified weight and cone angle with point without touching the surface is allowed to drop into the surface.

6.8.3   SOIL TESTING MATERIAL

Disturbed or remolded fine grained soil sample (silty soil or low plastic soil) of about 300 passing Sieve No. 40 (0.425-mm IS sieve) is taken for determination of consistency limits.

6.8.4   TESTING EQUIPMENT AND ACCESSORIES (IS: 11196 – 1985)

Perform the "LL" test in a laboratory environment using the following equipment and accessories.

1. A cup (50 mm in dia. & 50 mm deep)
2. Cone penetrometer device (Figure 6.12)
3. Glass Perspex for mixing soil
4. Stopwatch
5. Weighing balance accuracy 0.01 g
6. Measuring cylinder
7. Wash bottle
8. Spatula
9. Moisture content cans
10. Oven
11. Desiccator

12. 0.425 mm IS sieve
13. Disuntiled water
14. Duster and grease

6.8.5 Testing Program

The following procedure is followed for determination of liquid limit (Figure 6.12):

1. Take about 150 g of air-dried soil passing 0.425 mm (425 μm) sieve
2. Spread the soil sample on a glass Perspex sheet uniformly, sprinkle water on it, and mix it thoroughly. Add more water and keep mixing until a uniform paste is prepared.
3. Fill the soil paste in a cup of 50 mm internal diameter and 50 mm high in layers and tap it onto a rubber pad to remove entrapped air. Remove excess soil paste and level the top cup surface with a straightedge (Figure 6.12).

TABLE 6.8
Practice data sheet for liquid limit test by cone penetrometer method

Project/Site Name: Date:

Client Name: Job. No.   Sample No.

Sample Recovery Depth: Sample Recovery Method:

Sample Description:

Tested By:

Test Type: on gravimetric basis/on volumetric basis

Sl. No.	Observations and Calculations	Trial No.				
		1	2	3	4	5
Observation						
1	Penetration (mm)					
2	Moisture Container No.					
3	Weight of empty container, W_1 (g)					
4	Weight of container + wet soil, W_2 (g)					
5	Weight of container + dry soil, W_3 (g)					
Calculations						
5	Weight of water, $W_w = W_2 - W_3$ (g)					
6	Weight of solids, $W_s = W_3 - W_1$ (g)					
7	Water content, $w = (W_w)/(W_s)*100$ (%)					

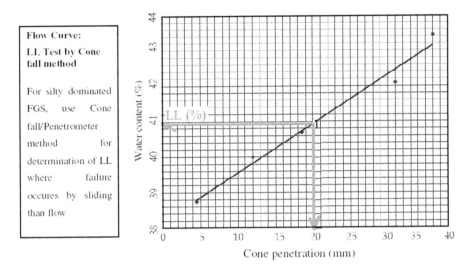

FIGURE 6.13 Flow curve for liquid limit test by cone penetrometer method.

4. Place the cup under the cone and lower the cone gradually so as to just touch the surface of the soil in the cup (Figure 6.12).
5. Set the initial dial gauge reading to zero and keep the stop watch ready for recording the time.
6. Release the cone to penetrate into the soil under its own weight for five (5) seconds only.
7. Record the penetration depth (mm) against the specified time of 5 seconds and remove the cup from the cone penetrometer device.
8. Transfer about 10 to 15 g of soil pat from the cup penetrated by the cone for water content determination by the oven-drying method.
9. Repeat the test at least up to 5 points for plotting the flow curve. The data observations and calculations are given in Table 6.8.

6.8.6   OBSERVATION DATA SHEET AND ANALYSIS

Based on test observations and data analysis, plot the flow curve representing cone penetration depth (mm) along the x-axis and water content (%) along the y-axis on normal graph paper (Figure 6.13). The water content against cone penetration depth of 20 mm is taken as liquid limit of the remolded soil. The test data is given in Table 6.8.

6.8.7   RESULTS AND DISCUSSIONS

From the test results from cone penetrometer test, the LL against 20 mm penetration, LL (%):

The cone is calibrated to read a penetration of 25 mm at the liquid limit. At any other water content, w_x, the penetration $P_x = x * (25 \text{ mm})$ is related with liquid limit as under:

$$LL = w_x + 0.01(25 - x) * (w_x + 15) \qquad (6.11)$$

Where: w_x = Water content (%) to give penetration "x" mm.

Equation (6.11) is valid if the depth of penetration (P) is between 20 and 30 mm.

A curve similar to the flow curve representing penetration versus water content may be drawn and "LL" corresponding to 25 mm penetration may be obtained.

Alternatively, "LL" can be computed using the following expression:

$$LL = \frac{w}{0.77 Log_{10} P} \qquad (6.12)$$

Where: P = cone penetration depth corresponding w (%).

6.8.8 ADVANTAGES OF CONE PENETROMETER METHOD

The cone penetrometer method is widely adopted for low plasticity soils. The cone penetrometer device is easier to maintain compared to the Casagrande device. The cone penetrometer device yields more reproducible results than the Casagrande device.

6.8.9 PRECAUTIONS

1. Soil paste should be prepared exactly in the same way as for the Casagrande method.
2. Check the cone penetrometer before the test.
3. Clean the cup and fill the soil pate gradually by giving simultaneous tapping so that no air is entrapped.
4. Level the top surface carefully and place the cup under the cone penetrometer centrally.
5. Touch the cone just with the top surface and check the dial gauge reading set to be zero.
6. Check the stop watch to be ready for test.
7. Release the cone to penetrate the soil for five (5) seconds only and record the final dial gauge reading carefully.
8. Clean the cup after the end of test and other accessories.

Test 6.9. To Determine Liquid Limit of a Remolded Soil Sample by ONE POINT Method using Casagrande Apparatus

References: IS: 2720-Part 5 & 6; IS: 11196-1985; ASTM D 427; BS 1377: Part 2 (1990)

6.9.1 Objectives

One-point method of determination of "LL" has the advantage of speed, since it involves only one determination of water content. However, it should not be used for soils of liquid limit above about 120% (BS: 1377).

Based on soil characterization under the influence of physico-chemical and rheological factors it has been concluded that the minimum shear strength of soils at LL is due to force field equilibrium and independent of soil type. This is also referred to as critical shear strength of soils at LL, which led to idea for determination no LL by one point method.

6.9.2 Definitions and Theory

Determination of LL either by the Casagrande or cone penetrometer method is a very tedious and time process. Since it has already been established that the shear strength of soils at LL is independent of soil type (SP 36-1987), therefore, LL can also be determined from a single test, generally referred to as one point method. Thus, the LL can be determined by the following equation:

$$LL = w_n \left(\frac{N}{25}\right)^{\tan \beta} \tag{6.13}$$

Where: w_n = Natural water content at the blow count-N of the test

β = Slope of the semi-log plot of water content versus $\log_{10} N$

For a good approximation liquid limit values for all soils, it was found that $\tan\beta = 0.121$. Therefore, Equation (6.13) can be rewritten as:

$$LL = w_n \left(\frac{N}{25}\right)^{0.121} \tag{6.14}$$

Equation (6.14) may yield good results for LL in a single test if N value lies in the range of 20 to 30, with an average value of $N = 25$ (with negligible error).

6.9.3 Soil Testing Material

Disturbed or remolded fine grained soil sample of about 300 passing Sieve No. 40 (0.425-mm IS sieve) is taken for determination of consistency limits.

6.9.4 Testing Equipment and Accessories

Perform the "LL" test in a laboratory environment using the following equipment and accessories and material.

1. Casagrande's liquid limit device (Figure 10.3)
2. 0.425 mm IS sieve
3. Grooving tolls of both standard and ASTM types (Figure 6.3)

4. Disuntiled water
5. Glass Perspex for mixing soil
6. Measuring cylinder
7. Porcelain dish
8. Moisture content cans
9. Wash bottle
10. Desiccator
11. Weighing balance accuracy 0.01 g
12. A plastic bag
13. Duster and grease
14. Spatula
15. Oven
16. A calculator

6.9.5 Testing Program

The liquid limit of a soil sample can be determined by one point method using the Casagrande apparatus (Figure 6.3). The following procedure is followed for determination of liquid limit:

1. Inspect the LL device and ensure that cups drops exactly by 10 mm onto a rubber base (Figure 6.4).
2. Take about 300 g of the air-dried soil sample passing through the 0.425 mm sieve. It is worth mentioning here that same soil paste should be used for all the three consistency limit tests (LL, PL, and SL).
3. Spread the soil sample on a glass Perspex sheet uniformly, sprinkle water on it, and mix it thoroughly. Add more water and keep mixing until a uniform paste is prepared such that materials in Casagrande's cup under repeated drops in the range of 10 to 15 blows.
4. Keep the soil paste in an airtight plastic bag for sufficient time until the water content is uniform in the whole soil paste.
5. Take out the soil paste from the plastic bag and remix it thoroughly before testing.
6. Take about 200 g of the soil paste for LL test and preserve remaining soil paste for plastic and shrinkage limit tests.
7. Fill the paste in Casagrande's cup as required with the help of spatula and level it carefully with a straightedge. Make sure that soil specimen has about 10 to 12 mm in the center of the cup (Figure 6.5) and excess soil paste is be transferred back to the glass Perspex sheet.
8. Now take either Casagrande's tool or ASTM tool as per requirement and cut a groove along the centerline through the soil pat in Casagrande's cup. Make sure that the groove is cut straight with at least 2 mm wide and 10 mm in height.
9. Now give repeated drops by turning the handle of Casagrande's cup at 2 revolutions per second and count the number of blows until the two sides of

the groove flow and join for a distance of about 12 mm at the bottom of the groove in the cup.

10. Record the number of blows for the first trial as N_1 (Table 6.3) when the two parts of the groove join together for about 12 mm length.

11. Take a representative sample of the soil paste (15–20 g) from the two parts of the groove, which joined together by repeated drops for determination of water content. Record this value of water content (as per codal procedure) as w_1 against the N-value of N_1 for the first trial of the test (Table 6.3).

12. Transfer the remaining soil paste from the cup to the glass Perspex sheet and clean the cup thoroughly for the next trial of the LL test.

13. Mix the remaining soil paste thoroughly for about 10 to 15 minutes (in some cases, it may take more time to dry the soil paste for next trial test) so as to change the water content.

14. Repeat steps 7 to 13 and record the number of blows (N) and determine the water content (w %) in each case for about three trials (for cross-check of moisture content).

15. Calculate LL using one-point method equation and semi-log plot for comparison.

6.9.6 Observation Data Sheet and Analysis/Results and Discussions

• **Computation of "LL"**

For the range of blows between 15 and 35, IS: 2720 specifies that "LL" shall be calculated using the expression:

$$LL = \frac{w}{1.3215 - 0.23Log_{10}(N)} \tag{6.15}$$

Where w = Water content of soil when groove closes for N blows.

Alternatively, LL is computed using one point method as (IS: 2720(Part-V):

$$LL = w\left(\frac{N}{25}\right)^n \tag{6.16}$$

Where $n = 0.092$ for soils with $LL < 50\%$
            $= 0.12$ for soils with $LL > 50\%$

Further, Equation (6.16) can be rewritten as:

$$LL = C * w_n \tag{6.17}$$

Where: C is a constant factor given as:

$$C = 0.98 \text{ for } N = 20, \text{ and } C = 1.02 \text{ for } N = 30$$

Thus, compute "LL" for all four tests using one-point method using the water content w_n and the corresponding blow count N for each test, obtaining four values of "LL."

Also, determine the "LL" using the semi-log plot for these four points and compare the resulting four values with the value obtained from semi-log plot.

Check: Do the one-point method equation values agree better or worse when $N = 25$?

6.9.7 General Comments

The one-point method of determination of liquid limit is only used for soils of similar geologic formation and that the slopes of the flow curves are constant on a logarithmic plot.

6.9.8 Precautions

Same as for "LL" by Casagrande apparatus.

6.10 DERIVED INDICES FROM ATTERBERG LIMITS

6.10.1 Index Properties of Fine Grained Soils

Index properties are generally indicative of engineering properties which have been developed by the researchers for wide use in various applications in geotechnical engineering analysis and design. Among various index properties, consistency limits, or Atterberg limits (liquid, plastic, and shrinkage limits) are known as index properties for fine grained soils, whereas PSD and relative density are known as index properties for coarse grained soils.

Thus, consistency limits (liquid, plastic, and shrinkage limits) and various derived indices from these consistency limits are not only used for soil identification and classification of fine grained soils, but are *indicative of engineering properties* of these soils.

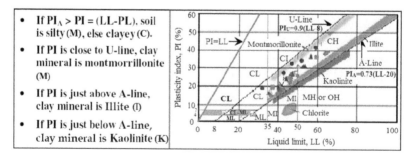

FIGURE 6.14 Relationship between "PI" and "LL" for classification of fine grained soils (ISSCS).

6.10.2 DERIVED INDICES FROM ATTERBERG LIMITS

The most widely used consistency limits defined by Casagrande (1932) in geo-technical engineering are:

1. Liquid limit, represented by: LL (%)
2. Plastic limit, represented by: PL (%)
3. Shrinkage limit, represented by: SL (%)

These consistency limits are known as direct parameters, which can be directly determined on soils by experimental work in the laboratory. However, various indices can be derived from these direct parameters, which are used for classification of in situ state of soils. These indices are:

1. **Plasticity index** represented by: PI = (LL − PL)
   Plasticity index (PI) is a measure of the plasticity of a soil.
2. **Plasticity index of A-line** of Plasticity Chart[*] (Figure 6.14) is represented by:

$$PI_A = [0.73(LL - 20)]$$

PI_A in association with PI is used to classify fine grained soils either silt dominated (e.g. silty soil) or clay dominated soils (Figure 6.14).

3. **Plasticity index of U-line** of Plasticity Chart (Figure 6.14) is represented by:

$$PI_U = [0.9(LL - 8)]$$

PI_U in association with PI is used to identify clay minerals such Kaolinite, Illite, Montmorillonite present in the fine grained soil.

4. **Shrinkage index**
   (SI) is a measure of the degree of swelling/thawing or heave/collapse properties of expansive soils. It is calculated as:

$$SI = [PL - SL]$$

5. **Liquidity index**
   (LI) or water-plasticity ratio is a measure of consistency of soil represented by:

$$LI = \frac{w - PL}{LL - PL} = \frac{w - PL}{PI}$$

Where: w = Natural water content (%).

6. **Consistency index** is represented by: $CI = \frac{LL - w}{LL - PL} = \frac{LL - w}{PI}$
   LI and CI indices are used to assess whether soil can be used either as construction material or as foundation medium to support structures. It may be noted that: LI + CI = 1.

7. **Flow index** represented by:

$$I_F = \frac{w_1 - w_2}{Log_{10}\left(\frac{N_2}{N_1}\right)}$$

Where w_1 = Water content (%) at N_1 blows, and
w_2 = Water content (%) at N_2 blows.
Flow index is the slope of flow curve (liquid limit test plot between "w and $\log_{10}N$") and indicates the rate of loss of shear strength of soils.
8. **Toughness index** is represented by: $I_T = \frac{PI}{I_F}$
Toughness index indicates the shear strength of soil at the plastic limit ($I_T \cong 0 - 3$).
9. **Activity is** represented by:

$$A = \frac{PI}{CF}$$

Where CF = %age of clay particles finer 2 μm IS sieve.
Activity is a measure of pozzolanic reactivity when imbibed with the water, e.g. how much a clayey soil specimen is highly active, active, or inactive, respectively.
Thus, index properties are not only used soil classification, but also used for various empirical correlations for some other engineering properties.

6.11 SIGNIFICANCE/APPLICATIONS OF ATTERBERG LIMITS AND INDICES

The Atterberg limits and derived indices are not only used for the identification and classification of soils, but also widely used by the researchers and engineers in the development of empirical correlations for assessing engineering properties of soils.

TABLE 6.9

Soil classification based plasticity index (PI) as per Indian Standard Soil Classification System (ISSCS)

Sr. No.	Liquid Limit, LL (%)	Soil Description	Plasticity Index, PI (%)	Soil Description or Degree of Plasticity
1	< 20	Non-plastic	0	Non-plastic
2	20–35	Silt (M) or clay (C) of low (L) plasticity (e.g. ML or CL)	<7	Low plastic
3	35–50	Silt or clay of medium (I) plasticity (e.g. MI or CI)	7–17	Medium plastic
4	> 50	Silt or clay of high (H) plasticity (e.g. MH or CH)	> 17	Highly plastic

The various indices derived from the consistency limits help in the identification of in situ state of soils in the field without regress. The significance of index properties and derived indices is explained as below.

6.11.1 IDENTIFICATION AND CLASSIFICATION OF FINE GRAINED SOILS

Since fine grained soils comprise of silt and clay, the consistency limits and derived indices play a vital role in identifying soil components (whether soil mass is silty or clayey), soil minerals (which clay mineral is dominantly present in the soil mass), and also classified based on plasticity behavior.

- *Based on plasticity behavior*

Liquid limit (LL) values indicate the water holding capacity and thereby compressibility behavior, whereas the plasticity index (PI) indicates the classes of plasticity of a soil. Therefore, soils are classified according their plasticity index and liquid limit values as given in Table 6.9.

6.11.2 CLASSIFICATION OF FINE GRAINED SOIL MASS USING INDEX PROPERTIES AND ALLIED INDICES

6.11.2.1 Identification of Constituents (e.g. silt and clay dominance) and Type of Clay Mineral

To classify fine grained soil either silt (M) or clay (C), plasticity chart is used, which is a graphical representation of "PI" (along y-axis) versus "LL" (along x-axis) on a normal graph. The main objective of using this chart is to identify the probable errors while computing LL or PL in terms of a graphical plot between "PI" and "LL" in the plasticity chart. In this chart, two lines, "A-Line- and U-Line," are used for the identification of constituents of fine grained soils and clay minerals present in the soil specimen (Figure 6.14). To use this chart for the identification of constituents of fine grained soils and the clay mineral type, the following parameters are calculated (e.g. calculates numerical values of PI, PI_A, and PI_U):

a. PI = [LL-PL]: computed from LL and PL values determined in the laboratory (direct values)
b. PI_A = [0.73(LL-20)]: computed from the equation of A-line from plasticity chart to help in the identification of fine grained soil either as silt dominated or clay dominated as per the following criteria:
   - If PI > PI_A, the soil sample is clay (C) dominated, e.g. symbol C is to be used
   - If PI < PI_A, the soil sample is silt (M) dominated, e.g. symbol M is to be used
   Check these values on the plasticity chart.

TABLE 6.10

Classification of Consistency of Fine Grained Soils Based on *LI* and *CI*

Sr. No.	Consistency	Description	*LI*	*CI*	q_u (kPa)	Remarks
1	Liquid	Liquid (slurry state)	>1	< 0	≈ 0	Soil not suitable material
2	Plastic	Very soft	0.75–1.00	0–0.25	< 25	
		Soft	0.50–0.75	0.25–0.50	25–50	
		Medium stiff	0.25–0.50	0.50–0.75	50–100	Soil suitable as const
		Stiff	0–0.25	0.75–1.00	100–200	material
3	Semi-solid	Very stiff or hard	< 0	>1	200–400	
4	Solid	Hard or very hard	< 0	>1	> 400	

c. $PI_U = [0.9(LL-8)]$: computed from the equation of U-line from the plasticity chart to help in the identification of clay mineral type present in the fine grained soil specimen as per the following criteria:

- If the value of *PI* $(= LL\text{-}PL)$ is near or close to U-line, then the clay mineral is montmorillonite (M)
- If the value of *PI* $(= LL\text{-}PL)$ is just above A-line, then the clay mineral is Illite (I)
- If the value of *PI* $(= LL\text{-}PL)$ is just below A-line, then the clay mineral is kaolinite (K)

Check these values on the plasticity chart.

6.11.2.2   To Check the In Situ State of Soil

To check the in situ state of soil mass and to ascertain whether the in situ soil deposit is suitable as an engineered construction material or as stable foundation medium to support superstructures, the following indices are used:

a. $PI = [LL\text{-}PL]$: computed from LL and PL values determined in the laboratory (direct values)

b. Liquidity index (LI): $LI = \frac{w_n - PL}{PI}$

c. Consistency index (CI): $CI = \frac{LL - w_n}{PI}$

Where: Natural water content (%) to be obtained from lab. test (practical No. 1) Liquid limit, *LL* (%) to be obtained from lab. Test (practical No. 6.5), and Plastic limit, *PL* (%) to be obtained from lab. Test (practical No. 6.6).

Once, the numerical values of the above parameters are calculated, then use the following Table 6.10 as a guideline to describe the class of consistency and suitability of fine grained soils based on "LI and CI."

The in situ state of a saturated soil deposit at its natural water content (w_n) may be studied by their liquidity index (LI) or consistency index (CI). LI is also equal to:

TABLE 6.11

Classification of Clays Based on Activity

Sr. No	Activity	Classification	Remarks
1	< 0.75	In-active clays	For a soil of specific origin, the activity is constant.
2	0.75–1.25	Normal clays	PI increases as the amount of clay fraction increases.
3	> 1.25	Active clays	Highly active minerals, e.g. montmorillonite can produce a large increase in the plasticity index even when present in small quantity.

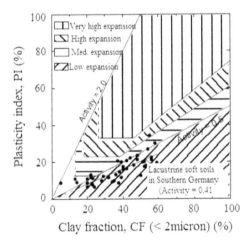

Activities of Various Clay Minerals	
Mineral	Activity
Smectites	1–7
Illite	0.5–1
Kaolinite	0.5
Halloysite (2H$_2$O)	0.5
Halloysite (4H$_2$O	0.5
Attapulgite	0.5–1.2
Allophane	0.5–1.2

FIGURE 6.15 Approximate values of the degree of expansiveness of clay soil (after Bell 1993).

$$LI = [1 - CI] \, or \, [LI + CI = 1]$$

Where: CI = Consistency index

When w_n = LL, CI = 0, when w_n = PL, CI = 1.

When w_n = PL, LI = 0, when w_n = LL, LI = 1.

If CI is negative, the soil mass behaves like a fluid under any given disturbance. If CI is greater than 1, the soil is in a semi-solid or solid state.

6.11.2.3 To Check the Water-Holding Capacity of Clays as Defined by "Activity"

The parameter "activity (A)" of a soil is defined as the ratio of PI to the percent of clay-sized particles (less than 2 μm) present, expressed as:

$$A = \left[\frac{PI}{CF} \right]$$

Where, PI = LL − PL (%), and

CF = Clay Fraction = %age of particles finer 2 μm (0.002 mm) IS sieve.

Further, the activity parameter is helpful in predicting the swell-shrink characteristics and to quantify the water holding-capacity expansive soils. Clays can be grouped into three qualitative categories (Table 6.11) depending upon their activity.

A clay specimen comprises three main minerals such as kaolinite, illite, and montmorillonite. Among these minerals, kaolinite has the lowest swelling potential of the clay minerals followed by the illite mineral which may swell by up to 15%. The montmorrillonite mineral has the highest swelling and water-holding capacity and may swell up to 100% or even more. Figure 6.15 shows the range of the degree of expansiveness of clay based on the activity and the approximate values for the activities of different clay minerals are listed in Figure 6.15.

6.11.3   USE OF CONSISTENCY LIMITS AND INDICES AS INDICATIVE OF ENGINEERING PROPERTIES

Soft soil deposits have a low bearing capacity and undergo large settlement over a long period of time and geotechnical engineers often face challenging problems for reliable measurement of engineering properties of soft soils. There are mainly four types of engineering properties of soils which need the utmost attention. Engineering properties along with their parameters are listed as below:

Sr. No.	Engineering Properties	Parameters	Remarks
1	Compaction characteristics	OMC (w_{omc}) and MDU (γ_{dmax})	MDU: Maximum dry unit weight
2	Coeff. of permeability	k	
3	Compressibility characteristics	C_c, C_s, e, p_c, d, p_s, m_v, C_v, C_h, etc.	
4	Strength characteristics	c_u and ϕ	

To obtain the above parameters for engineering properties, not only is a lot of time required, but it is expensive and a skilled staff and well-equipped laboratory is needed. Therefore, for an indicative assessment and for academic interest, index properties can be used. It has been seen that consistency limit and derived indices are somehow related with some of the above engineering parameters. However, for actual work/project execution, proper soil testing from experts may get done. The consistency limits and derived indices are used for assessment of various engineering properties as explained in next section.

6.11.4 Relationship between Consistency Limits and Derived Indices with Compaction Characteristics

Generally, the compaction is considered as the first foremost engineering property of soils. The compaction process densifies and enhances the compressibility and stability characteristics of marginal soils. The compaction characteristics parameters are measured in terms of optimum moisture content (w_{omc}) and maximum dry density (MDD: ρ_{dmax}) or maximum dry unit weight (MDU: γ_{dmax}), which can be determined in the soil testing laboratory either by standard proctor test or modified compaction test as per standard codal procedure. However, compaction parameters can be estimated from the liquid limit (LL) and the plasticity index (PI) as below:

a. For general soil specimen (fine grained dominated):

$$w_{omc}(std) = 6.77 + 0.4LL - 0.21PI\,(\%)$$

$$\gamma_{d\,max}(std) = 20.48 - 0.13LL + 0.05PI\,(kN/m^3)$$

b. For predominantly clayey and sandy gravels such as lateritic soils, "w_{omc}" can be estimated based on plastic limit as below:

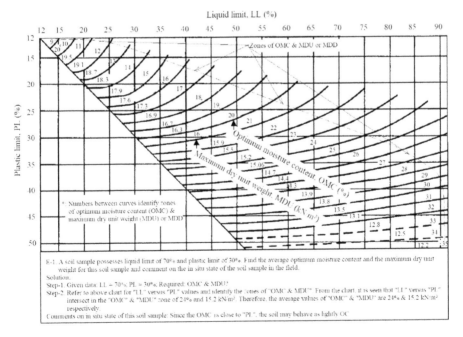

FIGURE 6.16 Chart for estimating average optimum moisture content based on LL and PI (Johnson and Sallberg 1962, p. 125).

$$w_{omc} = 0.42PL + 5(\%)$$

c. For clayey silty with sands such as micaceous soils (with consistency limits of the fines plotted below the A-line), "w_{omc}" can be estimated based on plastic limit and liquid limit as below:

$$w_{omc} = 0.45PL + 3.58(\%) \quad OR \quad w_{omc} = 0.5LL - 6(\%)$$

d. For black cotton soils (silty clays), "w_{omc}" can be estimated based on plastic limit as below:

$$w_{omc} = 0.96PL - 7.7(\%)$$

e. Other correlations of compaction characteristics with index properties are:

- OMC, $w_{omc} = 0:92 * PL$
- OMC (dry of optimum) $w_{omc} = \sqrt{S_r} * (9.46 + 0.2575\ LL)$
- OMC (wet of optimum) $w_{omc} = S_r^2 * (10.61 + 0.3615\ LL)$
- Max^m. dry unit wt. in terms of plastic limit, $\gamma_{dmax} = 0.23(93.3\text{-}PL)$
- CBR (soaked) $= 5.813 - 0.007826\ (LL) + 0.12097\ (PL)$

f. Average optimum moisture content is based on "LL" and "PI."

As per ASTM D 698-66T, the chart shown in Figure 6.16 (Johnson and Sallberg 1962, p. 125) is based on "LL" versus "PI" with different moisture curves and can be used for obtaining average optimum moisture content for first trial test in the standard compaction test. For known values of LL and Pl of a soil sample, this chart will help to find the approximate OMC ($\pm2\%$) of the soil sample and thereby helping in choosing the right choice for selecting the water contents for conducting the standard compaction test. Generally, in the case of clayey soils, starting initial water content for compaction test on oven-dried soil is about 10 to 12% followed by 3% increment up to point of optimum and 2% increment beyond point of optimum. It is recommended that at least 4 to 5 trials should be conducted for a compaction test (at least two points/trials after point of optimum).

6.11.5 Relationship between Consistency Limits and Derived Indices with Compressibility Characteristics

Most wide correlations of consistency limits and derived indices with compression index are:

a. Terzaghi and Peck (1967) gave an empirical correlation for C_c as:

$C_c = 0.009(LL - 10)$: For normally consolidated clays
$C_c = 0.007(LL - 10)$: For remolded clays

b. **Azzouz et al. (1976) gave an empirical correlation for C_c as:**

$C_c = 0004(LL - 9)$: For Brazilian clays

c. **Nacci et al. (1975) gave an empirical correlation for C_c as:**

$C_c = 0.02 + 0.014(PI)$: For natural deep-ocean soil samples

d. **By other Researchers an empirical correlation for C_c as:**

$Cc = 0.007 * (LL - 10)$	: For Remolded clays (Skempton 1944)
$Cc = 0.5 * PI * Gs$	: For all remolded normally consolidated Clays (Wroth and Wood 1978, Wood 1983)

$Cc = 0.543 * e_{LL} * 10^{(-0.168 \log 10 PI)}$: Natural soils in their NC uncemented state, for pressure range between 25 to 800 kPa (Nagaraj 1983)

e_{LL} = Void ratio at "LL"

$Cc = 0.007 * (LL - 7)$	: For Remolded clays (Bowles 1984)
$Cc = 0.0046 * (LL - 9)$	: For Brazilian clays (Bowles 1984)
$C_c = 0.2343 \left[\frac{LL(\%)}{100} \right] G_S$	: For NC clays (Nagraj and Murthy 1985)

$C_c = 0.37 * (e_o + 0.003 LL + 0.0004 w_n - 0.34)$: For all soils (Bowels 1986)

$Cc = PI/74$	: Data from different soils (Kulhawy and Mayne 1990)
$Cs = PI/370$	: Data from different soils (Kulhawy and Mayne 1990)

Where: PI = Plasticity index (%), LL = Liquid limit (%)

TABLE 6.12

Classification of Consistency of Fine Grained Soils Based on LI and CI

Sr. No.	Consistency	Description	LI	CI	q_u (kPa)	Remarks
1	Liquid	Liquid (slurry state)	>1	<0	≈ 0	Soil not suitable material
2	Plastic	Very soft	0.75 – 1.00	0 – 0.25	<25	
		Soft	0.50 – 0.75	0.25 – 0.50	25–50	
		Medium stiff	0.25 – 0.50	0.50 – 0.75	50–100	Soil suitable as const material
		Stiff	0 – 0.25	0.75 – 1.00	100–200	
3	Semi-solid	Very stiff or hard	<0	>1	200–400	
4	Solid	Hard or very hard	<0	>i	>400	

6.11.6 RELATIONSHIP BETWEEN CONSISTENCY LIMITS AND DERIVED INDICES WITH STRENGTH CHARACTERISTICS (UNDRAINED SHEAR STRENGTH)

a. To check loss of rate of shear strength of soils when inundated with water:
To check the rate of loss of shear strength of fine grained soils, the slope of the flow curve (I_F) is used as tentative criteria, which is given as:

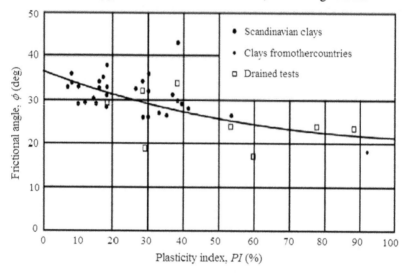

FIGURE 6.17 Variation of frictional angle with plasticity behavior for clayey soils.

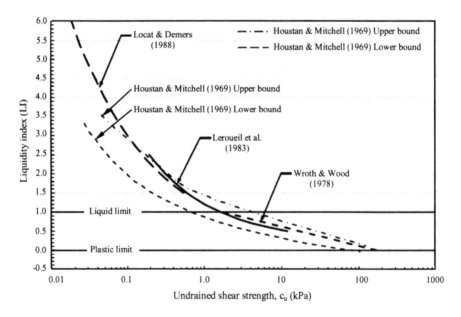

FIGURE 6.18 Variation of undrained shear strength with index properties.

$$I_F = \frac{w_1 - w_2}{Log_{10} \frac{N_2}{N_1}}$$

Where w_1 = Water content (%) at N_1 blows, and
w_2 = Water content (%) at N_2 blows.
Higher values of I_F indicate a higher (e.g. steeper slope of flow curve) rate of loss of shear strength (see Figure 10.6 for flow curve).

b. **To check in situ state of fine grained soil deposits:** To check the in situ state of fine grained soil deposits and to access their undrained shear strength, the consistency index or liquidity index in association with "LL" and natural water content are used as tentative criteria as below:

  i. Consistency index (CI): $CI = \frac{LL - w_n}{PI}$ or

  ii. Liquidity index (LI): $LI = \frac{w_n - PL}{PI}$
  iii. Natural water content (%) to be obtained from lab. test (practical No. 1)
  iv. Liquid limit, LL (%) to be obtained from lab. Test (practical No. 6.5)
  v. Plastic limit, PL (%) to be obtained from lab. Test (practical No. 6.6)
  vi. PI = LL-PL (%)

Once the numerical values of the above parameters are calculated, then use the following Table 6.12 as a guideline to describe the classes of consistency and suitability of fine grained soils based on LI and CI. The relationship between the

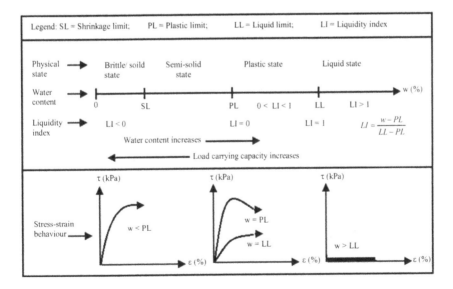

FIGURE 6.19   Stress-strain behavior of fine grained soils at various consistency limits.

plasticity index and friction is shown in Figure 6.17. The relationship between undrained shear strength and liquidity index is shown in Figure 6.18.

c. **Shear strength of fine grained soils at plastic limit:** To check the shear strength of fine grained soil deposits in the in situ state, use the toughness index, which is given as below:

$$I_T = \frac{PI}{I_F}$$

Where I_F is slope of the flow index.

Note: Toughness Index generally lies in the range of 0 to 3 for most soils. When I_T is less than 1, the soil is easily crushed (friable) at the plastic limit.

i. Two soils having the same plasticity index value possess toughness which varies in inverse proportion to their flow indices.
ii. It is, therefore, quite clear that different soils have different shear strengths at plastic limit.
iii. Further, shear strength of fine-grained soils at the plastic limit is tentatively equal to 100 times the shear strength of fine grained soils at the liquid limit.

d. **Stress-strain behavior of fine grained soils at various consistency limits**
e. **Other most widely used correlations between consistency limits and derived indices for assessment of undrained shear strength are:** The stress-strain behavior of fine-grained soils with increasing water content or at various consistency limits is illustrated in Figure 6.19. Figure 6.19 clearly explains the behavior of fine grained soils, which changes drastically from very non-linear behavior to brittle behavior with decreasing water content. This clearly shows that water content plays a very vital role in the behavior of fine grained soils.

Sr. No.	Equation	Reference	Region of Applicability
1	$c_u/\sigma'_{vc} \approx 0.23$	Mir and Juneja (2009)	NC clays
2	$c_u/\sigma'_{vc} \approx 0.30$	Mir and Juneja (2009)	Over consolidated clays
3	$c_u = 1/(LI - 0.21)^2$ (kPa)	Mitchell (1993)	From several clays (remolded strength)
4	$c_u/\sigma'_{vc} \approx 0.26$	Scherzinger (1991)	Constance lacustrine soft clays
5	$c_u/\sigma'_{vc} \approx 0.30$	Burland (1990)	Natural sensitive clays

(Continued)

Sr. No.	Equation	Reference	Region of Applicability
6	$c_u = 11.4 + 0.169 * \sigma'_{vc}$ (kPa)	Windisch and Yong (1990)	Barlow-Ojibway Lacustrine clays
7	$c_u = 2.32 + 0.260 * \sigma'_{vc}$ (kPa)	Windisch and Yong (1990)	East Canadian marine clays
8	$c_u = 3.05 + 0.260 * \sigma'_{vc}$ (kPa)	Windisch and Yong (1990)	Champlane sea clays
9	$c_u = 7.69 + 0.117 * \sigma'_{vc}$ (kPa)	Windisch and Yong (1990)	Scandinavian clays
10	$c_u/\sigma'_{vc} = -0.09 + 0.0092 * (PI)$ kPa	Windisch and Yong (1990)	Scandinavian clays
11	$c_u/\sigma'_{vc} = -0.18 + 0.0072 * (LL)$ (kPa)	Windisch and Yong (1990)	Scandinavian clays
12	$c_u/\sigma'_{vc} = 0.129 + 0.00435 * (PI)$ (kPa)	Wroth and Houlsby (1985)	NC clays
13	$c_u/\sigma'_{vc} = (0.23 \pm 0.04) * (OCR^{0.8})$	Jamiolkowski et al. 1985	All clays
14	$c_u/\sigma'_{vc} = 0.5743 * (3Sin\phi')/(3 - Sin\phi')$ (kPa)	Wroth and Houlsby (1985)	NC soils
15	$c_u/\sigma'_{vc} = 0.11 + 0.0037 * (PI)$	Skempton (1954) Bowles (1984)	NC soils, PI > 10%
16	$c_u/\sigma'_{vc} = 0.45 * (LL)$	Bowles (1984)	NC soils, LL > 40%
17	$\tau_f/\sigma'_{vc} = 0.2 + 0.0024 * (PI)$	Lerouneil et al. (1983)	Clays from eastern Canada, PI < 60%
18	$c_u/\sigma'_{vc} = 0.23 \pm 0.04$	Larsson (1980)	Soft sedimentary clays, PI < 60%
19	$c_u/\sigma'_{vc} = 0.33$	Larsson (1980)	Inorganic clays
20	$c_u = 170 * \exp(-4.6LI)$ (kPa)	Wroth and Wood (1978)	Remolded clays
21	$c_u/\sigma'_{vc} = 0.08 + 0.55 * (PI)$	Larsson (1977)	Scandinavian clays
22	$\tau_f/\sigma'_{vc} = 0.22$	Mesri (1975)	Soft clays
23	$c_u/\sigma'_{vc} = 0.14 + 0.003 * (PI)$	Lambe and Whitman (1969)	All clays
24	$c_u/\sigma'_{vc} = 0.45 * (PI/100)^{1/2}$	Bjerrum and Simons (1960)	NC clays
25	$c_u/\sigma'_{vc} = 0.45 * (LI)^{-1/2}$	Bjerrum and Simons (1960)	NC clays
26	$c_u/\sigma'_{vc} = 0.11 + 0.0037 * \log(PI)$	Skempton (1957)	NC soil, PI < 60%
27	$c_u/\sigma'_{vc} = 0.45 * (LL)$	Hansbo (1957)	Scandinavian clays

c_u = undrained shear strength, τ_f = undrained vane shear strength, σ'_{vc} = effective consolidation pressure, LL = liquid limit, PI = plasticity index (%), LI = Liquidity index, OCR = over consolidation ratio, ϕ' = angle of internal friction

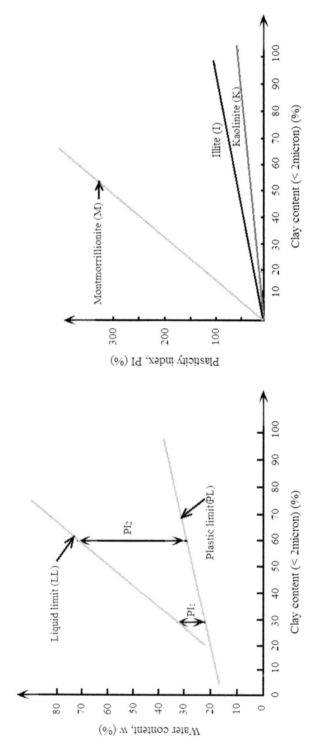

FIGURE 6.20   Theoretical relationships between the Atterberg limits and clay contents.

6.12 GENERAL COMMENTS

1. Atterberg limit are limiting water contents. Theoretical relationships between the Atterberg limits and clay contents are shown in Figure 6.20.
2. These limits are useful in selection of soils to be used for earth work construction.
3. Liquid limit depends on mineral composition. Minerals with high surface charge have high liquid limits.
4. High liquid limit means high compressibility and low permeability.
5. Liquid limit typically ranges anywhere from 20% for silts to over 100% for high plasticity clays.
6. Plasticity index typically ranges anywhere from near 0% (i.e. a non-plastic soil) for silts to over 50% for high-plasticity clays.
7. Plasticity index gives the range of water content within which the fine fraction is plastic.
8. Plasticity index is a qualitative measure of the swell potential of soil. Clays with high cation exchange capacity, including bentonite, montmorillonite, and smectite, have high swell potentials.
9. High compressibility coupled with high plasticity index is also indicative of potential swelling.
10. For the same liquid limit organic soils have low plasticity compared to inorganic clays.
11. The toughness index is a measure of shearing resistance at the plastic limit, and depends on colloidal content.
12. The in situ conditions of clays depend on the closeness of natural water content with plastic or liquid limit, as expressed by the liquidity index, and consistency index.
13. Shrinkage limit gives the water content up to which the soil can shrink. It ranges from 5 to 25%; the lower value indicating high shrink ability.
14. The Atterberg limits of a soil may be classified as the soil aggregate properties.

REFERENCES

ASTM D4318-10. "Standard Test Methods for Liquid Limit, Plastic Limit, and Plasticity Index of Soils." *Annual Book of ASTM standards*. American Society for Testing and Materials, Philadelphia, United States, www.astm.org.

ASTM D427-04. "Test Method for Shrinkage Factors of Soils by the Mercury Method (Withdrawn 2008)." *Annual Book of ASTM Standards*. American Society for Testing and Materials, Philadelphia. www.astm.org.

Bowles, J. E. 1970. *Engineering Properties of Soils and Their Measurements*. New Delhi:McGraw-Hill.

BS 1377-2. 1990. "Methods of Test for Soils for Civil Engineering Purposes-Part 2: Classification Tests for Determination of Water Content." British Standards, UK.

Casagrande, A. 1932. "Research on the Atterberg Limits of Soils." *Public Roads* 13(8): 121–136.

Holtz, R. D., and W. D. Kovacs. 1981. *An Introduction to Geotechnical Engineering*. Englewood Cliffs: Prentice Hall, p. 733.

IS: 2720 (Part 5). 1985. *"Method of Test for Soils: Determination of Atterberg Limits (Liquid and Plastic Limits)."* Bureau of Indian Standards, New Delhi.

IS: 11196. 1985. *"Method of Test for Soils: Specification for Equipment for Determination of Liquid Limit of Soils by Cone Penetration Method."* Bureau of Indian Standards, New Delhi.

IS: 2720 (Part 6). 1972. *"Method of Test for Soils: Determination of Shrinkage Factors."* Bureau of Indian Standards, New Delhi.

IS 10077. 1982. *"Specification for Equipment for Determination of Shrinkage Factors."* Bureau of Indian Standards, New Delhi.

Johnson, A. W., and J. R. Sallberg. 1962. "Methods for Estimating Moisture Content-Unit Weight Relationships." In *Factors Influencing Compaction Results*. Highway Research Board Bulletin No. 319. Washington, DC: National Academy of Sciences-National Research Council, pp. 120–135.

Karlsson, R. 1961. "Suggested Improvements in the Liquid Limit Test with Reference to Flow Properties of Remoulded Clays." Procc. 5th Int. Conf. SMFE, Paris, Vol. 1, pp. 171–184.

Norman, L. E. J. 1958. "A Comparison of Values of Liquid Limit Determined with Apparatus Having Bases of Different Hardness." *Geotechnique* 8(2): 79–84.

Reddy, Krishna R. 2000. *"Engineering Properties of Soils Based on Laboratory Testing."* Department of Civil and Materials Engineering University of Illinois at Chicago, USA.

Seed, H. B., R. J. Woodward, and R. Lundgren. 1964. *"Fundamental Aspects of the Atterberg Limits. Jl. Soil Mechanics and Foundation Divn."* ASCE, pp. 75–105.

Seed, H. B., R. J. Woodward, and R. Lundgren. 1964. *"Clay Mineralogical Aspects of the Atterberg Limits. Jl. Soil Mechanics and Foundation Divn."* ASCE, pp. 107–131.

Sherwood, P. T., and D. M. Ryley. 1968. *"An Examination of Cone Penetrometer Method for Determining the Liquid Limit of Soils."* RRL Report LR233, Road Res. Lab., Crownthrone, Berkshire, UK.

SP 36-1. 1987. *"Compendium of Indian Standards on Soil Engineering: Part-1 Laboratory Testing of Soils for Civil Engineering Purposes."* Bureau of Indian Standards, New Delhi.

7 Organic Matter in a Fine Grained Soil Sample

References: ASTM D 2974; BS 1377: Part 3 (1990)

7.1 OBJECTIVES

The main objectives of this test are:

a. To identify the amount of organic contents by dry weight of soil specimen and to ascertain the affect of organic matter on the mechanical properties of soils.
b. To select the appropriate soil improvement technique for stabilization of organic soils.
c. To assess the significance of organic soils for living organisms in the soils.

7.2 INTRODUCTION

Organic soil deposits are considered as most problematic soil deposits that exhibit very high water holding capacity, high compressibility, and very poor strength. High organic content has direct bearing on the properties of soils and have a significant effect on the index and engineering properties (Reddy 2000). The organic soils include different types of plant and animal materials, which decompose in the soil with time. The low contents by weight of organic matter (<20%) have no bearing on soil characteristics and the constituents of soil solid control the engineering behavior of soils. However, when the organic contents exceeds beyond 20%, it has direct bearing on engineering properties of soils such as coefficient of permeability, compressibility characteristics, and strength properties. The organic soils also play a vital role by providing nutrient support to living organisms in the soil and binding the soil particles into aggregates and enhance their water-retaining capacity. Therefore, it is mandatory to identify the amount of organic contents by dry weight of soil specimen and to ascertain the affect of organic matter on the mechanical properties of soils. This will help in selecting the appropriate soil improvement technique for stabilization of organic soils for sustainable development of various engineering structures built on or with such soil deposits.

7.3 DEFINITIONS AND THEORY

Organic matter constitutes a variety of decomposable materials such as dry plant roots, dead animals, wood ash, tree leaves, animal manures, bacteria,

DOI: 10.1201/9781003200260-7

earthworms, microorganisms, and crop residues, etc. These constituents when intermixed with the soil largely affect the top surface of the soil profile and the soil properties. The organic soils undergo high compressibility and have very low bearing capacity.

The organic content present in a soil is defined as ratio of weight organic content to the dry weight of soil. The weight of organic content can be determined in the soil testing laboratory by drying measured soil sample containing organic matter in a muffle furnace for 24 hours at a temperature of about >750 °C or as required per standard codal procedures. Generally, in geotechnical engineering applications, organic matter in soils is determined by the following two methods:

1. By direct method in which organic matter of a soil sample is determined using muffle furnace for determination of loss on ignition (LOI) as described in Part A below in Section 7.3 A (Method C or Method D as per ASTM and BS codes).
2. By indirect method in which the organic content in the soil sample is approximately estimated by comparing the liquid limits of air-dry and oven-dry soil samples. If the difference between these two values is more than 30%, then it is presumed that organic matter is present in the soil samples.

Organic matter in the soils is also determined as per IS: 2720 (Part 22) using reagents.

7.4   METHOD OF TESTING

Organic content is determined by the oven-drying method as a loss on ignition LOI by using a muffle furnace at a temperature of about 440 °C and above.

7.5   SOIL TESTING MATERIAL

Disturbed or remolded fine grained oven-dried soil sample of about 30 g passing through a Sieve No. 40 (0.425-mm IS sieve).

7.6   TESTING EQUIPMENT AND ACCESSORIES

The following equipment and accessories and soil sample is required in the soil testing laboratory for determining organic content in a given soil specimen:

1. Muffle furnace
2. 0.425 mm IS sieve
3. Chemical Balance—sensitive to 0.001 g
4. Porcelain dish
5. Moisture content cans
6. Oven, tongs, and a desiccator

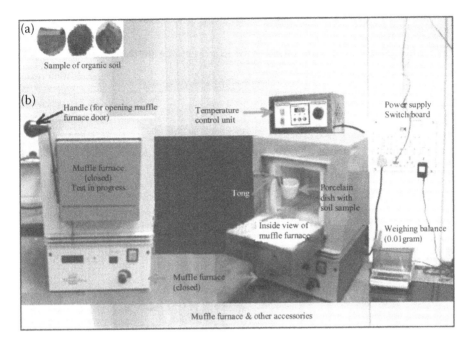

FIGURE 7.1   Test accessories for determination of organic content of soil.

Soil material and test equipment, along with other accessories, are shown in Figure 7.1.

7.7   TESTING PROGRAM PART A: DIRECT METHOD—DETERMINATION OF LOSS ON IGNITION

1. Prepare soil specimen as per IS: 2720 (Part I)-1983.
2. Take an empty, clean, and dry porcelain dish and determine and record its weight (W_1).
3. Take about 5 g soil sample passing through 0.425 mm IS sieve in the porcelain dish and record the weight of the porcelain dish and soil specimen (W_2).
4. Place the porcelain dish in a muffle furnace. Gradually increase the temperature in the furnace to $440 \pm 25\,°C$ (Method-C) or $750\,°C \pm 38\,°C$ Method-D) as required and leave the specimen in the furnace overnight or till constant mass is achieved.
5. Remove the porcelain dish using the tongs carefully from the muffle furnace and allow it to cool to room temperature in a desiccator.
6. Record the weight of the porcelain dish + dry soil (W_3).
7. Determine the organic content in the soil sample, as given in Table 7.1.
8. Repeat the test as per above procedures for the determination of average value of organic content present in the soil specimen.

TABLE 7.1
Organic Content Determination by Muffle Furnace method (as LOI)

Sl. No.	Observations and Calculations	Determination No.		
		1	2	3
Observation				
1	Porcelain dish No	D1	D2	D3
2	Weight of empty porcelain dish, W_1 (g)			
3	Weight of porcelain + oven dry soil, W_2 (g)			
4	Weight of porcelain + soil ash (burned soil), W_3 (g)			
Calculations: Loss on ignition, LOI				
5	Weight of dry soil, $W_S = W_2 - W_1$ (g)			
6	Weight of soil ash (burned soil), $W_{SA} = W_3 - W_1$ (g)			
7	Weight of organic matter, $W_{org} = W_S - W_{SA}$ (g)			
8	Organic matter content, O_M (%): $[W_{org}/W_S] * 100$			
9	Average Organic matter content, O_M (e.g. Loss on ignition as a percentage of the dry mass of soil passing a 0.425 mm test sieve) nearest to 0.1%. (%)			

Note: If soil is silty sand type, the sample passing 2 mm (instead of 0.425 mm) should be used in the test.

7.7.1 OBSERVATION DATA SHEET AND ANALYSIS

Test data analysis, observations, and calculations for determination of organic content are given in Table 7.1.

7.7.2 RESULTS AND DISCUSSIONS

Based on test results, organic matter content present in the soil sample = _____ %.

7.8 TESTING PROGRAM–PART B: INDIRECT METHOD

1. Prepare soil specimen as per IS: 2720 (Part I)-1983.
2. Sieve soil sample through 0.425 mm IS sieve (425 μm).
3. Divide the soil sample into two parts. Air-dry one part of the soil (W_{SA}) and oven-dry other soil sample (W_{SO})one as per codal procedure (IS:2720).
4. Determine the liquid limit of both samples using Casagrande apparatus as per procedure given in Practical No. 6 (LL_{WSA} and LL_{WSO}):
   ⇒ Liquid limit value for air-dry soil sample: $LL_{WSA} = A_{LL}$ (say = 25%)
   ⇒Liquid limit value for oven-dry soil sample: $LL_{WSO} = O_{LL}$ (say = 18%)
5. Find the difference between the liquid limit values as: $\Delta LL = 25 - 18 = 7\%$.

6. Find percentage difference between oven-dried and air-dried liquid limit values.
7. It has been reported by various researchers that in case the soil contains organic material, then the liquid limit value of the oven-dried soil sample will be less by more than 30% compared to the liquid limit value of the fresh or inorganic soil sample.
8. Thus, for given test data, find liquid limit values for the same soil in air-dried state and oven-dried state as per codal procedure (**refer to Practical No. 6 for details**) and complete data sheet accordingly.

7.9 RESULTS AND DISCUSSIONS

Based on liquid limit values and the %age difference between these values, it can be concluded that the soil sample contains organic matter:

a. If the liquid limit value for oven-dry soil sample (LL_{WSO}) is less by more than 30%.
b. That soil sample does not contain appreciable organic matter as the liquid limit value for oven-dry soil sample (LL_{WSO}) is not less by more than 30% (in this case, report value of "LL" of soil sample in both cases).

7.10 GENERAL COMMENTS

Organic soils or peaty soils are very soft and trouble soils with high compressibility and low bearing capacity. These soils are the end results of various decomposable materials thrown into the soil, which largely affect the top layer of the soil profile. Therefore, the top soil layer consisting of organic matter is not suitable either as construction material or as a foundation medium for various engineering structures. Presence of organic matter in the soil increases its plasticity and results in high plastic limit values. However, organic matter does not have any significant influence on the liquid limit. So, soils with organic matter have low plasticity indices with a high plastic limit. However, organic soils are useful for living organisms in the soil and binding the soil particles into aggregates and enhance their water retaining capacity.

7.11 PRECAUTIONS

1. Prepare the soil sample as per codal procedure
2. Sieve soil through 2 mm sieve down 425 µms
3. Classify the soil and use soil passing 2 mm in case soil is silt-sand dominated, else take soil passing 425-µm (0.425 mm) sieve.
4. Gradually increase the temperature of muffle furnace as required either by Method-C or Method-D
5. Make sure that desired temperature remains constant for more than 3 hours or until the weight of the material remains constant

6. Use tongs to remove porcelain dish from the muffle furnace as it will be very hot
7. Cool the sample to room temp. and record its weight nearest to 0.001 g
8. Calculate organic matter content as described in the above procedure

REFERENCES

ASTM D 2974. "Standard Test Methods for Moisture, Ash, and Organic Matter of Peat and Organic Soils." *Annual Book of ASTM Standards*, Vol. 04–08. American Society for Testing and Materials, Philadelphia, United States, www.astm.org.

BS 1377-3. 1990. *"Methods of Test for Soils for Civil Engineering Purposes-Part 3: Chemical and Electro-Chemical Tests."* British Standards, UK.

IS: 2720 (Part I). 1983. *"Method of Test for Soils: Preparation of Dry Soil Specimen for Various Tests."* Bureau of Indian Standards, New Delhi.

IS: 2720 (Part 22). 1973 (Reaffirmed 2006). *"Method of Test for Soils: Determination of Organic Matter."* Bureau of Indian Standards, New Delhi.

Reddy , Krishna R. 2000. *"Engineering Properties of Soils Based on Laboratory Testing."* Department of Civil and Materials Engineering University of Illinois at Chicago, USA.

Sehgal, S. B. 1992. *"A Text Book of Soil Mechanics."* CBS Publishers and Distributors, New Delhi.

8 Relative Density of a Soil Sample

References: IS: 2720 (Part 2) – 1973; ASTM 2049-64; ASTM D 4253; ASTM D 4254; BS 1377: Part 2, 1990

8.1 OBJECTIVES

The main objective of determining relative density is that it is used as a controlling parameter for compaction of granular soils in the field because the standard proctor compaction test does not produce a well-defined moisture-density curve for granular (sandy) soils.

8.2 INTRODUCTION

Relative density or density index is an index property of granular soils, which numerically compares density of in situ soil with the density represented by the same soil in the extreme states of compactness (e.g. either in densest or in loosest state). Since the standard proctor compaction test is not suitable for granular (sands) soils, therefore, relative density is used as controlling parameter to ensure that the in situ density of granular soils in the field is achieved as per standard codal procedures. The relative density and void ratio are widely used to predict the relative strength and compactability factor etc. in the construction of various engineering structures (Reddy 2000). The safe bearing capacity of sandy soils has a direct bearing on in situ soil conditions in the field, which is generally predicted by standard penetration number (SPT-N value) and static cone penetration resistance value (q_c). The relative density also depends upon the angle of internal friction (ϕ), which is also related with SPT-N value. Based on the SPT-N value and q_c (after Tang 1962), there exists an approximate relationship (Table 8.1) among relative density (R_D), SPT-N value, angle of internal friction (ϕ), static cone penetration resistance (q_c), and density for the granular soils (ρ_b), which can be used to assess the suitability of these soils for construction works.

From Table 8.1, it is seen that the angle of internal friction is significantly influenced by the relative density bearing capacity. This may also have direct bearing on settlement of footings on granular soils. Similarly, approximate correlation can be established between the relative density and the minimum and maximum in situ dry densities, as shown in Figure 8.1.

From Figure 8.1, the in situ dry density can be obtained for any desired value of relative density, which would be used to obtain the natural void ratio (e) against the same desired relative density by the equation:

DOI: 10.1201/9781003200260-8

157

TABLE 8.1

Relationship Between R_D, ϕ, and Density of Granular Soils Based on the SPT-N and q_c (after Tang 1962)

R_D (%)	SPT Value (N)	q_c (kg/cm²)	ϕ (deg)	Bulk Density, ρ_b (g/cc)	State of Compactness
0–15	0–4	< 20	25–32	1.12–1.60	Very loose
16–35	4–10	20–40	27–35	1.44–1.84	loose
36–65	10–30	40–120	30–40	1.76–2.08	Medium
66–85	30–50	120–200	35–45	1.76–2.24	Dense
86–100	>50	>200	>45	2.08–2.40	Very dense

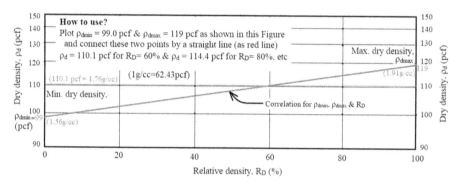

FIGURE 8.1   Relation between relative density and in situ dry density.

$$e = \left[\frac{G\rho_w}{\rho_d} - 1 \right] \qquad (8.1)$$

Where: G = Specific gravity of the soil specimen (standard value for granular soils = 2.65)

ρ_w = Density of water = 62.43 pcf (1 g/cm³ ≈ 62.43pcf ≈ 10 kN/m³), and

ρ_d = Dry density of water of soil specimen for desired relative density against which the in situ void ratio is required (from Figure 8.1).

Once the in situ or natural void ratio of the soil is obtained, the maximum and minimum void ratio of the soil specimen is obtained by rewriting the Equation (8.1) as:

$$e_{max} = \left[\frac{G\rho_w}{\rho_{d\,min}} - 1 \right] \quad and \quad e_{min} = \left[\frac{G\rho_w}{\rho_{d\,max}} - 1 \right] \qquad (8.2)$$

Alternatively, knowing the relative density and the maximum and minimum void ratio of a soil specimen, the natural void ration is obtained by the equation:

$$e_o = e = [e_{max} - (e_{max} - e_{min}) * R_D] \quad \therefore R_D = \left(\frac{e_{max} - e}{e_{max} - e_{min}} \right) \qquad (8.3)$$

It may be noted that the variation of angle of internal friction depends upon the relative density, particle size distribution, and shape of particles. Angular-shaped particles have a higher angle of internal friction value compared to sub-rounded or rounded particles. Thus, shape and soil grading of coarse grained soils have direct bearing on the shear strength of coarse grained soils. Approximate correlation between relative density, angle of internal friction, and dry density for various coarse grained soils are shown in Figure 8.2.

The relative density and the angle of internal friction sandy soils can be estimated by empirical correlations using the standard penetration test (ASTM D-1586). However, desired corrections are applied to minimize the effects of penetration resistance and in situ effective stress as proposed by Liao and Whitman (1986):

$$N' = [C_N * N_o] \qquad (8.4)$$

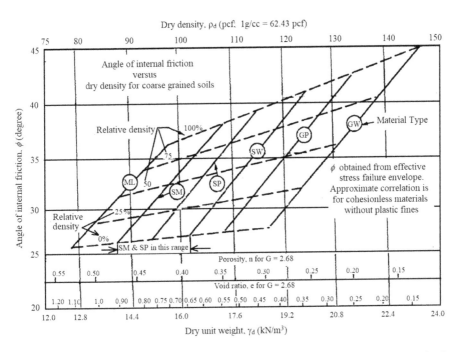

FIGURE 8.2 Approximate correlation between R_D, ϕ, and dry density for coarse grained soils.

Where: N' = Corrected resistance or SPT value
C_N = Correction factor = $[95.76/\sigma'_o]^{1/2}$
N_o = Measured resistance in the field during SPT test, and
σ'_o = Effective overburden pressure (kN/m^2).

8.3   DEFINITIONS AND THEORY

Relative density or density index is an index property of granular soils, which numerically compares the density of the in situ soil specimen with the density depicted by the same soil specimen in either a densest or a loosest state expressed as:

$$R_D = \left(\frac{e_{max} - e}{e_{max} - e_{min}} \right)(\%)$$
(8.5)

Where: e = Void ration in insitu state
e_{max} = Maximum void ratio of the soil in the loosest state determined in the laboratory
e_{min} = Minimum void ratio of the soil in the densest state determined in the laboratory, and
e_{max} = 0.87 - 0.94 and e_{min} = 0.29 - 0.34.
Relative density may also be expressed in terms of dry unit weight by the equation:

$$R_D = \left[\left(\frac{\rho_d - \rho_{d\,min}}{\rho_{d\,max} - \rho_{d\,min}} \right) * \left(\frac{\rho_{d\,max}}{\rho_d} \right) * 100 \right] (\%)$$
(8.6)

Where: ρ_d = Field dry density of the soil specimen in place, called "in-place dry density"
ρ_{dmin} = Minimum dry density of the soil specimen in the loosest state, and
ρ_{dmax} = Maximum dry density of the soil specimen in the densest state.

Relative density is used as controlling parameter to ensure that the in situ density of granular soils in the field is achieved as per standard codal procedures. The relative density and void ratio are widely used to predict the relative strength and compactability factor etc. in the construction of various engineering structures. Relative compaction (R_C) and relative density (R_D) are correlated as:

$$R_C = [80 + 0.2 * R_D]$$
(8.7)

e.g. R_C of 80% for loose soil corresponds to zero relative density.

8.3.1 MAXIMUM DENSITY

Maximum density predicts the densest state of compactness of a granular soil specimen (sand) in the field, which can be simulated in the laboratory by compacting a representative sand specimen in a calibrated mold of known volume under a surcharge of about 2 psi (1 psi = 6.895 kPa). In this test, the mold filled with the sand specimen is vibrated at a desired frequency for a specified time interval as per standard codal procedure (ASTM D4253) to achieve its densest state of compactness. The maximum density is then calculated from the compacted mass ($M_{s(max)}$) of sand specimen in known volume (V_C) as:

$$\rho_{max} = \left[\frac{M_{s\,(max)}}{V_C} \right] (g/csc \ \ or \ \ pcf) \tag{8.8}$$

Where: V_c = Volume of calibrated mold in which sand mass ($M_{s(max)}$) is compacted.

8.3.2 MINIMUM DENSITY

Unlike maximum dry density, minimum density predicts the loosest state of compactness of a granular soil specimen in the field, which can be simulated in the laboratory as per standard codal procedure (ASTM D4254). In this method, the representative dry sand specimen is poured into a calibrated mold of known volume through a funnel with a constant freefall of about 100–120 mm until the calibrated mold is filled. The minimum mass of sand specimen filled in the calibrated mold is weighed and the minimum density is then calculated from the known volume (V_C) as:

$$\rho_{min} = \left[\frac{M_{s\,(min)}}{V_C} \right] (g/cc \ \ or \ \ pcf) \tag{8.9}$$

Where: V_c = Volume of calibrated mold in which soil mass ($M_{s(min)}$) is filled under constant free fall as per standard codal procedures.

8.4 METHOD OF TESTING

The relative density of a sand specimen is determined by using a vibratory table in accordance with standard codal procedures.

8.5 SOIL TESTING MATERIAL

For determination of relative density, oven-dried sandy soil passing through a 4.75 mm sieve is taken for determination of minimum and maximum dry unit weight as per standard codal procedures.

8.6 TESTING EQUIPMENT AND ACCESSORIES

For determination of relative density of a cohesionless soil sample, the following equipments/accessories are required (Figure 8.3):

1. Representative of in situ soil sample collected from field
2. Vibratory table
3. Mould assembly consisting of standard mold
4. Dial gage indicator, balance, and scoop
5. Surcharge base-plate with handle
6. Surcharge weights
7. Guide sleeves
8. Straightedge

8.7 TESTING PROGRAM

1. Take a calibrated mold and weigh it (M_1 g)
2. Fill the mold by pouring the sand through a funnel under constant freefall
3. Trim off the excess sand on top of the mold and level the top surface carefully by a straightedge
4. Weigh the mold filled with sand in loose state (M_2 g)
5. Empty the mold and fill with new sand specimen as densely as possible and level the top surface by straightedge

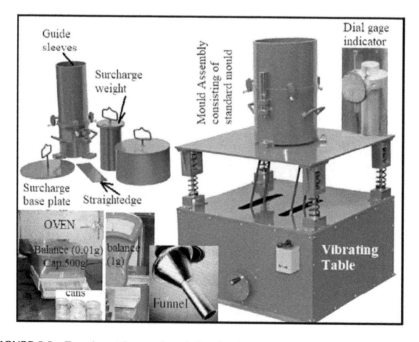

FIGURE 8.3   Experimental setup for relative density test.

6. Set vibratory table ready with all test accessories and attach the mold with filled with sand specimen (Figure 8.3)
7. Set the initial dial gauge reading and record initial reading as R_i to the nearest 0.025 mm
8. Vibrate the mold assembly and sand specimen for 10 minutes
9. Stop vibrating mold assembly and record the final dial gage reading R_f
10. Replace the mold assembly from the vibratory table and de-attach all accessories
11. Weigh the mold and compacted sand specimen, M_3 (in dense state in Figure 8.3)
12. Determine the volume of the calibrated mold for the compacted sand specimen in the densest state V_C
13. The observations and data analysis is given in Section 8.8 below

8.8   OBSERVATION DATA SHEET AND ANALYSIS

Observation data and necessary calculations are:

1. Weight of the of empty mold, M_1 (g):
2. Diameter of empty mold, d (cm):
3. Height of empty mold, h (cm):
4. Volume of empty calibrated mold, V_C (cm^3) = $[\pi/4*(d)^2*h]$:
5. Mass of mold and sand in loose state, M_2 (g) step-4:
6. Initial dial gauge reading before compaction of sand, R_i:
7. Final dial gauge reading after compaction of sand, R_f:
8. Thickness of surcharge base plate, t_P (mm):
9. Mass of mold and sand in dense state, M_3 (g):
10. Mass of sand in Loose state, $M_{sl} = M_2 - M_1$ (g):
11. Mass of sand in dense state, $M_{sd} = M_3 - M_1$ (g):
12. Minimum dry density, $\rho_{dmin} = M_{sl} / V_C$ (g/cm^3); 1 m^3 = 10^{-6}cm^3:
13. Maximum dry density, $\rho_{dmax} = M_{sd} / V_C$ (g/cm^3); 1 m^3 = 10^{-6}cm^3:
14. Specific gravity (based on expt. test for the same soil), G = (say 2.65):
15. Minimum void ratio, $e_{min} = \left[\dfrac{G*\rho_w}{\rho_{d\,max}} - 1 \right]$:
16. Maximum void ratio, $e_{max} = \left[\dfrac{G*\rho_w}{\rho_{d\,min}} - 1 \right]$:
17. Field dry density (based on expt. test for the same soil), $\rho_d = M_d/V_m$:
18. Void ratio in natural state, $e_o = e = \left[\dfrac{G*\rho_w}{\rho_d} - 1 \right]$:
19. Relative density, $R_D = \left[\left(\dfrac{e_{max} - e}{e_{max} - e_{min}} \right) * 100 \ (\%) \right]$:
20. Or cross check, $R_D = \dfrac{\gamma_d - \gamma_{d\,min}}{\gamma_{d\,max} - \gamma_{d\,min}} * \dfrac{\gamma_{d\,max}}{\gamma_d} * 100 \ (\%)$:

8.9   RESULTS AND DISCUSSIONS

Based on test results, relative density of the soil sample = _____%
   Soil type: ....
   ....
   ....

8.10   GENERAL COMMENTS

1. The relative density and void ratio are widely used to predict the relative strength and compactability factor etc. in the construction of various engineering structures. Relative density of at least 70% is recommended for construction of granular sub-bases for pavement design. It may be noted that foundations on sand with relative density between 40–70% may not show a sudden failure. However, as the settlement exceeds about 8% of the foundation width, bulging of sand starts at the surface. At settlements of about 15% of foundation width, a visible boundary of sheared zones at the surface appears. However, the peak of base resistance may never be reached. This type of failure is termed local shear failure, Figure 8.4(b), by Terzaghi (1943).
2. Vesic (1963) observed that foundations on loosely compacted ($R_D < 35\%$) exhibit punching failure (Figure 8.4c). However, the rate of settlement decreases with increasing base resistance.

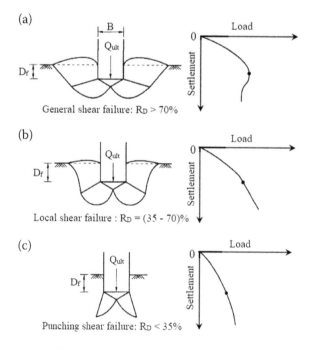

FIGURE 8.4   Modes of bearing capacity failure in soils (Vesic 1963).

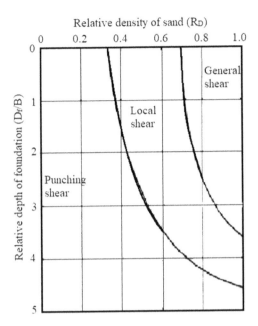

FIGURE 8.5 Types of failure of model footings in sand (Vesic 1963).

3. The types of mode of foundation failure of model footings on sand are shown in Figure 8.5 (Vesic 1963). From Figure 8.5, it is seen that the limiting relative densities increase with increase in the relative depth/width ratio. Also, it is seen that punching shear failure occurs below a critical depth ratio of about 8 without showing any sign of failure for rectangular footing. However, this ratio is limited to 4 for circular footings.

8.11 APPLICATIONS/ROLE OF RELATIVE DENSITY IN SOIL ENGINEERING

The relative density and void ratio are widely used to predict the relative strength and compactability factor etc. in the construction of various engineering structures. Approximate correlation between relative density, water content, and dry density is shown in Figure 8.6.

8.12 SOURCES OF ERROR

1. If type of soil is not appropriate. The relative density is meaningful only for cohesionless materials.
2. Material segregated while being processed
3. Molds not accurately calibrated
4. Gain in moisture of oven-dried material before or during testing. A small amount of moisture in the soil can cause erroneous measurements of the minimum density and, to a much lesser degree, of the maximum density.

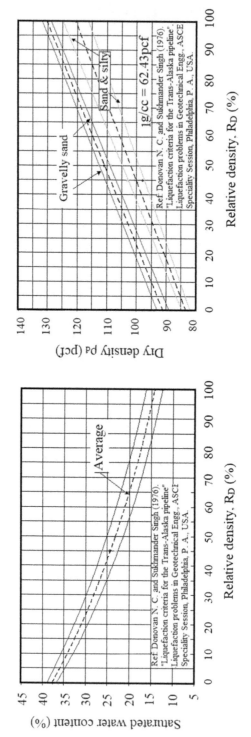

FIGURE 8.6 Approximate correlation between relative density, water content, and dry density.

5. Disturbance of mold during filling. Inadvertent jarring of the mold or impact of the falling particles will increase the measured minimum density.
6. Segregation of material while filling mold
7. Loss of material from mold before weighing. Insufficient amplitude of vibratory table under load.
8. Misalignment of guide sleeve with mold

8.13 PRECAUTIONS

1. Check the requirements of vibratory table to be fit for test.
2. Check if the soil sample is appropriate.
3. Check that sand is poured with constant freefall through the funnel.
4. Check that all the accessories and dial gauge readings are set in place before vibrating the assembly mold.

REFERENCES

ASTM D 4253. "Standard Test Methods for Maximum Index Density and Unit Weight of Soils Using a Vibratory Table." *Annual Book of ASTM Standards*. American Society for Testing and Materials, Philadelphia, United States, www.astm.org.

ASTM D 4254. "Standard Test Methods for Minimum Index Density and Unit Weight of Soils and Calculation of Relative Density." *Annual Book of ASTM Standards*. American society for testing and materials, Philadelphia, United States, www.astm.org.

ASTM D 2049-64. "Standard Test Methods for Calculation of Relative Density." *Annual Book of ASTM standards*. American society for testing and materials, Philadelphia, United States, www.astm.org.

BS 1377-2. 1990. "*Methods of Test Sor soils for Civil Engineering Purposes-Part 2: Classification Tests for Determination of Water Content*." British Standards, UK.

IS: 2720 (Part 1). 1980. "*Indian Standard Code for Preparation of Soil Samples*." Bureau of Indian Standards, New Delhi.

IS: 2720 (Part I). 1983. "*Method of Test for Soils: Preparation of Dry Soil Specimen for Various Tests*." Bureau of Indian Standards, New Delhi.

Reddy, Krishna R. 2000. "*Engineering Properties of Soils Based on Laboratory Testing*." Department of Civil and Materials Engineering University of Illinois at Chicago, USA.

Terzaghi, K. 1943. *Theoretical Soil Mechanics*. New York:Wiley & Sons.

Vesic, A. 1963. "Bearing Capacity of Deep Foundations in Sand." *National Academy of Sciences, National Research Council, Highway Research Record* (39): 112–153.

9 Compaction Characteristics of a Soil Specimen

References: IS: 2720-Part 7 (1980); IS: 2720-Part 8 (1980); ASTM D698-12e2; ASTM D1557-12e1; BS 1377: Part 4 (1990)

9.1 OBJECTIVES

Compaction test is mandatory:

1. For determination of compaction parameters (OMC and MDD or MDU) for soils in the laboratory.
2. The compaction parameters are used to establish the controlling parameters for achieving the desired degree of compaction in the field.
3. To compact the loose soil fills, un-engineered fills, earth-fill dams, embankments, road construction works etc., for sustainable development.
4. To reduce compressibility and achieve maximum shear strength at minimum void ratio.

9.2 INTRODUCTION

The compaction method is generally suitable for fine grained soils (cohesive soils) conceived by R.R Proctor (1930), which is usually known as standard Proctor test. In this method, soil specimen about 3,000 g passing 20 mm, retaining and passing 4.75 mm sieves is taken in proportion to the percentage retaining and passing 4.75 mm sieve for compaction test (for example, percentage retained on 4.75 mm sieve = 40%, percentage passing 4.75 sieve = 60%, then weight of 3,000 g of soil specimen for compaction test is taken in proportion as: $3,000 * 0.40 + 3,000 * 0.60 = 1.200 + 1.800 = 3,000$ g). A 1,000 cm^3 capacity compaction mold is used when percentage passing through 4.75 mm sieve is more than 80%. However, or use a compaction mold of 2,250 cm^3 capacity if the %age passing through 4.75 mm sieve is less than 80%. Also, reject the soil retained on 20 mm sieve. Replace this soil by equivalent amount passing 20 mm sieve if necessary. The prominent variables influencing the compaction are dry density (ρ_d), moisture content (w), compactive effort (CE), and soil type.

Dry density and water content are determined for each compaction trial test. About four to six trials are required for compaction to determine OMC and MDD. Compactive effort is the measure of mechanical energy applied for compacting the soil specimen in the compaction test, which is about 593.7 kJ/m^3 in a standard

DOI: 10.1201/9781003200260-9

Proctor test as per standard codal procedure (ASTM D 698, BS 1377-4 and IS 2720-7). The compaction test is mostly suitable for cohesive soils (fine grained soils). However, it may be noted that the compaction test is not suitable for cohesionless (sands) soil, as it does not produce a proper compaction curve for these soils. In the field compaction test, various types of rollers are used to compact the soil depending on the soil type and the compactive effort is measured in terms of passes or "coverages" by a roller.

For a standard Proctor compaction test in the soil testing laboratory, compactive effort is applied by compacting the soil specimen in three equal layers by dropping a hammer of 2.6 kg from a freefall of 30 cm in a compaction mold of 10 cm diameter and 12.7 cm height (nearly 1,000 cm³ capacity mold). It may be noted that each layer is tamped 25 times by the 2.6 kg hammer from a freefall of 30 cm. Therefore, CE is computed by the following equation:

$$CE = \left[\frac{Hammer\ wt.\ (kg) * Ht.\ of\ free\ fall\ (m) * No.\ of\ layers\ (n) * No.\ of\ drops\ (N)}{Volume\ of\ compacting\ mould\ (m^3)} \right] \left(\frac{kg.\ m}{m^3} \right)$$

(9.1)

Likewise, for a modified compaction test in the soil testing laboratory, compactive effort is applied by compacting the soil specimen in five equal layers by dropping a hammer of 4.89 kg weight with *25 evenly distributed blows* from a freefall of 45 cm in each layer for *1,000 cm³* mold and *56 evenly distributed blows for 2250 cm³ diameter mold*. Since more compactive effort is applied in the modified compaction test, therefore, the compaction curve will move towards left side of the curve for standard Proctor compaction test, as illustrated in Figure 9.1.

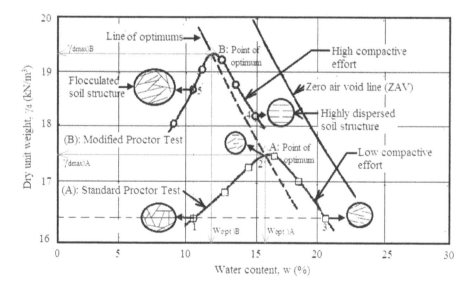

FIGURE 9.1 Standard and modified compaction curves (Mir and Sridharan 2013).

Therefore, for a compaction test, about four to six trials of compaction tests are conducted on a fresh sample for each trial with increasing water content (for clay dominated soils, first trial of compaction test may be started with water content of 10% followed by 4% increment of water to be added for each trial up to point of optimum and 2% increment for trials beyond point of optimum). It may be ensured that each compaction trial test is given uniform compactive effort.

Note: The point of optimum is the point on the compaction curve representing the compaction test parameters (OMC and MDD or MDU). It may be considered when the volume of compacted soil mold starts decreasing. Generally, point of optimum is reached at after three trials of compaction test.

For each trial of compaction test, water content and dry density is determined as per standard codal procedures (Test No 1 & 2). Each compaction trial test with water content and dry density represent a single point on the compaction curve. When a complete compaction test is completed (e.g. test with at least four to six trials), then a dry density-water content curve is plotted between water content and dry density on a natural graph paper as shown in Figure 9.1. The point on the peak of the curve is known as the point of optimum and the water content corresponding to optimum point is known as optimum moisture content (OMC) and the dry density corresponding to OMC is known as maximum dry density (MDD) or maximum dry unit weight (MDU).

From Figure 9.1, it is seen that dry density increases up to point of optimum and decreases beyond the point of optimum. However, the water content keeps increasing through out the test. Therefore, the dry density on point of optimum is taken as "MDD" and the corresponding water is known as "OMC," which yields maximum dry density. It may be noted that the compaction curve is unique for a given soil specimen, type of compaction method (e.g. either standard or modified or custom designed as required) and the mechanical energy imparted during the test. It may also be noted that the maximum dry density obtained in the laboratory test necessarily does not represent the maximum dry density of the same soil in the field. Therefore, relative compaction is used to ensure that the maximum dry density is achieved in the field as per specified specifications.

Figure 9.1 also illustrates the influence of compactive effort on the formation of structural arrangement of compacted soil particles. It is clearly seen that the soil particles exhibit highly flocculated structure (point 3 in curve-B) in a modified compaction test (high effort) as compared to just flocculated structure (point 1 in curve-A) in the standard Proctor test (low effort) on the dry side of optimum. Similarly, it is seen that the soil structure tends to be highly dispersed (point 4 in curve-B) for high effort test and just dispersed (point 3 in curve-A) for low effort test on the wet side of optimum. Further, among the different methods of compaction, the kneading compaction (such as the one induced by sheepsfoot rollers) produces a more oriented structure as compared to either static or impact type compaction. Thus, soil fabric becomes more dispersed with increase in water content and the "CE" by kneading compaction method, which induces more shearing strains in the soil. For general guidance, "OMC" and "MDD or MDU" for normal cohesive soils vary in the range of about 15 to 20% and 14 to 20 kN/m^3, respectively.

9.3 DEFINITIONS AND THEORY

Generally, the moist or bulk or total density of a soil ample (ρ_b) is defined as:

$$\rho_b = \rho_t = \left[\frac{M}{V}\right] (g/cc) \ or \ \gamma_b = \left[\frac{M}{V} * g = \frac{W}{V}\right] (kN/m^3) \qquad (9.2)$$

Where: M = Mass of moist soil sample (g)
  V = Volume of the soil sample encompassing soil mass (M)
  g = Gravitational acceleration constant ($g = 32.2 \ ft/s^2 = 9.81 \ m/s^2 = 981 \ cm/s^2$)
  $\gamma_b = \rho * g$ = Bulk unit weight of soil sample, and
  W = weight of wet soil sample.

Similarly, for a known volume of the soil sample, the dry density a of soil sample determined as:

$$\rho_d = \left[\frac{M_s}{V}\right] (g/cc) \ or \ \gamma_d = \left[\frac{M_s}{V} * g = \frac{W_s}{V}\right] (kN/m^3) \qquad (9.3)$$

Where: M_s = Mass of dry soil sample
  V = Volume of the soil sample encompassing soil mass (M_s)
  g = Gravitational acceleration constant ($g = 32.2 \ ft/s^2 = 9.81 \ m/s^2 = 981 \ cm/s^2$)
  $\gamma_d = \rho * g$ = Dry unit weight of soil sample, and
  W_s = Weight of dry soil solids, respectively.

However, in a compaction test, for known water content of the soil sample, the dry unit weight is calculated as:

$$\gamma_d = \left[\frac{\gamma_b}{1 + w/100}\right] (kN/m^3) \qquad (9.4)$$

Where: w = Water content (%) of soil sample taken from compaction test.

9.3.1 Zero Air Curve and Theoretical Dry Density

A zero air-void line or a zero air curve delineates the boundary beyond which soil sample achieves full saturation. A zero air void line is a graphical representation of theoretical dry density for given water content for a saturated soil sample, which is mathematically expressed as:

$$\gamma_{d\,(theoretical)} = \left[\frac{\gamma_w}{\frac{w(\%)}{100} + \frac{1}{G}}\right] = \left[\frac{G * \gamma_w}{(1 + \frac{w(\%) * G}{100})}\right] \cdot \left(\because \gamma_d = \frac{G * \gamma_w}{(1+e)} \ \& \ wG = Se, \atop when \ S = 1 \Rightarrow e = wG\right) \qquad (9.5)$$

Where: $\gamma_{d(theoretical)}$ = Theoretical dry density representing zero air void line.

However, Holtz et al. (2011) proposed that $\gamma_{d(theoretical)}$ can be related with degrees of saturation and the Equation (9.5) may be rewritten as:

$$\gamma_{d\,(theoretical)} = \left[\frac{\gamma_w * S}{\left[\left(\frac{w(\%)}{100} \right) + \left(\frac{\gamma_w}{\gamma_s} \right) * S \right]} \right] \quad \because \left(G = \frac{\gamma_s}{\gamma_w}, \quad e = \frac{w * G}{S} \right) \quad (9.5a)$$

When degree of saturation is 100% (e.g. S = 1), the Equation (9.5a) is the same as Equation. (9.5):

$$\gamma_{d\,(theoretical)} = \left[\frac{G * \gamma_w}{\left[1 + \frac{w(\%) * G}{100} \right]} \right] \quad \because \left(S = 1, \quad \frac{1}{G} = \frac{\gamma_w}{\gamma_s} \right)$$

In the compaction test, value of specific gravity (G) is known. Therefore, for different intervals of water content (w %) and assuming that the degree of saturation is 100%, the theoretical dry density can be computed using Equation (9.5). About three points are required for superimposing the zero air void line or zero air void curve onto the compaction curve as shown in Figure 9.1. It may be noted that a soil sample tested in a compaction test never achieves full saturation (e.g. S = 1), therefore, the compaction curve always lies towards the left side of the zero air void curve. Thus, the zero air void curve helps in assessing the reliability of compaction test data and it can be concluded that some air is still present in some of the soil void space between the soil particles. However, about more than 90% degree of saturation is presumed to be achieved on the wet side of optimum.

9.4 METHOD OF TESTING

Compaction parameters (optimum moisture content (OMC) and maximum dry unit weight (MDU) or maximum dry density (MDD)) of a soil sample are determined by the standard Proctor or by a modified compaction test.

9.5 SOIL TESTING MATERIAL

About 20–30 kg of an oven-dried soil sample passing through a 20 mm sieve (¾ in.) and retaining and passing through a 4.75 mm sieve (No. 4) is taken for the compaction test.

9.6 TESTING EQUIPMENT AND ACCESSORIES

The following equipment and accessories are required for a compaction test:

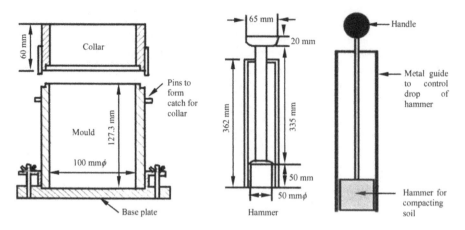

FIGURE 9.2 Schematic of a compaction test and allied accessories.

1. Soil sample passing 20 mm sieve (¾ in.) and retaining and passing 4.75 mm sieve (No. 4)
2. IS sieves 4.75 mm and 20 mm
3. Weighing balances, accuracy 1 g and 0.01 g
4. Compaction mold, cap. 1,000 ml (2,250 ml for modified test)
5. Detachable base plate
6. Weighing balance of 1 g accuracy to weigh up to 10 kg
7. Dessiccators, oven
8. Weighing balance of 0.01 g accuracy to weigh moisture cans
9. Moisture content cans
10. Collar, 60 mm high (should conform to IS 10074:1982)
11. G.I. trays (600 × 500 × 80 mm)
12. Compacting hammers of weight 2.6 kg and 4.89 kg
13. Mixing tools, spoons, trowels, straightedge, duster, and grease, etc.
14. Distilled water, graduated jar, plastic squeeze bottle with water
15. Extractor jack
16. A calculator

A compaction test setup with accessories is shown in Figure 9.2.

9.7 TESTING PROGRAM

The necessary equipment and process of mixing of soil in a compaction test is illustrated in Figure 9.3. The step-by-step procedure is given as follows:

1. Collect the soil sample from the field and pulverize and air-dry it in the soil testing laboratory.
2. Take about 20 kg of the oven-dried soil sample for the standard compaction test.
3. Take about 45 kg of the oven-dried soil sample for the heavy or modified compaction test.

FIGURE 9.3 Standard Proctor compaction equipment and test procedure.

4. Sieve the soil sample through 20 mm and 4.75 mm sieves (Figure 9.3).
5. Determine the percentage passing through 20 mm and 4.75 mm sieves and retained on the 20 mm sieve. Reject the soil retained on the 20 mm sieve. Replace
   this soil by the equivalent amount passing through the 20 mm sieve if necessary.
6. Select the compaction mold of 1,000 cm^3 capacity if the percentage passing
   through 4.75 mm is more than 80% or take a mold of 2,250 cm^3 capacity if
   the percentage retained on the 4.75 mm sieve is greater than 20%. Clean the
   mold thoroughly and apply a thin layer of grease to its inner surface uniformly and record empty weight of the mold (W_1 g).
7. Now, take a soil sample in proportion to the percentage retaining and passing the 4.75 mm sieve in step 6 for the compaction test. For example:
   Let total weight of sample taken for compaction test (passing 20 mm
   sieve) = M = 10 kg
   Weight of soil sample retaining on 4.75 mm sieve = M_1 = 2.5 kg (2,500g)
   Weight of soil sample passing through 4.75 mm sieve = M_2 = 7.5 kg (7,500g)
   %age retaining on 4.75 mm sieve = P_1 = (M_1/M) * 100 = (2.5/10) * 100 = 25%
   %age passing through 4.75 mm sieve = P_2 = (M_2/M) * 100 = (7.5/10) *
   100 = 75%
   or %age passing through 4.75 mm sieve = 100 – P_1 = 75%
   Let weight of soil sample to be taken for compaction test = M_{ct} = 3 kg = 3,000 g (a)
   Therefore, based on above soil gradation, soil sample weight of 3 kg will be
   computed as:
   Mct = 0.25 * M_{ct} + 0.75 * M_{ct} = 0.25 * 3 + 0.75 * 3 = 0.75 + 2.25 = 3 kg (same
   as above)
8. For a standard compaction in 1,000 cm^3 capacity mold, take about 3 kg soil
   sample for each compaction trial test. It may be noted that the same soil
   sample can also be reused for next trial of compaction if required; however,
   water content increment has to be added to this moist soil and mixed
   thoroughly. But this practice may be avoided if sufficient quantity of soil
   sample is available for the test.
9. Similarly, for a heavy or modified compaction in 2,250 cm^3 capacity mold,
   take about 6 kg soil samples for each compaction trial test.
10. For a modified compaction in 2,250 cm^3 capacity mold, take about 8.5 kg
    soil sample for each compaction trial test.
11. A water content of about 10% is suggested as the first or starting increment to be
    added to fine grained soils and mixed thoroughly for the first trial of the com-
    paction test.
    Note: The moist soil prepared in step 11 should be placed in an airtight container
    for about 16 hours or overnight so that the moist soil attains equilibrium or the
    water added is distributed uniformly. The soil sample may be thoroughly mixed
    again before conducting the test. Generally, the moist soil sample is allowed to
    mature in an airtight container for the minimum period of time as given below:
    i.   About 3–4 hours standing time is recommended if the soil sample is silty
         sand dominated with appreciable gravel content.
    ii. About 16–20 hours standing time is recommended if the soil sample is
         clay dominated.

12. In a standard compaction, the thoroughly mixed moist soil is compacted in three equal layers by compacting each layer by dropping a hammer of 2.6 kg weight 25 times (e.g. 25 no. of blows) from a freefall of 30 cm in 1,000 cm^3 capacity mold. It may be noted that the each layer is scratched before compacting the next layer and a collar is added before compacting the third layer. The soil sample is also uniformly distributed into three parts so that at the end of the third soil layer, the soil should extend slightly above the top of the rim of the compaction mold.

13. When the first trial is completed by compacting the soil in three equal layers in the mold of known volume, remove the collar by rotating it to break the bond between the collar and the soil before lifting it off from the mold.

14. Trim off the extra soil flush with the top and bottom of the mold and clean the outer surface with a neat piece of cloth.

15. Record the weight of the mold filled with moist soil (W_2). Then find weight of moist soil sample, $W_s = W_2 - W_1$.

16. Determine the bulk or total unit weight of the moist soil sample compacted in known volume of mold (V) given by $\gamma_b = [W_s/V]$.

17. Now, extract the soil sample from the mold and clean it thoroughly. Take a representative soil sample for the determination of water content ($w_1\%$). Once the water content is known, the dry unit weight for the first trial of compaction is determined as $\gamma_{d1} = [\gamma_{b1}/(1 + (w_1/100))]$. Thus, γ_{d1} and w_1 represent the first point on to the compaction curve (Figure 9.1).

18. Steps 12 to 17 are repeated for remaining trials for the compaction test (at least 5 to 6 trials are required for plotting the compaction curve) on the next soil sample. If the soil sample from step 17 is reused, then additional water increment of about 4% is further added to the soil (now total water added is 14%) and thoroughly mixed so as the water content is evenly distributed in the soil sample and the test is completed. Likewise, add further increment of 4% to each next trial and determine the water content and dry unit weight for plotting the compaction curve.
    Note: Fresh sample is recommended for each trial test if the size of sample is small. However, for large size samples, the large quantity of soil sample is required for a test using fresh soil samples. Therefore, the same sample is used to obtain all the density-water measurements.

19. Likewise, a water content of about 4% is suggested as the first or starting increment to be added to coarse grained soils thoroughly for the first trial of compaction test and the procedure from steps 11 to 18 is repeated to complete the compaction test.

Note: Johnson and Sallberg (1962) developed a chart (Figure 6.16: Practical No. 6) between "PL" and "LL" for representing the optimum moisture content curves, which helps in choosing an approximate values of "OMC and MDU or MDD" from LL and PL. This water content is taken as the starting water content for the first trial test, followed by the desired increments of water content for subsequent trials. Generally, in the case of clayey soils, starting initial water content for compaction test on air-dried soil is about 8 to 10% followed by 3% increment upto point of optimum and 2% increment beyond point of optimum.

20. Similarly, the above procedure may be repeated for a heavy or modified compaction test. However, the moist soil is compacted in five equal layers by compacting each layer by dropping a hammer of 4.89 kg weight 25 times (e.g. 25 no. of blows) from a freefall of 45 cm in 1,000 cm^3 capacity mold and 56 times (e.g. 56 no. of blows) from a freefall of 45 cm in 2,250 cm^3 capacity mold, respectively.

Note: It may be noted that at least five test trials (at least two points either side of point of optimum) will produce a well defined compaction curve. Therefore, to avoid a large number of test trials, it is recommended to start the first test trial with a water content of about 5 to 7% from the dry side of OMC (e.g. lower than OMC). Then add 3 to 4% moisture content (by weight of dry soil) on successive trials (2–3% if same sample is extruded from the mold and re-used for next trial).

9.8   OBSERVATION DATA SHEET AND ANALYSIS

1. Prepare the observation and data sheet as shown in Table 9.1 and plot the compaction curve with between (w %) and (γ_d), as shown in Figure 9.4.
2. From the compaction curve, find "OMC" and "MDU" or "MDD" as desired.
3. Plot the zero air void line as in usual manner ($\gamma_d = G\gamma_w/(1+w_G)$).

9.9   RESULTS AND DISCUSSIONS

Optimum moisture content (OMC) = _____%
    Maximum dry unit weight (MDU) = _____ kN/m^3
    Students are advised to complete the compaction test and record all observations and then calculate water content and dry unit weight, which will result in a plot of water content versus dry unit weight. Once the compaction curve is plotted, plot the ZAV line, and this completes the compaction test. From the compaction curve, find OMC and γ_{dmax} and comment on the type and suitability of the soil sample. A sample of the compaction curve is given below for guidance only. However, complete data for this figure is not given here.

Data sheet for sample for compaction curves for soil samples A and B							
Test-A		G = 2.69 $\gamma_w(kN/m^3) = 10$		Test-B		G = 2.54 $\gamma_w(kN/m^3) = 10$	
w (%)	Dry unit wt. (kN/m^3)	w (%)	Dry unit wt. (kN/m^3)	w (%)	Dry unit wt. (kN/m^3)	w (%)	Dry unit wt. (kN/m^3)
9.36	17.05	16	18.81	12.12	14.85	18	17.430
12.72	17.92	18	18.121	15.37	16.00	20	16.840
15.63	18.53	20	17.491	18.17	16.83	22	16.30
18.90	17.42	22	16.90	22.09	15.92	24	15.78
21.61	16.56	NA	NA	25.01	15.16	26	15.30

TABLE 9.1

Data observation sheet for compaction test

Diameter of the mold	= _____ (cm or m)
Height of the mold	= _____ (cm or m)
Volume of the mold, V	= _____ (cm^3 or m^3)
Specific gravity of solids, G	= _____

Project/Site Name: **Date:**

Client Name: **Job. No.**   **Sample No.**

Sample Recovery Depth: **Sample Recovery Method:**

Sample Description:

Tested By:

Test Type: on gravimetric basis/on volumetric basis

Sl. No.	Observations and Calculations	Trial No.				
		1	2	3	4	5
A	Calculation of total unit weight, γ_b(kN/m^3)					
1	Wt. of empty mold, W_1 (g)					
2	Wt. of mold + compacted soil, W_2 (g)					
3	Wt. of moist soil, W_s (g) [1 gram = 10^{-5}kN]					
4	Volume of mold, V (m^3)[1 cm^3 = 10^{-6}m^3]					
5	Total unit weight, γ_b (kN/m3) = W_s/V					
B	Calculation of water content, w (%)					
1	Container No.					
2	Weight of empty container, W_1 (g)					
3	Weight of container + soil, W_2 (g)					
5	Weight of water, $W_w = W_2 - W_3$ (g)					
6	Weight of solids, $W_s = W_3 - W_1$ (g)					
7	Water content, $w = (W_w)/(W_s) * 100$ (%)					
C	Calculation of dry unit weight, γ_d(kN/m^3)					
1	Dry unit weight, γ_d (kN/m^3), $\gamma_d = \gamma_b/ (1 + w/100)$					
8	Void ratio, $e = \dfrac{G\gamma_w}{\gamma_d} - 1$					
9	Degree of saturation, $S = (wG/e) * 100$					
10	Theoretical dry unit wt., $\gamma_{dth} = G\gamma_w/(1+w_G)$					

9.10   GENERAL COMMENTS

The standard Proctor compaction or the modified compaction test is suitable only for fine grained dominated soils, which produces a curve of well-defined points. However, a well-defined curve is not produced in the case of cohesionless soils (e.g. sandy soils). Therefore, the relative density test is recommended for cohesionless soils. The values of OMC and MDU in a standard compaction test will be always on the higher side compared to the modified test due to higher compactive effort in the modified compaction test.

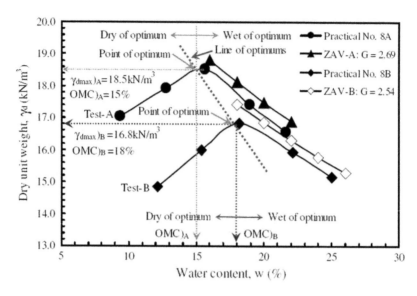

FIGURE 9.4   Compaction curves for soil samples A and B (Mir 2020).

9.10.1 Why (OMC) Against (MDD)?

In a compaction test, soil is compacted by expulsion of air from soil pores and adding water content, which brings soil particles close together by means of lubrication effect. Since the dry unit weight increases with increase in water content up to the peak point only and then decrease on further increase in water content. Therefore, the water content corresponding the peak point, which produces (MDD) is termed "OMC."

9.10.2 Other Important Comments

Compaction is first engineering property among others and the parameters obtained in this test such as "OMC" and "MDU or MDD" are referred to as the compaction characteristics or compaction parameters which are used for preparation of remolded soil samples for conduct of other tests such as coefficient of permeability, consolidation test, and shear strength tests. Therefore, while conducting the compaction test (especially on particularly fine grained soils), the researchers/students/technicians have to be very careful about the out-come behaviour of compaction behaviour of soils, particularly fine grained soils for construction of various types of projects or to establish the project quality control criterion. Based on experience, following are some of the very important comments regarding the compaction test, which will be of great interest to readers.

1. Fresh samples may yield less dry unit weight if not mixed properly and cured for some time as per codal procedures.

2. It may be ensured that the compaction mold is placed on a stiff base so that it does not vibrate while giving repeated drops and the hammer energy is fully imparted to the soil as compaction effort.

3. It has been conceded now that the *structural arrangement of the compacted soil mass* (especially when fine-grained soils are present) is intimately associated with the compaction process and the water content at which the soil mass is compacted (whether on the dry or wet side of the OMC). This concept is extremely important for compacting the clay cores of dams where settlements could cause cracks in the core. It has been found that the dispersed soil structure (parallel particle orientations) obtained by compacting these soils on the wet side of optimum results in a soil mass that has somewhat lower shear strength but that can undergo larger settlements without cracking and causing a large leakage and/or actual dam failure. Compaction of soil on the wet side of optimum also reduces the permeability, as compared with compacting the soil on the dry side of optimum. In this case, the dispersed soil structure has fewer interconnected pores to produce continuous flow paths so is more resistant to water transmission.

4. However, the *flocculated structure* of a compacted clayey soil on the dry side of optimum is less susceptible to shrinkage on drying, but is more susceptible to swelling on saturation. This is attributed to a combination of the flocculated structure, sensitivity of additional water at the contact points, and the lower reference (w at compaction) water content. For soils compacted on the wet side of the OMC, the reference water content is already closer to saturation ($S = 100\%$) so swelling would be less and shrinkage as the soil dries would be more since the reference point is higher.

5. Repercussions of water content on the dry density of the compacted soil.

It may be noted that there is a negligible increase in soil's dry unit weight during the initial hydration phase as the water being added to the soil is absorbed on the surface of the soil particles. However, adding more water content results in a large increase in the soil particle's packing and increases the unit weight on the dry side of optimum due to the lubrication effect as illustrated in Figure 9.5.

However, Figure 9.5 also demonstrates the repercussions of water content on the dry unit weight of the compacted soil during the compaction process on the wet side of optimum. When the water content is increased beyond the OMC, it displaces the soil particles apart and occupies the space between the soil particles, which reduces the dry unit weight. This phenomenon is generally known as swell phase on wet side of OMC. However, on further increasing water content, the water fills the void pores between the soil grains completely and hence reduces the dry unit weight further. The phenomenon is known as saturation phase of wet side of optimum. In both these cases, the free water occupies the pore void space by displacing the soil solids apart, which results in the reduction of dry unit weight of the soil sample on wet side of optimum (Fig. 9.5).

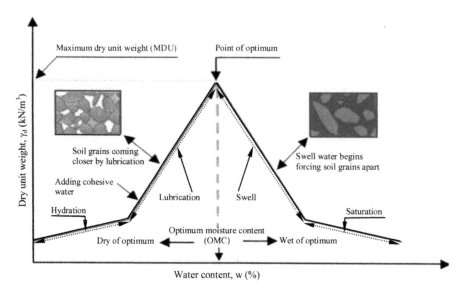

FIGURE 9.5   Repercussions of water on density (dry unit weight).

6. It may be noted that the soil compacted on the dry side of optimum possess high strength under low stress level and fails abruptly under high level stress. However, the soil compacted on the wet side of optimum exhibits plastic failure and non-linear behavior. Also, it is noted that the residual strength of the compacted soil is almost the same on either side of the point of optimum (e.g. dry of OMC and wet of OMC).

9.11   APPLICATIONS

The compaction test plays a vital role in the design of building structures, design of earthen dams, highway pavements, and airfields. The compaction parameters are used to establish the controlling parameters for achieving the desired degree of compaction in the field and to reduce compressibility and achieve maximum shear strength at minimum void ratio. Therefore, the laboratory compaction test parameters are used to write the compaction specification for field compaction of soils.

Relative compaction (RC) and density index or relative density (R_D) are correlated as:

$$RC = [80 + 0.2R_D] \tag{9.6}$$

RC of 80% for loose sand corresponds to zero relative density.

Where relative compaction or degree of compaction is given by:

$$RC\,(\%) = \left[\frac{Field\ \ dry\ \ density}{MDD} * 100 \right] \tag{9.7}$$

Field compaction is also controlled by another factor known as compatibility (C) given by:

$$C = \left[\frac{\gamma_{max} - \gamma_{min}}{\gamma_{min}} \right] = \left[\frac{\rho_{max} - \rho_{min}}{\rho_{min}} \right] \qquad (9.8)$$

- **Field compaction criteria**

Compaction is necessary to ensure that engineering properties assumed in the design of soil structures are actually accomplished uniformly during construction. Quality control is exercised by controlling:

a. Thickness of compacted layer, dry density, and water content, besides involving choice of quarry material, thickness of lifts (layers), no of passes, choice of rollers (heavy or light, kneading of vibratory).
b. The in situ oil should be compacted to 95% of MDU, at a water content of ±2% of OMC as obtained in a standard compaction test.
c. Relative compaction (RC) is yet another controlling factor that may be specified. It is also known as degree of compaction. *The RC values should range from 90% to 105%.*
d. Rolling is done in layers of 400 to 250 mm, 150 mm of compacted thickness being common.
e. *Number of passes range from 6 to 16* depending on the soil type.
f. The OMC serves as a guide for placement water content for soils with an appreciable amount of fines, but has little meaning in the case of gravelly soils. For small works, two to three tests per lift are necessary.

9.12 SOURCES OF ERROR

1. The main source of error in the compaction is if the specified compactive effort is not imparted to the soil in the compaction mold. This is produced in erroneous test parameters.
2. Further, if the additional increment of water on reused water is not thoroughly mixed and given adequate time for hydration.
3. The dry unit weight may be overestimated by taking a re-used soil sample for subsequent test trials. In this case, the soil sample is repeatedly mixed, rehydrated, and compacted, which results in a higher dry unit weight compared to a fresh soil sample.

9.13 PRECAUTIONS

1. Make sure that the soil sample is properly and carefully sieved through designated sieves to select the right compaction mold for the compaction test.

2. Make sure that the soil sample is thoroughly mixed with water and an adequate time period is allowed for hydration before conducting the test.
3. Make sure that the compaction mold is placed on a stiff base so as to avoid any sort of vibrations during the compaction process.
4. Make sure that the specified number of blows are evenly distributed over the surface of each layer.
5. Make sure that each layer is scratched before adding another layer so as to get proper bonding between the layers.
6. Make sure that the collar is rotated before removing it from the mold to break the bond of the collar and the soil.

REFERENCES

ASTM D698-12e2. 2012. "Standard Test Methods for Laboratory Compaction Characteristics of Soil Using Standard Effort (12 400 ft-lbf/ft3 (600 kN-m/m3))." ASTM International, West Conshohocken, PA, www.astm.org.

ASTM D1557-12e1. 2012. "Standard Test Methods for Laboratory Compaction Characteristics of Soil Using Standard Effort (56,000 ft-lbf/ft3 (2,700 kN-m/m3))." ASTM International, West Conshohocken, PA, www.astm.org.

BS 1377-4. 1990. *Methods of Test for Soils for Civil Engineering Purposes-Part 4: Compaction-Related Tests.* British Standards, UK.

Holtz, R. D., W. D. Kovacs, and T. C. Sheahan. 2011. *An Introduction to Geotechnical Engineering.* 2nd ed. New York:Pearson, 853 p.

IS: 2720 (Part 1). 1980. *"Indian Standard Code for Preparation of Soil Samples."* Bureau of Indian Standards, New Delhi.

IS: 2720 (Part 7). 1980. *"Method of Test for Soils: Determination of Water Content-Dry Density Relation Using Light Compaction."* Bureau of Indian Standards, New Delhi.

IS: 2720 (Part 8). 1980. *"Method of Test for Soils: Determination of Water Content-Dry Density Relation Using Heavy Compaction."* Bureau of Indian Standards, New Delhi.

Johnson, A. W., and J. R. Sallberg. 1962. "Factors Influencing Compaction Results." *Highway Research Board Bulletin* (319): 125–148.

Mir, B. A., and A. Sridharan. 2013. "Physical and Compaction Behavior of Clay Soil-Fly Ash Mixtures." *J. Geotech. Geol. Eng* 31(4): 1059–1072.

Proctor, R. R. 1933. "Fundamental Principles of Soil Compaction." *Engineering News-Record* 111(9, 10, 12, and 13): 372.

Reddy, Krishna R. 2000. "Engineering Properties of Soils Based on Laboratory Testing." Department of Civil and Materials Engineering, University of Illinois at Chicago, USA.

10 Coefficient of Permeability or Hydraulic Conductivity of Soils

References: IS: 2720-Part 1 (1980); IS: 2720-Part 17 (1980); BS 1377: Part 5 (1990)

10.1 OBJECTIVES

The main objective of determination of the coefficient of permeability is to ascertain the suitability of soils for its desired function either as a barrier material (e.g. used for clay liners or in core dams to avoid seepage loss etc.) or as draining material (e.g. used as fill material behind retaining walls to avoid development of pore water pressure).

The coefficient of permeability defines the quality of a soil material that allows the flow of water through it. It classifies the soil either as pervious material (e.g. having high permeability such as coarse grained soils) or as an impervious soil material (e. g. having very low permeability such as clays).

10.2 DEFINITIONS AND THEORY

Permeability, generally termed as the coefficient of permeability, is an intrinsic property of a porous material (i.e. pore size, tortuosity, degree of compaction and surface area and pozzolanic activity). The coefficient of permeability is an engineering property of soils and plays a vital role in the identification and characterization of soils, which are used as construction material in various engineering structures such as earthen dams (e.g. dam cores), clay liners beneath landfills, as fill material behind retaining walls (Reddy 2000). The permeability is an engineering property of soils, which defines the rate of laminar flow of water through the continuously connected soil voids of unit cross section and unit hydraulic gradient. This is generally quantitatively expressed in terms of coefficient of permeability or hydraulic conductivity, k (m/s). It can vary over a wide range, from 10^{-3} m/s for sands to as low as 10^{-11} m/s for clays. However, "k" is affected by various factors such as soil particle size, pore void size, fluid viscosity, temperature, etc. The maximum particle size permitted in the soil testing laboratory procedure is 10 mm (with 90% or more percent of soil aggregation passing through the 4.75 mm IS sieve). The concept of flow of fluids through soil pores was conceived by French scientist H. Darcy (1856), which is known as Darcy's Law. During his investigation of water flow through sand layers of varying thicknesses and

DOI: 10.1201/9781003200260-10

pressures, he found that the rate of flow remained proportional to the hydraulic head under laminar flow. However, he also found that Darcy's Law is not valid for turbulent flow of high velocities. Thus, Darcy's Law plays a vital role in the analysis and design of various practical problems in geotechnical engineering provided that the flow through soils remains laminar. Based on water flow through soils (e.g. discharge velocity) and hydraulic gradient, the validity of laminar flow through the soils is depicted in Figure 10.1, which is generally known as Reynold's number (Taylor 1948).

From Figure 10.1, it is seen that at a relatively low-flow velocity, the flow remains laminar; however (e.g. v ∞ i) when the velocity of flow is increased, the flow becomes turbulent and Darcy's Law is no longer valid (Reynolds 1883). Thus, for a laminar flow through homogeneous soils, Darcy's Law states that the rate of water flow is directly proportional to the pressure or hydraulic gradient expressed as:

$$[v \infty i] \tag{10.1}$$

Where v = Rate of flow of water through the soils
 i = Hydraulic gradient = Δh/L
 L = Length of soil sample (cm or m), and
 Δh = Head difference to create flow (cm or m).

$$\therefore v = [ki] = \left[k * \frac{\Delta h}{L} \right] = k * \Delta h/L \tag{10.2}$$

Where k = Coefficient of permeability (m/s) or constant of proportionality.
 Therefore, the discharge of water per unit time (cm^3 or m^3) is computed as:

$$q = [kiA] = \left[k * \frac{\Delta h}{L} * A \right] \tag{10.3}$$

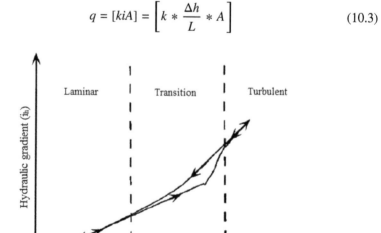

FIGURE 10.1   Schematic diagram of the laminar and turbulent flows (Taylor 1948).

Where: A = Total cross-sectional area of the soil sample (cm^2 or m^2), which includes the area of soil solids (A$_S$) and the area of soil voids (A$_V$), e.g. A = A$_S$ + A$_V$.

If i = 1 (e.g. Δh = L), k = v, then from Equation (10.3), we have:

$$\Rightarrow \quad q = [v * A = v_D * A] \quad or \quad \left[v_D = \frac{q}{A}\right] \qquad (10.4)$$

Where v_D = v = Discharge velocity through soil mass.

However, it may be noted that the velocity of flow through soils, e.g. the discharge velocity is always less than the actual or true seepage velocity (v_S). From the definitions of the discharge velocity and seepage velocity, we have:

$$q = [v * A = v_S * A_V]$$
$$OR \quad v_S = \left[v * \frac{A}{A_V}\right] \qquad (10.5)$$

But we know that: $\left[\frac{A_V}{A} = \frac{V_V}{V} = n \ (known \ as \ porosity)\right]$. Therefore, Equation (10.5) can be rewritten as:

$$v_S = \left[v * \frac{A}{A_V}\right] = \left[v * \frac{1}{n}\right] = \left[\frac{v}{n}\right] \qquad (10.5a)$$

Thus, rearranging Equation (10.5a), we have the desired relationship between the discharge velocity ($v = v_D$) and the true seepage velocity (v_S) expressed as:

$$v = v_D = [n * v_S] \qquad (10.6)$$

Where: n = Porosity of the soil sample = [e/(1 + e)].

Equations (10.4) through (10.6) are valid only if the flow of water through the soil mass is laminar within the hydraulic gradient range of 0.2 to 0.5. It may be noted that the lower values of the hydraulic gradient refer to coarse grained soils, which are highly pervious and the higher values of hydraulic gradient refer to fine grained soils, which possess low permeability and need a high-pressure gradient for flow to take place between tow points. In soil engineering practice, the approximate order of the values of k is given in Table 10.1 (Terzaghi and Peck 1967).

Also, the U.S. Bureau of Reclamation (USBR 1985) classifies the soils according to their coefficients of permeability as below:

1. Impervious: k < 10^{-08} m/s
2. Semi-pervious: k between 10^{-08} to 10^{-06} m/s
3. Previous: k > 10^{-06} m/s

TABLE 10.1

Coefficients of permeability for various types of soils (Terzaghi and Peck 1967)

Sr. No.	Coef. of Permeability, k (m/s)	Soil Description	Degree of Permeability
1	$> 10^{-3}$	Gravel	Highly pervious
2	10^{-3}–10^{-5}	Sandy gravel, clean sand, fine sand	Medium
3	10^{-5}–10^{-7}	Sand, dirty sand, silty sand	Low
4	10^{-7}–10^{-9}	Silt, silty clay	Very low or impermeable
5	$< 10^{-9}$	Clay	Practically impermeable

10.3 METHOD OF TESTING

The following permeability tests are conducted on soil samples in the laboratory for determination of "k":

1. The constant head test
2. The falling head test

It may be noted that the hydraulic gradient is always constant for the constant head method while it is changing in the falling head method.

10.4 SOIL TESTING MATERIAL—PREPARATION OF SOIL SPECIMEN

Since the coefficient of permeability is an engineering property of soils, therefore, a proper care has to be taken for preparing the test specimens in undisturbed state, in remolded state or in reconstituted state of soil sample. However, reconstituted samples can be prepared from clayey soils in the laboratory only. Remolded soil samples can be prepared in the laboratory for coarse grained soils, which cannot be collected in an undisturbed state from the field due to lack of cohesion.

a. **Undisturbed samples:** Undisturbed soils are generally collected from the field at the desired locations in a core cutter or SPT tubes presuming that their in situ properties are intact. However, undisturbed soil samples can be collected only in the case of clayey soils.

b. **Remolded/Disturbed samples:** A remolded or disturbed soil sample is collected from the field at a desired location by excavating the soil when an undisturbed soil sample cannot be collected in a core cutter of SPT tube. The disturbed soil samples are collected in the polythene of cement bags and transported to the soil testing laboratory. These samples are generally coarse grained soils, which lack cohesion and cannot be collected in undisturbed state in the field.

These are referred to as disturbed soil samples because their in situ properties (e.g. density, structure, etc.) are highly disturbed while excavating in the field. Therefore, these soil samples are prepared in the laboratory at a desired water content and unit weight for laboratory testing. Thus, these soil samples are also known as remolded soil samples.

c. **Reconstituted samples:** Reconstituted samples can be prepared using disturbed soil samples (clayey soils) collected from the field. These samples are prepared by consolidating the soil slurry under desired stress intensity on the laboratory floor. After consolidation, the soil sample is trimmed to the desired sample size by using a soil lathe. However, it may be noted that these soil samples lose their stress history and their properties are known as "intrinsic" (Burland 1990). The main objective of these intrinsic properties is that these are used to establish a reference model for behavior simulation of in situ soils in the field.

10.4.1 Preparation of Undisturbed Soil Specimens

Undisturbed test specimen can be prepared in two ways:

1. By core cutter of the same size testing mold (Figure 10.2):
   a. Take a core cutter of the same size of testing mold and apply a thin layer grease inside the cutter.
   b. Go to the field site and clean and level it by removing the top soil by at least 1 ft or up to the depth at which the test specimen is to be taken.
   c. Place the core cutter on the leveled surface and keep a dolly or a wooden block on its top.
   d. Drive the core cutter vertically straight into the ground by a lightweight hammer until the dolly or wooden blocks reach the leveled ground surface.
   e. Remove the dolly or wooden block from the top of the core cutter and excavate the surrounding soil carefully without disturbing the core cutter.
   f. Collect the core cutter carefully and also collect some disturbed soil with it in an airtight plastic bag and transport it to lab.
   g. Level the top and bottom surfaces of the core cutter and extrude the soil sample from it using the sample extruder carefully.
   h. Check the dimensions (dia. and height) of the test specimens carefully and take its weight before the test to the nearest gram (M_s). Thus, the undisturbed test specimen is ready for the permeability test.
   i. Take some soil from the remaining soil sample for water content determination.
2. Using a core cutter of different (bigger) size than the testing mold (Figure 10.3):
   a. Take a core cutter bigger than the testing mold and apply a thin layer grease inside the cutter.
   b. Repeat sub-steps 'b' to 'g' as in Step 1 above.
   c. Trim the extruded soil cylinder using the soil lathe. A soil lathe comes with different sizes of trimming accessories, as shown in Figure 10.3.

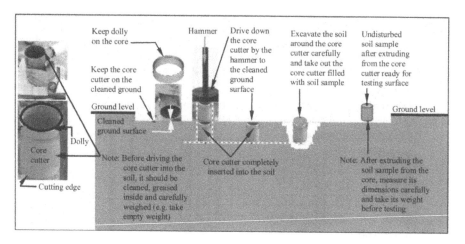

FIGURE 10.2   Collection of undisturbed soil sample from the field.

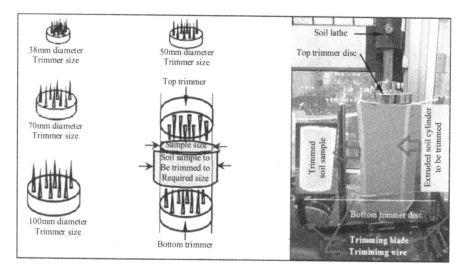

FIGURE 10.3   Preparation of undisturbed soil sample of desired specimen size.

    d. Check the dimensions (dia. and height) of the test specimens carefully and take its weight before the test to the nearest gram (M_s). Thus, the undisturbed test specimen is ready for the permeability test.

    e. Take some soil from the remaining soil sample for water content determination.

Note: It may be noted that the undisturbed soil samples from the field can be collected in cohesive (soft clayey soil) soils only. Undisturbed samples cannot be collected in the case of cohesionless soils. Therefore, remolded or disturbed samples are prepared in the laboratory.

10.4.2 PREPARATION OF REMOLDED SOIL SPECIMENS

1. Clean the compaction testing split mold and take its internal dimensions (int. dia. "D" and effective height "h") for determination of its volume or capacity, V (cm^3).
2. Weigh the split mold, W_1 (g).
3. Check the "OMC" and "MDD" for the soil sample to be used for the permeability test (OMC and MDD can be determined by the standard Proctor compaction).
4. Assume the density of soil sample is 0.9MDD and water content is OMC, e.g.

$$\gamma_d = 0.9 * \gamma_{dmax} \, (kN/m^3) \qquad\qquad (a)$$

Note: Ninety percent density is chosen for preparation of the test specimen because it is presumed that 100% compaction is not achieved in the field. Or, any other value of dry density may be chosen as desired. Further, depending on the nature of problem, water content may be chosen either dry side or wet side of OMC.

5. Now, calculate the dry weight of soil sample (W_d) and water content (w) as:

$$W_d = \left[0.9 * \gamma_{d\,(max)} * V \right] \left(\because \gamma = \frac{W}{V} \right) \qquad\qquad (b)$$

Where: V = Volume of the mold = $(\pi D^2 * h)$.

$$w = \left[\frac{OMC}{100} * W_d \right] (ml) \qquad\qquad (c)$$

6. Now, mix the dry sample (W_d) and water (w) thoroughly in a tray. Use hand gloves and assure that water is not wasted while mixing the soil (it is recommended that at least 10 ml extra may be taken in addition to calculated water content to compensate for any loss of water during mixing).
7. Apply a thin layer of grease inside the split mold surface and place it on the bottom spacer.
8. Fill the mold with mixed soil and place the top spacer on it.
9. Compact the soil sample in the split mold under the static compression machine up to its effective height (h), carefully, as shown in Figure 10.4.
   Note: Spacers should be chosen such that the desired size of the compacted soil sample is obtained. It should be noted that the soil sample should be compressed slowly until the effective height is achieved. After achieving the effective height of the compacted sample, wait for a few minutes and then remove the split mold from the machine to avoid any swelling of the sample if removed immediately.
10. In the absence of a static compression machine, prepare the soil sample at OMC and 0.9MDD using a suitable compacting device (static or dynamic device).

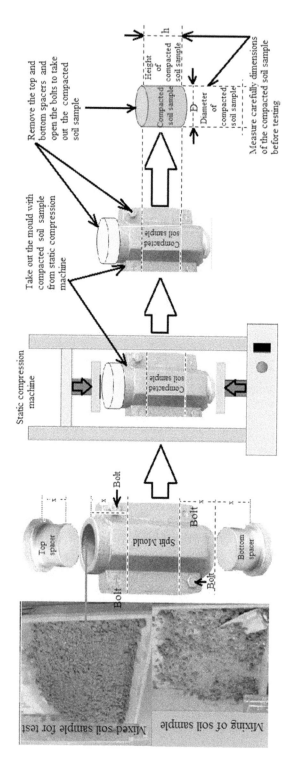

FIGURE 10.4 Preparation of remolded soil sample in the laboratory.

Note: A static compaction device is more convenient and accurate to prepare the soil sample at the desired density and water content. Dynamic compaction is effective to compact the soil to the maximum dry density at the optimum moisture content.

11. After completion of the compaction, take out the split mold from the compression machine and remove the spacers. Trim both ends flush with the top and bottom surfaces of the split mold.
12. Weigh the split mold with the compacted soil specimen to the nearest gram (W_2).
13. Now open the bolts of the split mold and take out the compacted or remolded soil specimen and cross-check its dimensions (height and diameter) carefully.
14. The remolded soil sample is ready for the permeability test as per test procedure.

10.4.3 PREPARATION OF RECONSTITUTED SOIL SPECIMENS

Fundamental research on soils can either be completed on natural, compacted, or reconstituted soils. Natural soils are not routinely tested because of difficulties in obtaining identical specimens. Compacted or remolded soil specimens are prepared either by kneading, dynamic (e.g. standard Proctor or modified Proctor tests) or by static compaction (e.g. by compression in a compression machine, generally known as compaction by press technique to avoid dynamic effect). The adaption of these techniques for preparation of soil samples can affect the results and interpretation of the test results.

Collecting an undisturbed soil sample is essential to determine the engineering properties like shear strength and compressibility of cohesive soil. However, it is always challenging to collect soil samples in an undisturbed condition from a significant depth below the existing ground surface. Further, compacted samples of cohesive soils are generally unsaturated and have lot of time for saturation due to low permeability ($k < 10^{-9}$ m/s). Also, collecting undisturbed samples for sandy soils is impossible. Therefore, reconstituted soil samples are preferred to avoid the above-cited difficulties. The necessary setup for preparation of the reconstituted soil samples has been designed and fabricated in the Geotechnical Engineering Laboratory, National Institute of Technology Srinagar.

a. **Preparation of reconstituted clay specimens:** Reconstituted clay specimens are those prepared at water contents in excess of the liquid limit (Burland 1990). Reconstituted samples can be prepared using disturbed samples collected from the similar depth and could provide an insight on the compressibility and shear strength behavior of cohesive soil. Limited research has been reported on preparation of reconstituted soil samples in the laboratory that could simulate better in situ condition in the literature. Therefore, a custom-designed setup has been designed and fabricated for preparing reconstituted clay specimens for laboratory testing, which is outlined briefly below.

Reconstituted specimens of clayey soil in the form of clay slurry are prepared at an initial water content of about 1.5 to 2 times the liquid limit (e.g., 1.5LL – 2LL) in a motor-operated slurry mixer, as shown in Figure 10.5a. After thorough mixing, the clay slurry is de-aired (Figure 10.5b) and consolidated (1-D consolidation) in 450 mm long and 250 mm diameter cylindrical mold, first under its own self-weight

(a)   (b)   (c)   (d)

Clay slurry mixer   Clay slurry being de-aired   Clay slurry consolidated   Reconstituted specimen

FIGURE 10.5 Reconstituted specimen from clay slurry sample (Mir and Juneja 2010): (a). Clay slurry mixer, (b). Clay slurry being de-aired, (c). Clay slurry consolidated, and (d). Reconstituted specimen. Source: Mir, B. A. (2010). Study of the influence of smear zone around sand compaction pile on properties of composite ground. Ph.D. Thesis, IIT Bombay under supervision of Prof. A. Juneja.

and later under desired stress intensity (e.g. choose pressure intensity, 100 kPa, 150 kPa, or 200 kPa as required) applied in stages on top of the clay surface using a custom-designed pneumatic load frame, as shown in Figure 10.5c (Mir 2010). After the slurry is consolidated under the desired stress intensity, the consolidated clay sample is extruded from the cylindrical mold and trimmed into desired specimen sizes using a soil lathe (Figure 10.5d).

Depending upon the sample requirement (specimen required for permeability, consolidation, DST, or triaxial tests) number of reconstituted clay specimens can be prepared for laboratory tests. It has been seen that test parameters obtained from reconstituted test results are in good agreement compared to those of undisturbed samples. Therefore, it is essential to use uniform reconstituted specimens prepared under controlled conditions in the laboratory (Mir 2010; Mir and Juneja 2010).

b. **Preparation of reconstituted sand samples:** Collecting undisturbed samples for sandy soils from the field is impossible due to lack of cohesion between sand particles. Though there are techniques such as ground freezing for collecting high-quality undisturbed sand samples from the field, but it turns to be very expensive and hence not preferred. Therefore, reconstituted representative samples of sandy soils by various techniques such as slurry deposition, dry or wet pluviation, vibrations, or moist-tamping in layers by under-compacting each layer to its succeeding layer are prepared in the laboratory (Ladd 1974; Vaid and Negussey 1988; Raghunandan et al. 2012). However, the detailed procedure is beyond the scope in this manual.

10.5 TESTING EQUIPMENT AND ACCESSORIES (FOR BOTH TEST METHODS)

The following equipment and accessories are required for the permeability tests:

1. Coarse grained soil passing Sieve No. 4 (4.75 mm IS sieve), 4.75 mm IS sieve
2. Permeability cell with top and bottom with controlled outlets/inlets

3. Permeability mold, internal dia. = 80 mm, eff. height = 60 mm, capacity = 300 ml
4. Detachable collar, 80 mm diameter, 40 mm height
5. Drainage base, having porous disc, dummy plate, 108 mm diameter, 12 mm thick
6. Water supply reservoir with constant head, constant head chamber, two manometers, and a graduated jar
7. De-aired water, graduated jar
8. Stopwatch, thermometer
9. Filter paper, supporting frame for the stand pipe and the clamp
10. Oven, desiccator
11. Weighing balance accuracy 0.1 g, large funnel and scoop
12. Grease, moisture content cans, straightedge
13. A static/dynamic compaction device for preparation of test specimen from disturbed soil

10.6 TESTING PROGRAM

10.6.1 Coefficient of Permeability of Soil by Constant Head Test Method

The constant head test is primarily suitable for coarse grained soils ($k > 1.0 * 10^{-6}$ m/sec) in which a constant supply of water flow is passed through the saturated soil sample under a constant head. Following is a step-by-step procedure followed during this method (Figure 10.6):

1. Take a dry and cleaned permeability mold of known dimensions (e.g. internal diameter "D" and effective height or length "L").
2. Apply a thin layer of grease inside the permeability mold surface weigh it, W_1 (g).
3. Now, place the permeability mold on the base plate and put the extension collar on its top and clamp the assembly properly.
4. Prepare the remolded soil sample (passing 4.45 mm sieve) at OMC and 0.9 MDU using a suitable compacting device (static or dynamic device) in the permeability mold.
5. Unclamp the permeability mold and trim the excess soil flush with top and bottom surfaces of the permeability mold.
6. Clean the outside surface of the permeability mold with a neat piece of cloth and weigh it, W_2 (g).
7. Now determine the net weight of the wet compacted soil sample in the permeability mold, $W_s = W_2 - W_1$ (g).
8. Determine the bulk unit weight ($\gamma_b = W_s/V$) of the moist soil sample compacted in the permeability mold.
9. Also, take about 2 to 30 g of the wet mixed soil sample in a container to determine its water content before the permeability test as per standard codal procedure (see Test No. 1).
10. Now place a saturated porous stone on the bottom base of the permeability cell with a controlled drain valve.

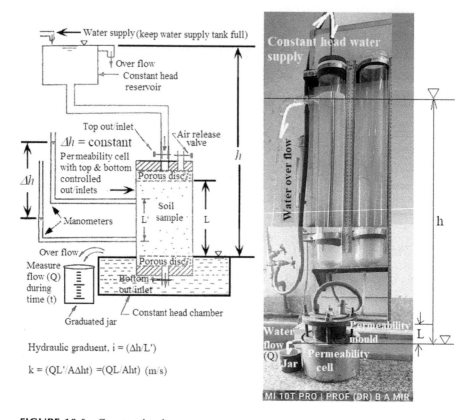

FIGURE 10.6   Constant head permeameter.

> **Note:** The porous discs used shall be de-aired and saturated before assembling in the mold. The air content of the porous discs can be eliminated by boiling in water.

11. Keep a saturated filter paper on the saturated porous stone and place the permeability mold with compacted soil sample. Care should be taken while placing the mold on the porous stone.

12. Likewise, place a saturated filter paper on the top surface of the soil sample.

13. Now keep the saturated porous stone over the filter paper and place the top drainage cap on the permeability cell and fix it properly using washers.

> **Note:** Filters are required to be provided between the specimen and porous disc to prevent the migration of the soil grains from the former to the latter due to seeping water.

14. Place the permeability cell in the constant head chamber (Figure 10.6) and connect it with the water reservoir at the bottom outlet and open the air valve at the top of the permeability cell.

15. Saturate the soil sample from bottom to top such that all the entrapped air is removed (e.g. water should flow from bottom to top while saturating the soil sample).

Note: For the fine grained soil of medium to high plasticity may be saturated by applying the vacuum gradually so that all the entrapped air is removed. The vacuum pressure may be gradually increased to 700 mm of mercury column height and should be maintained constant for about 20 minutes until full saturation is achieved.

16. After saturating the soil sample, disconnect the reservoir from the bottom outlet and connect the constant head water reservoir to the drainage cap inlet at the top cap (Figure 10.6).

17. Now connect the top inlet of the permeability cell with the constant head water reservoir and allow the water to flow downward through the soil sample until the steady state flow is constant.

18. Record the room temperature in the soil testing laboratory during the test and note the viscosity of water against the room temperature. The values of absolute viscosity are given in Table 10.2.

19. Now measure the discharge (Q) for a desired time interval (t) in a graduated jar.

20. Also, record the head difference (h) between the top of the constant head reservoir and the bottom constant head chamber (e.g. outflow of water), as shown in Figure 10.6.

21. Find the coefficient of permeability, k = [QL/Aht].

22. Repeat the above test procedure for the average value of "k" and complete the observation and data analysis sheet shown in Table 10.3.

23. Dismantle the permeability cell assembly and take out the soil sample after completing the test.

24. Also, determine the water content of the soil sample post-permeability test.

25. Calculate the void ratio (e) and hence find porosity of the soil sample, n = [e/(1 + e)] at the start and end of the test.

TABLE 10.2
Absolute values of viscosity for distilled water (μ)

Temperature°C	Density[#] (g/cm³)	Viscosity (Poise*)	Temperature°C	Density[#] (g/cm³)	Viscosity (Poise*)
4	1.00000	0.01567	23	0.99757	0.00936
16	0.99897	0.01111	24	0.99733	0.00914
17	0.99880	0.01083	25	0.99708	0.00894
18	0.99862	0.01056	26	0.99682	0.00874
19	0.99844	0.01030	27	0.99655	0.00855
20	0.99823	0.01005	28	0.99627	0.00836
21	0.99802	0.00981	29	0.99598	0.00818
22	0.99780	0.00958	30	0.99568	0.00801

Notes

\# For conversion of density into unit weight (CGS system to SI system of units): 1 g/cm³ = 10 kN/m³

* 1 poise = 10^{-3} gm-sec/cm² = dyne second/cm² = 10^{-4} kN-s/m² = 1,000 millipoises

TABLE 10.3
Data sheet for constant head permeameter

L=_____ (cm), D=_____ (cm), A=_____ (cm^2)

Sl. No.	Observations and Calculations	Trial No.		
		1	2	3
1	Hydraulic head, h (cm)			
2	Time interval, t (s)			
3	Total Quantity of flow Q (cm^3)			
	(a) First trial t_1 (s)			
	(b) Second trial t=(s)			
	(c) Third trial t_3 (s)			
	Average Q (cm^3)			
4	Lab. Or room temperature, T ($^\circ$C)			
5	Viscosity of water at Lab. or room temp., μ (poise)			
6	Viscosity of water at 20°C., μ (poise), 1 poise = 10^{-3} g-s/cm^2	0.01005		
Calculations				
7	Coeff. of permeability at any temp. T$^\circ$C, $kT = (QL/Aht)$			
8	Coeff. of permeability at standard temp. 20°C, $k_{20} = k_T * \frac{\mu_T}{\mu_{20}}$			
9	Average permeability, k (cm/s) $= k = (QL/Aht)$			
10	Coefficient of permeability, k (m/s)			
11	Degree of permeability			

26. Find seepage velocity, $v_s = v/n$ (where: $v = k = Q/At$).

27. The observation and data sheet are shown in Table 10.3.

10.6.1.1   Observation Data Sheet and Analysis

- **Determination of "k"**

Based on the test results and known test data, thye value of "k_T" at room temperature (T$^\circ$C) is computed as:

$$k_T = \left[\frac{qL}{Ah} = \frac{QL}{Aht} \right] \tag{10.7}$$

Where: q = Discharge per unit time (cm^3/s)

Q = Total volume of discharge (cm^3, assume 1ml = 1 cm^3, and 1 cm^3 = 10^{-6} m^3)

t = Time period (second)

h = Head difference (cm)

L = Length or effective height of specimen

$A = \pi * (D^2/4)$ = Cross-sectional area of soil sample (cm^2, take 1 cm^2 = 0.0001 m^2), and

D = Diameter of the soil sample (cm).

Once the "k_T" at the test temperature is known, it can be standardized with the standard temperature (T = 20°C) as below:

$$k_{T=20°C} = \left[\frac{\mu_T}{\mu_{20}} * k_T \right]$$ (10.8)

Where: μ_T and μ_{20} are the viscosities (in poise and 1 poise = 1,000 millipoises = 10^{-4} kN-s/m^2 = 10^{-3} gm-sec/cm^2 = 1 dyne-s/sq.cm) at any temperature T°C of the test and at standard temperature of 20°C, respectively.

The test observations and data analysis are given in Table 10.3.

Determination of unit weight, water content, and the void ratio of the soil specimens.

Given data: G = 2.65, L = _____ (cm), D = _____ (cm), A = _____ (cm^2), V = A * L = _____ (cm^3)

The observations and data analysis are given in Table 10.4.

10.6.1.2 Results and Discussions

Coefficient of permeability of the given soil, k = _____ m/s

    Void ratio of the soil sample at the start of test, e_o = _____

    Void ratio of the soil sample at the end of test, e_f = _____

Type of soil: ....

10.6.2 Coefficient of Permeability of Soils by Falling Head Test Method

Falling head test method is generally suitable for fine grained soils (e.g. clays) which require high hydraulic gradient for water flow due to their low permeability (k < 1.0 * 10^{-6} m/s). Therefore, in case of cohesive swoils, "k" is computed as (Figure 10.7):

$$k = \left[\frac{2.303aL}{At} \log_{10} \left(\frac{h_1}{h_2} \right) \right]$$ (10.9)

Where: h_1 = Initial head (cm)

    h_2 = Final head (cm)

    t = Time interval (s)

    a = Cross-sectional area of the liquid stand pipe (cm^2)

    A = Cross-sectional area of the specimen, (cm^2), and

    L = length of soil specimen (cm).

The principle of the falling-head test is illustrated in Figure 10.7. In this method, the constant head is not maintained as in the case of constant head test method.

TABLE 10.4

Determination of unit weight, water content, and the void ratio

Project/Site Name: **Date:**

Client Name: **Job. No.   Sample No.**

Sample Recovery Depth: **Sample Recovery Method:**

Sample Description:

Tested By:

Test Type: on gravimetric basis/on volumetric basis

Sl. No.	Observations and Calculations	Trial No.		
		1	**2**	**3**
A	Calculation of total unit weight, γ_b(kN/m³)			
1	Wt. of empty mold with dummy plate, W_1 (g)			
2	Wt. of mold + soil and dummy plate, W_2 (g)			
3	Wt. of moist soil, W_s (g) [1 g = 10^{-5}kN] = $W_2 - W_1$			
4	Volume of mold, V (m³) [1 cc = 10^{-6} m³]			
5	Total unit weight, γ_b (kN/m3) = W_s/V			

Sl. No.	Observations and Calculations	@ start of test			@ end of test		
B	Calculation of water content, w (%)						
6	Container No.	**C1**	**C2**	**C3**	**C1**	**C2**	**C3**
7	Weight of empty container, W_1 (g)						
8	Weight of container + soil, W_2 (g)						
9	Weight of water, $W_w = W_2 - W_3$ (g)						
10	Weight of solids, $W_s = W_3 - W_1$ (g)						
11	Water content, w = $(W_w)/(W_s) * 100$ (%)						
C	Calculation of dry unit weight, γ_d(kN/m³)						
12	Dry unit weight, γ_d (kN/m³), $\gamma_d = \gamma_b/1 + w/100$)						
13	Void ratio, $e = \frac{G\gamma_w}{\gamma_d} - 1$						
14	Porosity, $n = \frac{e}{1+e}$						
15	Degree of saturation, $S = (wG/e)_*100$						
16	Discharge velocity, $v_D = v = k^* = Q/At$ (cm/s)						
17	Seepage velocity, $v_s = v_D/n = k/n$ (cm/s)						

Note

* It is presumed that h ≅ L, i = 1 ⇒ v = k, ∴ k = QL/Aht = Q/At for i = 1.

The following procedure is followed during this method:

1. Follow the test procedure for preparation of saturated soil sample given in the constant head test method from step 1 to step 17.
2. Record the initial head difference (h_1) between the top of the stand pipe and the top of the constant head water chamber, as shown in Figure 10.7 and allow the water to flow downward through the soil sample into the constant head water chamber.

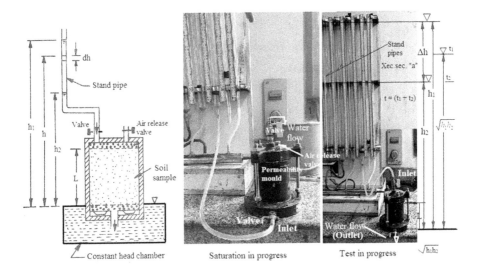

FIGURE 10.7   Falling head permeameter.

3. Stop the test after desired time interval (t) and record the final head difference (h_2).
   Note: Ensure that the values of h_1 and h_2 are recorded from the center of the outlet such that ($h_1 - h_2$) is about 30cm.
4. Mark the level of $\sqrt{h_1 h_2}$ from the center of the outlet on the stand pipe.
5. Record the room temperature in the soil testing laboratory during the test and note the viscosity of water against room temperature. The values of absolute viscosity are given in Table 10.2.
6. Now open the valve and start the stopwatch. Allow water to flow for a desired time interval for head to fall from h_1 to h_2.
   Note: It may be noted that if the time interval for head to fall from h_1 to $\sqrt{h_1 h_2}$ and from $\sqrt{h_1 h_2}$ to h_2 is equal, then only the laminar flow is established.
7. Repeat the procedure for different values of h_1 and h_2 for an average value of "k."
8. Dismantle the permeability cell assembly and remove the soil sample.
9. Also, determine the water content of the soil sample post-permeability test.
10. Calculate the void ratio (e) and hence find porosity of the soil sample, $n = [e/(1 + e)]$ at the start and end of the test.
11. Find the seepage velocity, $v_s = v/n$ (where: v = k).
12. The observation and data sheet shown, as given in Table 10.5.

10.6.2.1   Observation Data Sheet and Analysis

- **Determination of "k"**

In the falling head test method, the coefficient of permeability is given by:

TABLE 10.5

Data sheet for falling head permeameter test

Sl. No.	Observations and Calculations	Determination No.		
		1	2	3

Observation

1 Initial head, h_1 (cm)

2 Final head, h_2 (cm)

3 Head $\sqrt{h_1 h_2}$ (cm)

4 Time interval, t (s)

5 a. From h1 to $\sqrt{h_1 h_2}$ (cm)

   b. From $\sqrt{h_1 h_2}$ to h_2 (cm)

   c. from h_1 to h_2 = (a)+(b)

6 $Log_{10}(h_1/h_2)$

7 Viscosity of water at Lab. or room temp., μ (poise)

8 Viscosity of water at 20°C., μ (poise), 1 poise $=10^{-3}$g – s/cm^2

Calculations

7 Coefficient of permeability at temp. T°C $k_T = \frac{2.303aL}{At} \log_{10}\left(\frac{h_1}{h_2}\right)$

   (cm/s)

8 Coefficient of permeability at temp. 20°C $k_{20} = k_T * \frac{\mu_T}{\mu_{20}}$, μ = coeff.

   of viscosity of water (poise, 1 poise = 10^{-3} g-s/cm^2)

9 Average permeability, k (cm/s)

10 Degree of permeability

$$k = \left[\frac{2.303aL}{At} \log_{10}\left(\frac{h_1}{h_2}\right) \right]$$

Where: h_1 = Initial head (cm) = _____

h_2 = Final head (cm) = _____

t = Time interval (s) = _____

a = Cross-sectional area of the liquid stand pipe (cm^2) = _____

A = Cross-sectional area of the specimen, (cm^2) = _____

L = Length of soil specimen (cm) = _____

Other given data is:

G = Specific gravity of the soil sample, = 2.69

T = Temperature of water (°C) = _____

μ = Viscosity of water at temp. T (°C) = _____

Observation and data analysis is given in Table 10.5.

Determination of unit weight, water content, and the void ratio of the soil specimen

Given data: G = 2.65, L = _____ (cm), D = _____ (cm), A = _____ (cm^2),

$V = A * L = \underline{\hspace{2cm}}$ (cm^3)

The observations and data analysis is given in Table 10.6.

10.6.2.2 Results and Discussions

Coefficient of permeability of the given soil, k = \underline{\hspace{2cm}} m/s

Void ratio of the soil sample at the start of test, e = \underline{\hspace{2cm}}

Void ratio of the soil sample at the end of test, e = \underline{\hspace{2cm}}

Type of soil: ....

TABLE 10.6

Determination of unit weight, water content, and the void ratio

Project/Site Name: Date:

Client Name: Job. No.   Sample No.

Sample Recovery Depth: Sample Recovery Method:

Sample Description:

Tested By:

Test Type: on gravimetric basis/on volumetric basis

Sl. No.	Observations and Calculations	Trial No.		
		1	2	3
A	**Calculation of total unit weight, γ_b(kN/m^3)**			
1	Wt. of empty mold with dummy plate, W_1 (g)			
2	Wt. of mold+soil and dummy plate, W_2 (g)			
3	Wt. of moist soil, W_s (g) [1 g = 10^{-5} kN] = $W_2 - W_1$			
4	Volume of mold, V (m^3)\1 cm^3 = 10^{-6} m^3]			
5	Total unit weight, γ_b (kN/m^3) = W_s/V			

Sl. No.	Observations and Calculations	@ start of test			@ end of test		
B	**Calculation of water content, w (%)**						
6	Container No.	C1	C2	C3	C1	C2	C3
7	Weight of empty container, W_1 (g)						
8	Weight of container + soil, W_2 (g)						
9	Weight of water, $W_w = W_2 - W_3$ (g)						
10	Weight of solids, $W_s = W_3 - W_1$ (g)						
11	Water content, w = $(W_w)/(W_s) * 100$ (%)						
C	**Calculation of dry unit weight, γ_d (kN/m^3)**						
12	Dry unit weight, γ_d(kN/m^3), $\gamma_d = \gamma_b/(1 + w/100)$						
13	Void ratio, $e = \frac{G\gamma_w}{\gamma_d} - 1$						
14	Porosity, $n = \frac{e}{1+e}$						
15	Degree of saturation, $S = (wG/e) * 100$						
16	Discharge velocity, $v_D = v = k$ (cm/s)						
17	Seepage velocity, $v_s = v_D/n = k/n$ (cm/s)						

10.7   GENERAL COMMENTS

It may be noted that the coeff. of permeability is dependent on the void ratio of the soil material. Therefore, the void ratio should also be determined for each test value of "k." Variation of coeff. of permeability with void ratio for granular soils is shown in Figure 10.8.

10.7.1   ADDITIONAL COMMENTS

Based on research inputs, it has been observed that laboratory methods' constant head and falling head methods do not provide a reliable value "k" because of varied reasons and major issues as enlisted below:

1. The remolded soil sample prepared in the laboratory does not represent the in situ state conditions of the field soils.
2. Boundary conditions may influence the water flow paths. There is a vast difference between the field and laboratory boundary conditions. The actual boundary conditions in the field cannot be truly reproduced in the remolded soil sample prepared in the laboratory. Thus, there will be variation in the values of "k."
3. The laboratory tests may not establish steady state flow (e.g. laminar flow) because of high hydraulic head compared to field conditions. Thus, Darcy's Law may not be valid if turbulent flow is established due to high hydraulic head.

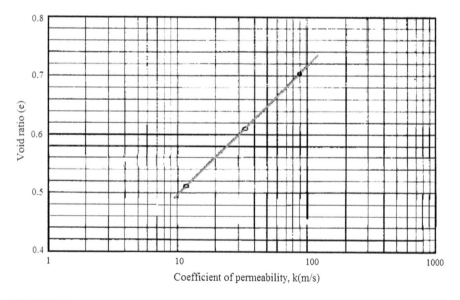

FIGURE 10.8   Variation of coeff. of permeability with void ratio for granular soils.

10.8 APPLICATIONS

In soil engineering practice, the coefficient of permeability is used to ascertain the suitability of soils for its desired function either as barrier material (e.g. used for clay liners or in core dams to avoid seepage loss, etc.) or as draining material (e.g. used as fill material behind retaining walls to avoid development of pore water pressure).

The coefficient of permeability defines the quality of a soil material which allows the flow of water through it. It classifies the soil either as pervious material (e.g. having high permeability such as coarse grained soils) or as an impervious soil material (e. g. having very low permeability such clays). The coefficient of permeability helps in the development of various empirical correlations with other soil parameters such as:

- **Relationship between effective particle size and coefficient of permeability**

The most commonly used empirical relationship to determine the coefficient of permeability from the index property is (Equation 4.6-Chapter 4 is):

$$k = CD_{10}^2 \quad [holds \ good \ for \ k \le 10^{-5} \ m/s]$$

Where: k = Coefficient of permeability (cm/s)

C = 100 = Constant (cm^{-1}/s) valid for clean sands (with less than 5% fines), and

D_{10} = Effective particle size corresponding to 10% by weight passing (mm).

The above equation is given by Alan Hazen. The coefficient of permeability depends on shape and soil of soil particles and varies with square of particle size diameter.

Thus, reverse is the case, e.g. if the coefficient of permeability is known, the mean effective particle size can ne calculated using above empirical equation tentatively.

Correlation has also been found to exist between the coefficient of permeability, porosity, and effective size, as shown in Figure 10.9.

The effective diameter of grains (D_{10}) is related to permeability and porosity for spherical-shaped soil grains as follows (Kotyakho 1 949):

$$D_{10} = \sqrt{\frac{72(1 - n)^2}{n^3}} * k \quad OR \quad k = \left[\frac{D_{10}^2 * n^2}{72(1 - n)^2} \right] \qquad (10.10)$$

Where: n = Coefficient of porosity

Similarly, permeability may increase if the voids are more as illustrated by the expression:

$$k = \left[\frac{e^3}{1 + e} \right] \qquad (10.11)$$

Where: e = Void ratio.

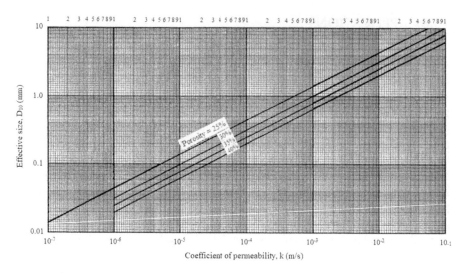

FIGURE 10.9 Correlation between effective size, porosity, and coefficient of permeability (Turneaure and Russell 1947).

- **Coefficient of vertical consolidation (C_v)**

Since the consolidation test time consuming (at least 15 days test), therefore, the coefficient of vertical consolidation (C_v) is indirectly obtained from permeability test given by:

$$C_v = \left[\frac{k}{\gamma_w * m_v} \right]$$

(10.12)

Where: m_v = Coefficient of volume change or coefficient of volume compressibility
k = Coefficient of permeability (in vertical direction), and
γ_w = Unit weight of water.
Similarly, the coefficient of horizontal or radial consolidation (C_h or C_r) is given by:

$$C_r = C_h = \left[\frac{k_r}{\gamma_w * m_v} = \frac{k_h}{\gamma_w * m_v} \right]$$

(10.13)

Where: C_r = Coefficient of radial consolidation
C_h = Coefficient of horizontal consolidation
k_r = Coefficient of radial permeability, and
k_h = Coefficient of horizontal permeability.

10.9 SOURCES OF ERROR

The various sources of error in a permeability test may affect the value of "k" as given below:

1. Remolded soil sample in the laboratory may not simulate the field conditions
2. If the constant head water supply is not maintained throughout the test
3. If the room temperature is not correctly recorded during the test in the laboratory
4. If the air bubbles are not completely removed during the saturation stage

10.10 PRECAUTIONS

1. The permeability mold may be cleaned and greased before the test.
2. Saturated porous stones and filter paper may be used.
3. Remolded soil sample may be prepared by static compaction rather than by tamping.
4. De-aired and distilled water must be used to saturate the soil sample.
5. High hydraulic head may be avoided, which may establish turbulent flow.
6. Constant head water supply may be maintained in the constant head test method.

REFERENCES

ASTM D2434. 2012. "Standard Test Method for Permeability of Granular Soils (Constant Head) (Note: The Falling Head Test Method Is Not standardized)." ASTM International, West Conshohocken, PA, www.astm.org.

BS 1377-5. 1990. "Methods of Test for Soils for Civil Engineering Purposes-Part 5: Compressibility, Permeability and Durability Tests." British Standards, UK.

Bowles, J. E. (1973). "Permeability Coefficient Using a New Plastic Device." *Highway Research Record No. 431*, pp. 55–61.

Burland, J. B. 1990. "On the Compressibility and Shear Strength of Natural Clays." *Géotechnique* 40(3): 329–378.

Darcy, H. 1856. *The Public Fountains of the City of Dijon* (English Translation by Patricia Bobeck). Paris: Dalmont.

IS: 2720 (Part 1). 1980. "Indian Standard Code for Preparation of Soil Samples." Bureau of Indian Standards, New Delhi.

IS: 2720 (Part 17). 1986. "Method of Test for Soils: Determination of Permeability." Bureau of Indian Standards, New Delhi.

Johnson, A. I. 1963. "Application of Laboratory Permeability Data. Open File Report, Water Resources Division Denver." United States Department of the Interior Geological Survey Colorado, USA.

Kotyakho, F. I. 1949. "Interrelationship Between Major Physical Parameters of Sandstones." *Neft. Khoz.* 12: 29–32.

Ladd, R. S. 1974. "Specimen Preparation and Liquefaction of Sands." *J. of Geotech. and Geoenv. Eng.* ASCE 100(10): 1180–1184.

Mir, B. A. 2010. *Study of the Influence of Smear Zone Around Sand Compaction Pile on Properties of Composite Ground.* Ph.D. Thesis, IIT Bombay.

Mir, B. A., and A. Juneja. 2010. "Some Mechanical Properties of Reconstituted Kaolin Clay." Procc. of the the17th Southeast Asian Geotech. Conference, Taipei, Taiwan, pp. 145–148.

Raghunandan, M., A. Juneja, and B. Hsiung. 2012. "Preparation of Reconstituted Sand Samples in the Laboratory." *International Journal of Geotechnical Engineering* 6(1): 125–131. doi:10.3328/IJGE.2012.06.01.125-131.

Reddy , Krishna R. 2000. "Engineering Properties of Soils Based on Laboratory Testing." Department of Civil and Materials Engineering University of Illinois at Chicago, USA.

Taylor, D. W. 1948. *Fundamentals of Soil Mechanics*. New York: John Wiley.

Terzaghi, K., and R. B. Peck. 1967. *Soil Mechanics in Engineering Practice*. New York:Wiley.

Turneaure, F. E., and H. L. Russell. 1947. *Public Water Supplies*. 4th ed. New York: John Wiley & Sons, p. 704.

U.S. Bureau of Reclamation. 1985. "Permeability and Settlement of Soils. Test Designation #E-13, Earth Manual." Second Edition, 1974, fourth printing.

Vaid, Y. P., and D. Negussey. 1988. "Preparation of Reconstituted Sand Specimens." *Advanced Triaxial Testing of Soil and Rock*. ASTM STP 977, R. Donaghe, R. Chaney, and M. Silver (Eds.). West Conshohocken, PA: ASTM International, 405–417.

11 Consolidation Test of a Soil Sample

References: IS: 2720 (Part 1); 2720 (Part 15); 2720 (Part 40); 2720 (Part 41); ASTM D2435; ASTM D4546; BS 1377: Part 5 (1990); BS 1377: Part 6 (1990)

11.1 OBJECTIVE

The main objective of conducting a consolidation test is to determine the various consolidation/compressibility parameters (e.g. e_o, C_c, C_s, p_c, C_v, a_v, m_v, δc, etc.). These compressibility parameters are used in the settlement analysis of various structures on compressible soils.

11.2 DEFINITIONS AND THEORY

Consolidation is a slow and time-dependent process of compression of saturated fine grained soils (e.g. clays) under static loading due to expulsion of water. However, the degree of saturation remains constant. Generally, a soil mass is a very complex, heterogeneous, and triparticulate system. However, a saturated soil mass is a two-phase soil system in which the void pores in the fine grained soil mass are completely filled with water. When a saturated soil sample is subjected under static long-term loading ($\Delta\sigma$), firstly the load will be borne by water, thereby generating excess pore water pressure (Δu, e.g. $\Delta\sigma \approx \Delta u_w$) in the soil mass in addition to its hydrostatic pore water pressure. Since the fine grained soils, especially clays, have low permeability, the excess pore water pressure dissipates slowly under applied loading and volume decreases. When the pore water pressure drains out completely (e.g. $\Delta u_w \approx 0$), the applied load is resisted by soil solids (e.g. $\Delta\sigma \approx \Delta\sigma'$), known as intergranular stress or effective stress. The change in volume is generally referred to as settlement and the gradual process of compression with time due to expulsion of water under gradually applied static load is called consolidation.

11.2.1 COMPRESSION OF SOILS DUE TO EXPULSION OF WATER: VOLUME CHANGE BEHAVIOR IN SOILS

When a saturated soil sample is subjected under static long-term loading, the following two types of volume changes are observed in soil mass:

1. Volume change due to physico-chemical factors, which occurs due to change in the internal pressure, e.g. swelling, shrinkage (Figure 11.1).

FIGURE 11.1   Volume change in saturated soils due to physico-chemical factors.

    2. Volume change due to rheological factors, volume change due to external loading e.g. elastic compression, compaction, and consolidation.

In civil engineering, consolidation refers to compression of saturated clayey soils under static long-term loading conditions. According to Karl von Terzaghi (1923), "Consolidation is a slow process of compression of a saturated soil under a static load, associated with a change of water content due to squeezing out of pore water, involving development and dissipation of pore pressure and transfer of external load to soil solids" (Figure 11.2), e.g.:

- First, to pore water in the form of pore water pressure (e.g. $\Delta\sigma = \Delta u_w$)
- Secondly, to soil grains in the form of effective stress ($\Delta\sigma = \Delta\sigma'$ as $\Delta u_w = 0$)

It may be noted that the volume change in the soil mass takes place under applied static or dynamic loading due to compression of void pores and not due to deformation of soil solid grains. Therefore, the soil mass may compress, consolidate, collapse of compact under applied loading due to reduction in volume of voids. Thus, there should be clear distinction between the above-mentioned four volume change mechanisms, which ultimately lead to reduction in the volume of voids in a soil mass.

- **Compression:** defined as the mechanism of rapid decrease in volume of voids in a partially saturated soil mass by expulsion of air from the void pores under applied loading.
- **Consolidation:** defined as the process of volume change in a saturated soil mass due to expulsion of water from soil voids under applied loading. This is also known as time-dependent reduction in volume of a saturated soil mass under static loading.
- **Collapse:** defined as the process of rapid decrease in volume due to collapse of structural arrangement of soil particles with increased water content under applied loading. Collapsible soils exhibit high apparent strength when dry, but exhibit a sudden decrease in volume when imbibed with water and subjected to low-level stress intensity. This process of volume reduction in pore voids in a soil mass is also known as hydro-consolidation or hydro-compression.
- **Compaction:** defined as the process of volume change in a soil mass due to expulsion of air under kneading or impact type of loading.

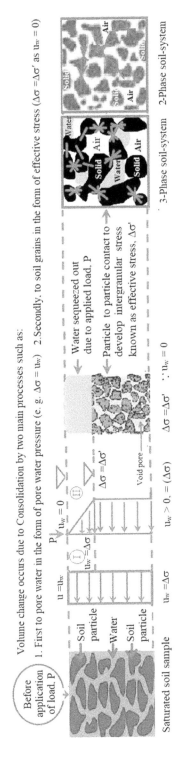

FIGURE 11.2 Volume change due to consolidation of a saturated soil.

Note: It may be noted that the water content in the soil mass is at par with plastic limit of the soil mass during compaction process (e.g. less than the saturation water content), and hence the decrease in volume is due to expulsion of air from soil voids.

Further, it may also be noted that in some cases, there will be increase in saturated or partially saturated soil volume under static or dynamic loading, which is referred to as dilation process (e.g. soil mass dilates in volume). Furthermore, black cotton soils or expansive soils expand (e.g. swell) when imbibed with water wherein volume increases due to inter-particle repulsions between clay particles under externally applied loading (e.g. swelling pressure of expansive is higher than applied loading). However, when these soils dry up, these shrink and there is a decrease in volume. Thus, consolidation tests play vital role in understanding the swell-shrink behavior of such soils.

11.2.2 CONSOLIDATION: TIME-DEPENDENT LOAD-DEFORMATION PROCESS

Soil mass is a heterogeneous and very complex material comprised of soil solids and fluid (water and gas (air)) filled in the void pores in between soil solids, which is generally referred to as a three-phase soil system. When the void pores are completely filled with water, the soil mass is referred to as a saturated two-phase soil system (e.g. soil solids + water filled in void pores). Since the clayey soils have low permeability, the pore water pressure (Δu_w) dissipates very slowly under gradually applied static loads and the deformation due to expulsion of water continues slowly with time. However, if the load is rapidly increased, there is a buildup of excess pore water pressure, Δu_e ($\Delta u_e > \Delta u_w$), which is initially carried by the pore water filled in the void pores. When more loads are applied, the pore water pressure starts dissipating and the load is gradually transferred from pore water to soil solids due to particle-to-particle contacts. When there is complete dissipation of pore water pressure from void pores, the soil solids realign themselves into a denser configuration under externally applied loads, which is now resisted inter-granular stress or effective stress. Thus, the time-dependent deformation process is known as consolidation, which results in the reduction of void ratio and increases shear strength. This time-dependent deformation process for a saturated clayey sample is explained by means of a spring piston analogy, as shown in Figure 11.3.

The spring-piston analogy for a saturated clay sample under externally applied static loading involves the following stages (Figure 11.3):

1. Application of external load, P ($\Delta\sigma = P/A$)
2. Development of excess pore pressure: ($\Delta\sigma = \Delta u_i = \Delta u_e$)
3. Flow of water out of soil skelton: (expulsion of water)
4. Dissipation of pore pressure from soil sample ($\Delta u_i \rightarrow 0$)
5. Transfer of pore pressure to soil grains as effective stress ($\Delta u_i \rightarrow \Delta\sigma'$)
6. Compression of soil skelton framework: (Deformation—ΔV)

Note: To ensure free drainage at top and bottom, porous material layers (e.g. sand) are provided.

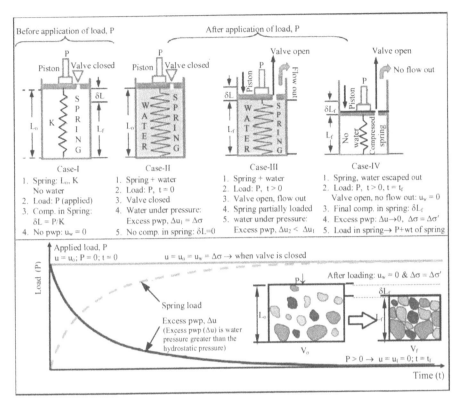

FIGURE 11.3   Schematic diagram of the spring-piston analogy for a saturated clay sample.

However, the coarse grained soils have high permeability and allow water to drain out quickly compared to fine grained soils. Thus, the applied load is quickly transferred onto the soil solids, resulting in rapid reduction in volume. The dissipation of pore water pressure under applied static load in a given time (t_1) for coarse grained and fine grained soils is shown in Figure 11.4.

11.2.3   CONSOLIDATION PARAMETERS

The consolidation or compressibility parameters are determined by the consolidation tests, which play a vital role in the settlement analysis of various structures on compressible soils (Reddy 2000). These parameters are determined in the laboratory using Terzaghi's 1-D consolidation theory. The test results are generally plotted between volume change (e.g. void ratio, e) versus effective stress (σ') as shown in Figure 11.5ab. This plot between (e-logσ' plot) or (e-σ' plot) is generally known as a consolidation curve. Various parameters obtained from the consolidation curve are:

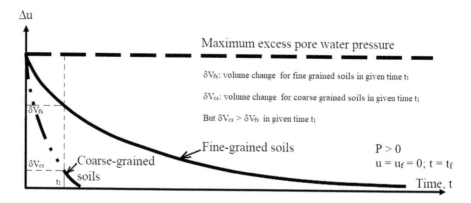

FIGURE 11.4 Schematic diagram of the dissipation of pore water pressure for coarse grained and fine grained soils.

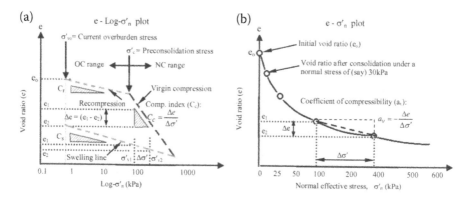

FIGURE 11.5 Consolidation test parameters.

11.2.3.1 Compression Index: C_c

The compression index is the slope of a virgin or normally consolidated line (NCL) consolidation curve (e-logσ_v) computed from Figure 11.5a as:

$$C_c = \frac{-\Delta e}{\Delta \log_{10}\sigma'_v} = \frac{\Delta e}{\Delta \log_{10}\sigma'_v} = \frac{(e_1 - e_2)}{(\log_{10}\sigma'_{v2} - \log_{10}\sigma'_{v1})} \quad (11.1)$$

C_c is used for computation of settlement for NC soils.

11.2.3.2 Re-Compression/Swelling Index: C_r or C_s

The re-compression/swelling index (C_r or C_s) is the slope of the re-compression line or unloading/re-loading line (URL) of e-logσ_v plot, which is computed as (Figure 11.5a):

$$C_r = C_s = \frac{-\Delta e}{\Delta \log_{10} \sigma'_v} = \frac{\Delta e}{\Delta \log_{10} \sigma'_v} = \frac{(e_1 - e_2)}{(\log_{10} \sigma'_{v2} - \log_{10} \sigma'_{v1})} \quad (11.2)$$

C_r or C_s is used for computation of settlement of OC soils.

11.2.3.3   Initial Void Ratio (e_o)

From the consolidation curve, the initial void ratio can be computed by rearranging Equation (11.1). The initial void ratio is generally determined either by "height of solids method" or by the "change in void ratio method."

The initial void ratio (e_o) is calculated from Figure 11.5a for the soil sample tested for a desired or given loading, which represents the natural void ratio of the soil sample in the field under that loading.

11.2.3.4   Coefficient of Compressibility: a_v

The coefficient of compressibility is the slope of virgin or NCL of the e-σ'_v plot (Figure 11.5b), which is computed as:

$$a_v = \frac{-\Delta e}{\Delta \sigma'_v} = \frac{(e_1 - e_2)}{(\sigma'_{v2} - \sigma'_{v1})} \quad (m^2/kN) \quad (11.3)$$

a_v is used for computation of modulus of volume change, settlement, and compression index.

11.2.3.5   Modulus of Volume Change: m_v

The compressibility is also defined in terms of modulus of volume change, m_v, as:

$$m_v = \frac{\Delta \varepsilon_V}{\Delta \sigma'_v} = \frac{\Delta V/V_o}{\Delta \sigma'_v} = \frac{\Delta e}{(1 + e_o)\Delta \sigma'_v} = \frac{a_v}{(1 + e_o)} = \frac{1}{E} \quad (m^2/kN) \quad (11.4)$$

The modulus of volume change measures the resistance to volume change of a material when compressed from all sides.

11.2.3.6   Preconsolidation Pressure: p'_c

The pre-consolidation pressure (p'_c) is defined as the maximum past effective stress experienced by a soil mass at a desired location ($p'_c > \sigma'_o$, where σ'_o is the present effective overburden or vertical stress at a desired location where p'_c is determined). In 1936, Casagrande proposed a classical method, which is most widely adopted method in conventional soil mechanics for determination of pre-consolidation pressure from the consolidation curve (e.g. from $e - \log \sigma'$ plot), as shown in Figure 11.6 (see Section 11.6.1 for determination of preconsolidation pressure).

To determine the Preconsolidation stress, p'_c, the following steps are taken (refer to Figure 11.6):

1. Identify the maximum curvature point-A on e-log s plot and draw a horizontal line A-B as shown in Figure 11.6.

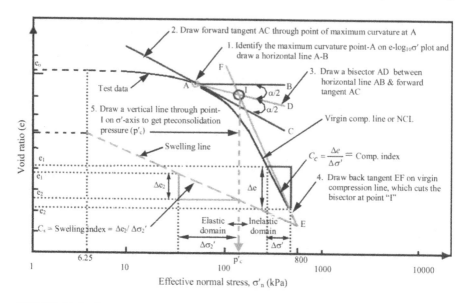

FIGURE 11.6 Preconsolidation stress from Casagrande's method (1936).

2. Draw a tangent AC (generally known as forward tangent) through point of maximum curvature at A.
3. Draw bisector AD between horizontal line AB and forward tangent AC.
4. Draw back tangent EF on virgin compression line, which cuts the bisector at point "I", which is known as point of intersection or preconsolidation pressure point.
5. Draw a vertical line through point-I on σ'-axis to get preconsolidation pressure (p'_c).

The pre-consolidation pressure parameter is used in the settlement analysis and to classify the soils either as normally consolidated (NC) or over-consolidated (OC) in its in situ state with the aid of OCR and present effective over-burden stress/pressure (σ'_{vo}), e.g.: If OCR = 1 (e.g. $p'_c = \sigma'_{vo}$), the soil is normally consolidated (NC). The OCR is defined as:

$$OCR = \frac{p'_c}{\sigma'_{vo}} \tag{11.5}$$

- If OCR > 1–3 (e.g. $p'_c > \sigma'_{vo}$), soil is lightly over-consolidated (OC)
- If OCR > 3–8 or higher, the soil is heavily over-consolidated and soil is in compression zone
- If OCR < 1 (e.g. $p'_c < \sigma'_{vo}$), soil is under-consolidated (UC–very rare case)
- If $(\sigma'_{vo} + \Delta\sigma_z) = \sigma_T < p'_c$, the soil is still in the compression zone (OC), where $\Delta\sigma_z$ is the applied loading intensity.

Thus, it is seen that OCR has a very significant effect in the behavior of clayey soils though its effect is marginal in sandy soils.

11.2.3.7 Coefficient of Consolidation (1-D vertical): C_v or (radial: C_h or C_r)

Coefficient of Consolidation is defined as the rate of compression of a saturated soil sample under increased static load in the vertical direction (e.g. 1-D consolidation), generally expressed as C_v (m²/sec). If loads are applied in the horizontal or radial direction, it is represented by C_h or C_r, respectively. There are various methods available for determination of C_v, as shown below (as illustrated in Figure 11.7):

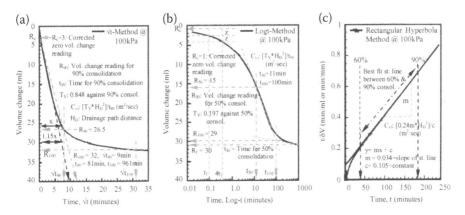

FIGURE 11.7 Determination of coeff. of consolidation by various graphical methods.

1. Taylor's square root of time fitting method (1948)
2. Casagrande's log time fitting method (1940)
3. Rectangular hyperbola method (1987)

11.2.3.8 Coefficient of Permeability: k

Coefficient of permeability can also be determined indirectly from compressibility parameters by using the following empirical correlation:

$$k = \frac{C_v * a_v * \gamma_w}{1 + e_o} = C_v * m_v * \gamma_w \tag{11.6}$$

Where: γ_w = Unit weight of water
C_v = Coeff. of consolidation
e_o = Initial void ratio, and
$m_v = \frac{a_v}{1+e_o}$ = Modulus of volume change or constrained modulus = 1/E.

11.2.3.9 Settlement

Settlement is defined as the vertical deformation or decrease in volume of a saturated soil mass under external loading, which is generally referred to as consolidation settlement (δ_c). Clays and organic soils are highly prone to consolidation settlement. Consolidation settlement constitutes primary consolidation (due to removal of water) and secondary settlement (due to rearrangement of particles). Primary consolidation is the major component and it can be reasonably estimated by using consolidation parameters.

11.2.3.10 Field Consolidation Curve

Laboratory and field consolidation curves illustrate the degree of sample disturbance on "e vs. log σ'_v" curve. Field consolidation curve can be constructed with the lab. consolidation curve, as shown in Figure 11.8. For normally consolidated soil, the field consolidation curve is constructed for known values of initial void ratio and preconsolidation pressure by following the procedure:

1. A horizontal line is drawn from a point "a" (Figure 11.8a) on the ordinate axis corresponding to initial void ratio (e_o = wG assuming that S = 1) of the soil specimen up to point "b" corresponding to preconsolidation pressure ($p'_c = \sigma'_o$).
2. Locate point on the ordinate axis corresponding to the void ratio value of $0.42e_o$.
3. Draw a horizontal line from this ordinate point to cut the lab. consolidation curve at point "c." If necessary, extend lab. consolidation curve to point "c."
4. Join point b and c. From Schmertmann's (1955) empirical observation, the field consolidation curve for NC soils will be defined by the line "bc."

Similarly, for normally consolidated soil, the field consolidation curve is constructed for known values of initial void ratio, present overburden stress, and preconsolidation pressure by following the procedure:

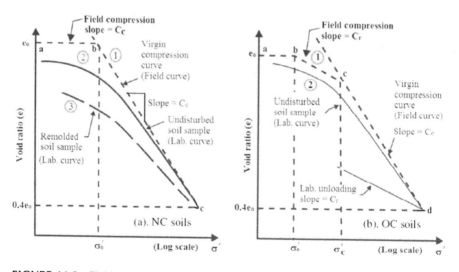

FIGURE 11.8 Field and laboratory consolidation curves.

1. A horizontal line is drawn from a point "a" (Figure 11.8b) on the ordinate axis corresponding to the overburden stress (σ'_o) at a depth from where the soil specimen has been collected.
2. Through point "a," draw a line parallel to the average slope of the unloading-recompression curve up to point "b," corresponding to preconsolidation pressure, p'_c ($p'_c > \sigma'_o$ for OC soils).
3. Locate a point on the ordinate axis corresponding to the void ratio value of $0.42e_o$.
4. Draw a horizontal line from this ordinate point to cut the lab. Consolidation curve at point "d." If necessary, extend lab. consolidation curve to point "c."
5. Join points c and d. From Schmertmann's (1955) empirical observation, the field consolidation curve for OC soils will be defined by the curve "bcd."

The slope of the line bc gives the compression index (C_c) for NC soils (Figure 11.8a). The slope of the line bc gives the recompression or swelling index, C_r, while the slope of the line cd gives the compression index (C_c) in the case of OC soils (Figure 11.8b).

It has been reported that $C_{c\ field} > C_{c\ lab}$ by about 15%. In a given problem, appropriate regions of the field consolidation curve can be used, depending on the range of the pertinent stress increment.

11.3 METHOD OF TESTING

1-D consolidation test is conducted on a saturated soil sample in the laboratory for determination of various compressibility parameters. In this test, a soil sample is laterally confined and vertically loaded with increment ratio of 1. The free drainage is allowed from top and bottom surfaces. After completing the test, the soil sample is unloaded by one-forth of the loading applied on the sample.

11.3.1 Pre-Requisite for One-Dimensional Consolidation Test

Since consolidation is an engineering property and some of the physical and index tests such as soil grading, specific gravity and consistency limit tests are pre-requisite for conducting this test. Consolidation test is conducted either on un-disturbed (in situ) fine grained soil specimen or remolded or reconstituted soil specimen as per requirement.

11.3.2 Soil Testing Material

Fine grained soil passing through a 0.425 mm IS sieve is taken for remolded soil sample preparation in the laboratory as per desired OMC and MDD. The soil sample is saturated before consolidation.

11.3.3 SIZE AND PREPARATION OF SOIL SPECIMEN

For a consolidation test on clayey soils, a soil specimen passing through a 425 µm IS sieve may be used. For silt-dominated soils, a soil specimen passing through a 2 mm IS sieve may be used. For a specimen of 75 mm diameter and 20 mm high in the consolidation test, the maximum size of soil particle should not exceed 2 mm. For other test specimens, as a general rule, the particle size should not exceed one-tenth of the specimen height.

The quantity of soil specimen depends upon the method of soil specimen preparation. If remolded soil specimen based on OMC and MDD (compaction test) is to be prepared, then weight of soil mass may be calculated as per consolidation ring size chosen. If reconstituted soil specimen is to be used, then about 10 kg of soil may be taken for preparation of slurry and consolidated on the laboratory floor up to the desired water content or stress/consistency of soil mass. However, it may be noted that the soil slurry is prepared at a water content of about 1.5 * LL to 2 * LL (LL-liquid limit of soil mass), thoroughly mixed and de-aired before consolidation to desired consistency (in situ state of soil). For soil specimen preparation, refer to Section 10.4 (Test No. 10).

As a general guide, the following quantities are recommended for general laboratory use (Table 11.1).

TABLE 11.1

Recommended Quantity of Soil Sample for General Laboratory Use

Sr. No.	Size of particles more than 90% passing	Minimum quantity of soil sample to be taken for test (g)	Remarks
1	0.425 mm IS Sieve	25	1. For sieve sizes, refer IS:
2	2 mm IS Sieve	50	460 (Part-I)-1978.
3	4.75 mm IS Sieve	200	2. The drier the soil, the
4	9.50 mm IS Sieve	300	greater shall be the
5	19 mm IS Sieve	500	quantity of soil taken.
6	37.5 mm IS Sieve	1000	

11.4 TESTING EQUIPMENT AND ACCESSORIES

For determination of compressibility characteristics of soil samples, the following equipment and accessories are required.

1. Consolidation cell including porous stones, filter papers, dial gauge, and load plate
2. Consolidation ring, 75 mm diameter, 20 mm high
3. Weighing balance accuracy 0.01 g
4. Adjustable counterbalance weights on beam
5. Loading frame yoke assembly
6. Cutting tools, straightedge, spatula, etc.
7. Calibrated weights (with accuracy of 1%)
8. Glass Perspex sheet for mixing soil

9. Constant head water reservoir
10. De-aired water
11. Squeeze bottle
12. Desiccator
13. Stopwatch
14. Moisture content cans
15. Oven
16. 2.0 mm sieve
17. 0.425 mm sieve
18. Permanent marker
19. Grease
20. Thermometer
21. Scale

Note: As of now, there are various types of loading categories applied such as using deadweight with a mechanical advantage or apply load using hydraulic or pneumatic pressure. Some systems record data automatically using computer-driven logging systems and electronic deformation indicators, while others rely on manual data recording using analog dial gauges. However, the basic principles are the same for conducting the test. Consult with your instructor for details on type of equipment available and how to operate the equipment in your laboratory.

11.5   TESTING PROGRAM

1. Check and prepare apparatus (in this test, a manually operated fixed ring consolidation cell is used).
2. Clean the consolidation ring and weigh it to the nearest 0.01 g (W_1).
3. Measure dimensions (internal diameter and height) of consolidation ring and find its volume, V (cm^3).
4. Prepare a soil specimen of 75 mm diameter and 20 mm high for the test. Soil specimen may be either undisturbed sample prepared by trimming natural soil sample taken from field in a Shelby tube or core cutter or remolded sample prepared in the laboratory based on ring dimensions and compaction characteristics (e.g. using OMC and MDD; refer to Section 9.4, Test No. 9 for details). For a remolded specimen, the dry weight of soil sample is taken as: W_s = 0.9MDD ∗ V (g) and water content is taken as: w = (OMC/100) ∗ W_s (ml). A remolded soil sample is prepared by mixing dry soil with desired water content and compacted in the consolidation ring by static compression (Figure 11.9).
5. After sample preparation in step 3, above, take weight of ring plus moist weight nearest to 0.01 g (W_2).
6. Determine the water content and specific gravity from trimmings or left-out soil sample in step 3.
7. Assemble the consolidation ring with the soil specimen into the consolidation cell. Place the consolidation ring with the soil specimen on the lower porous stone. Be sure that porous stones are saturated and kept in distilled/de-aired water bath before use and saturated filter papers are placed below and on top of the soil specimen. Also, put a saturated filter paper on the top

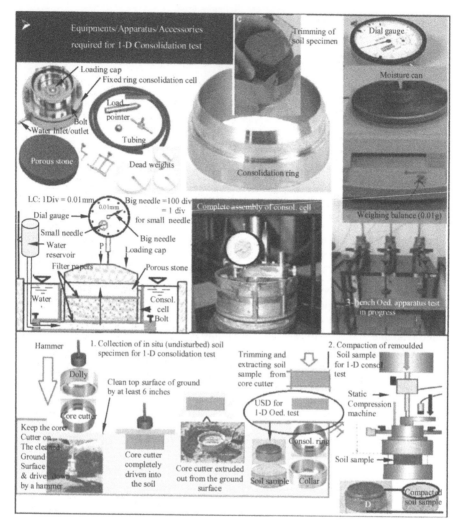

FIGURE 11.9   1-D consolidation testing apparatus and sample preparation.

surface of the soil sample in the consolidation ring overlaid by a saturated porous stone and top-loading cap carefully.

8. Place the consolidation cell on the loading frame and set up the loading yoke, adjust the beam to horizontal, and set the dial gauge reading carefully.

9. Connect the consolidation cell in a constant head de-aired water tank. However, ensure that the bottom level of water in the de-air tank is at about the same level as that of the soil specimen in the consolidation cell and saturate the soil specimen.

10. Apply an initial seating load (5 kPa) to the hanger to check the swelling.

Note: It may be noted that expansive soils expand when applied load is less than their swelling pressure. Therefore, for expansive clays, higher seating pressure (e.g. higher than standard seating load) is applied so that the soil sample does not swell. However, it is decreased (e. g. lower than standard seating load) for soft soils to minimize consolidation under the seating load (e.g. for very soft soils, a seating load of 2.5 kPa or less may be applied). For most fine grained soils, a load of 25 kPa is usually enough to prevent swelling, but if swelling occurs, apply additional load increments until the swelling ceases. Were the specimen permitted to swell, the resulting void ratio-pressure curve would have a more gradual curvature and the preconsolidation pressure would not be well defined.

11. After complete saturation of the soil specimen, note the initial reading of the dial gauge or set it to zero again for further test readings.
12. Apply a normal load increment (6.25 kPa, 12.5 kPa, 25 kPa, 50 kpa, 100 kpa, 200 kPa, 400 kPa, and 800 kPa—with loading increment ratio of 1, as shown in Figure 11.10) to the specimen and record the final settlement readings in dial gauge at time intervals given in Table 11.2 for determination of the void ratio and other consolidation parameters.

Note: Above-mentioned loads are standard loads; however, any desired load can be calculated from the known details of oedometer apparatus as illustrated in Figure 11.10.

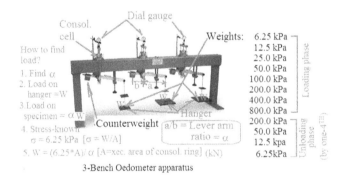

FIGURE 11.10 Incremental loading on 3-Bench oedometer apparatus.

13. Maintain each loading increment for 24 hours or until the settlement becomes constant or until 95% consolidation has reached. Add more load to the specimen next day such that the load increment ratio is one. Here we have $\Delta\sigma/\sigma = 1$ (where $\Delta\sigma$ = increase in pressure and σ = the pressure before applying the loading increment, $\Delta\sigma$).
14. Similarly, apply remaining load increments until the total load on the specimen becomes 800 kPa or as desired by the client.

Note: Ensure that each load increment is maintained for 24 hours as in step 13.

Ensure that the final dial reading of the preceding loading increment becomes the initial reading for the next loading increment and so on. Record initial and

TABLE 11.2

Determination of Void Ratio, Compression Index, and Preconsolidation Pressure by Height of Solids Method (Figure 11.11)

Initial ht. of soil sample, $H = 20$ mm, Initial xec. sec. area of soil sample, $A = 44.1844$ cm², Sp. gravity, $G = 2.63$, Least count of dial gauge, $LC = 0.01$ mm, Dry weight of soil sample, $W_d = 133.45$ g, Unit weight of water, $\gamma_w = 10$ kN/m³, Height of solids, $= H_s = \left[\dfrac{W_d}{G*A*\gamma_w}\right] = 11.484$ mm

Date & time	Stress, σ' (kPa)	Dial gauge readings	Change in sample ht, Δh (mm)	Ht. of sample, $H = H_o - H_s$ (mm)	Void ratio, $e = (H-H_s)/H_s$	$\Delta e = e_1-e_2$	$C_c = \dfrac{\Delta e}{\Delta \log \sigma'}$	$a_v = \Delta e/\Delta \sigma'$ (m²/kN)	$m_v = \dfrac{a_v}{1+e_o}$
………	0	2180	0	20	0.742	—			
	6.25	2174.5	-0.055	19.945	0.739	-0.003			
	12.5	2158	-0.165	19.78	0.725	-0.014			
	25	2122	-0.36	19.42	0.704	-0.020			
	50	2045	-0.77	18.65	0.633	-0.071			
	100	1968	-0.77	17.88	0.557	-0.076	0.23	$3*10^{-4}$ kPa^{-1}	$2.03*10^{-4}$ kPa^{-1}
	200	1899	-0.69	17.19	0.488	-0.069			
	400	1827	-0.72	16.47	0.425	-0.063			
	800	1730	-0.97	15.5	0.350	-0.076			
………	200	1790	0.6	16.1	0.402	0.052		NA	NA
	50	1851	0.61	16.71	0.455	0.053			
	12.5	1899	0.48	17.19	0.497	0.042			
………	6.25	1921	0.22	17.41	0.516	0.019			

final dial gauge readings for each loading increment of 6.25 kPa, 12.5 kPa, 25 kPa, 50 kpa, 100 kpa, 200 kPa, 400 kPa, and 800 kPa, respectively.

15. Also record dial gauge readings at elapsed times of 0 minutes (initial reading), 0.25 minutes, 1.0 minutes, 2.25 minutes, 4 minutes, 6.25 minutes, 9 minutes, 12.25 minutes, 20.25 minutes, 25 minutes, 36 minutes, 60 minutes, 120 minutes, 240 minutes, 480 minutes, and 1,440 minutes (24 hours) for determination of coefficient of consolidation.

16. Unload the soil specimen with one-fourth loading increment after reaching total pressure of 800 kPa (e.g. unload soil specimen from 800 kPa to 200 kPa (800/4) and wait until the dial gauge reading becomes constant. Record the final dial gauge reading and unload for the second unloading increment to 50 kPa (200/4) and further to 12.5 kPa and finally to 6.25 kPa. Be careful that initial and final dial gauge readings are recorded for each unloading increment once the dial gauge readings becomes constant before unloading.

17. Repeat Steps 11 to 14 for different preselected pressures, if any.

18. After completion of the test, remove the soil specimen, and take the entire soil specimen for water content measurements.

19. Plot readings of e versus σ' as desired (either as e-logσ or e-σ plots), and calculate the consolidation parameters from graph and/or calculations.

However, for a reliable time-consolidation curve, it should be ensured that the room temperature should not vary $\pm\ 2\,°C$ during the test.

11.6   OBSERVATION DATA SHEET AND ANALYSIS

11.6.1   COMPRESSIBILITY PARAMETERS (C_C, ϵ, P_c) BY HEIGHT OF SOLIDS METHOD

Test observations, data analysis, and calculations for void ratio, compression index, and preconsolidation pressure are given in Table 11.2.

Before test—Measured data:

Description of soil sample: _____

Specific gravity of soil sample, G: _____, Dia. of the soil sample, d: _____ (cm)

Initial height of the soil sample, H: _____ (cm), Area of C/S, A = π/4 $*$ (d)^2: _____(cm^2)

Dry wt. of the soil sample, W_d: _____ (g), Initial water content, w- ____(%)

Vol. of the consolidation ring, V = $A_o * H$: ___ (cm^3), Ht. of solids, $H_s = W_d/(G * A * \rho_w)$(cm)

Consolidation ring No.: _____, Wt of empty ring, W_1: _____ (g)

Initial bulk unit weight, γ_b: _____ (kN/m^3), Dry unit weight, γ_d: $W_d/(H * A)$ (kN/m^3)

Initial void ratio before test, (e_o): _____ $= \dfrac{G\gamma_w}{\gamma_d} - 1 = \dfrac{H-H_s}{H_s}$

Initial degree of saturation, S_o = [H_{wo} /(H-H_s)] $* 100$ (%)

H_{wo} = Original height of water = $W_{wo}/(A * \rho_w)$ and W_{wo} = Wt. of water in specimen before test

Wt. of Ring + wt. of wet the soil sample, W_2: _____ (g) [check for water content]

Initial dial gauge reading at seating load: _____, Least count of dial gauge: _____

After test—Measured data (final water content determination):

Weight of ring + glass plate, W_{gp}= _____ (g)
Weight of wet sample + ring + glass plate, W_1= _____ (g)
Weight of moisture can, W_c= _____ (g)
Weight of moisture can + wet soil, W_2 = _____ (g)
Weight of wet specimen, $W_s = W_1 - W_{gp}$= _____ (g)
Weight of moisture can + dry soil, W_3 = _____ (g)
Weight of dry specimen, $W_d = W_3 - W_c$= _____ (g)
Weight of water lost, $W_w = W_s - W_d$ = _____ (g)
Final moisture content of specimen, $w_f = (W_w/W_d) * 100$ =_____ %
Final void ratio, $e_f = w_f * G$ = =_____
Final degree of saturation, $S_f = [H_{wf} /(H_f - H_s)] * 100$ (%)
H_{wf} = Final height of water = $W_{wf}/(A * \rho_w)$ and W_{wf} = Wt. of water in specimen after test

Preconsolidation pressure is determined using Casagrande's method by the following steps:

1. Choose a point of maximum curvature on e-logσ' plot (point C in Figure 11.11a).
2. Draw a horizontal line through point C.
3. Draw a forward tangent through point C.
4. Draw a bisector between the horizontal line and the forward tangent.
5. Draw a back tangent on the virgin NC line, which cuts the bisector at point I.

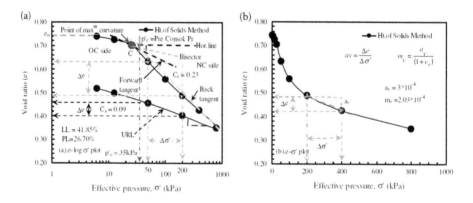

FIGURE 11.11 Consolidation parameters by height of solids method: (a) e – logσ' plot and (b) e – σ' plot.

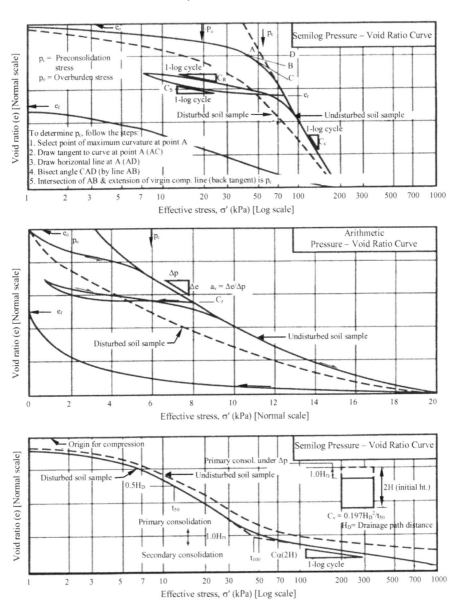

FIGURE 11.12 Consolidation test parameters and relationships.

6. Draw a vertical line through point I on the logσ′ axis, which represents the preconsolidation pressure. From Figure 11.11a, preconsolidation pressure, $p'_c = 35$ kPa.

Similarly, parameters a_v and m_v are calculated from Figure 11.11b (e – σ′ plot). The consolidation test parameters and relationships for both disturbed and undisturbed soil samples is also shown in Figure 11.12. The advantage of the "Height of Solids" method is that it can be used even for unsaturated soils. If the void ratio and water content at the beginning and the end of the test are known, the degree of saturation values can be calculated.

11.6.2 DETERMINATION OF VOID RATIO, COMP. INDEX, AND PRECONSOLIDATION PRESSURE BY CHANGE IN VOID RATIO METHOD

In the "change in void ratio" method, the volume change in a saturated soil sample is determined by computing the final void ratio as:

$$e_f = [w_f * G] \begin{pmatrix} \because & Se = wG \\ \& & S = 1 \end{pmatrix} \tag{11.7}$$

Where: w_f = Final water content post consolidation test.

For a laterally confined 1-D consolidation test, any change in vertical height (ΔH) per unit of initial height (H_o) of soil specimen is always equal to change in volume (ΔV) per unit of initial volume. Therefore, change in volume is a consequence of a decrease in void ratio (Δe) per unit of initial void ratio (e_o), which can be expressed the following relationship:

$$\frac{\Delta H}{H} = \frac{Change\ in\ volume}{Original\ volume} = \frac{\Delta e}{1 + e_o} \tag{11.8}$$

Substituting the known values of e_f and H_f at the end of the test in Equation (11.8), we get:

$$\Delta e = \frac{\Delta H}{H_f}(1 + e_f) \tag{11.9}$$

The change in void ratio (Δe) under each stress increment is calculated from Equation (11.9). The equilibrium void ratio at the end of each stress increment can be then calculated by working backwards from the known value of e_f at the end of test. Test observations, data analysis, and calculations for void ratio, compression index, and preconsolidation pressure are given in a data sheet presented in tabular form as below (Table 11.3).

The pressure-void ratio curve is shown in Figure 11.13. For the same test data, it is seen that the preconsolidation pressure increases from 35 kPa to 50 kPa and other consolidation parameters are almost the same.

11.6.3 DETERMINATION OF COEFF. OF VERTICAL CONSOLIDATION (C_v)

The coefficient of vertical consolidation (C_v) measures the rate of compression of saturated soils. C_v is not really a constant parameter, but is a function of the stress increment for a given soil. Since both coefficient of permeability (k) and coefficient of

TABLE 11.3

Determination of Void Ratio by Change in Void Ratio Method (Figure 11.13)

Initial ht. of soil sample, H_o = 20 mm, Initial xec. sec. area of soil sample, A = 44.1844 cm^2, Sp. gravity, G = 2.63, Dry weight of soil sample,

W_d = 133.45, Unit weight of water, γ_w = 10 kN/m^3, ρ_w = 1 g/cc

Date & time	Stress, σ' (kPa)	Dial gauge readings	Change in sample ht, ΔH (mm)	Ht. of sample, H $= H_o - \Delta H$ (mm)	$\Delta e = \frac{\Delta H}{H_f}(1+e_f)$	Void ratio, e $= (e_f \pm \Delta e)$	$C_c = \frac{\Delta e}{\Delta \log \sigma'}$	$a_v = \Delta e / \Delta \sigma'$ (m^2/kN)	$m_v = \frac{a_v}{1+e_0}$
....	0	2180	0	20	—	0.699			
....	6.25	2174.5	-0.03	19.97	-0.002	0.696			
	12.5	2158	-0.165	19.805	-0.011	0.685			
	25	2122	-0.235	19.57	-0.016	0.669			
	50	2045	-0.82	18.75	-0.059	0.610	0.22	$3*10^{-4}$ kPa^{-1}	$2.14*10^{-4}$ kPa^{-1}
	100	1968	-0.87	17.88	-0.066	0.545			
	200	1899	-0.79	17.09	-0.062	0.482			
	400	1827	-0.72	16.37	-0.059	0.423			
....	800	1730	-0.87	15.5	-0.076	0.347			
	200	1790	0.6	16.1	0.050	0.398	$C_s = \frac{\Delta e}{\Delta \log \sigma'} 0.08$	NA	NA
	50	1851	0.61	16.71	0.049	0.447			
	12.5	1899	0.48	17.19	0.038	0.484			
....	6.25	1921	0.22	17.41	0.017	0.501			

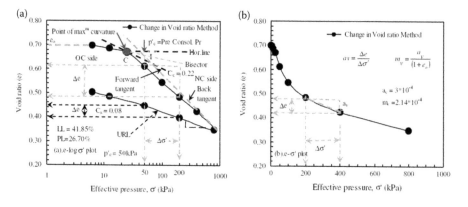

FIGURE 11.13 Consolidation parameters by change in void ratio method: (a) e-logσ'plot and (b) e-σ' plot.

compressibility (a_v) decrease with a decrease in void ratio, C_v is a function of the ratio $[k\{1 + e_o\}/a_v]$, and remains more or less the same within a considerable pressure range. There are various methods available for determination of C_v as follows:

1. Taylor's square root of time fitting method (1948)
2. Casagrande's log time fitting method (1940)
3. Rectangular hyperbola method (1987)

11.6.3.1 Taylor's Root of Time Fitting Method

Taylor (1942) suggested that the time factor (T_V) corresponding to 90% consolidation (T_{90}) is equal to 1.15 times the factor for 60% consolidation (T60), as shown in Figure 11.14a.

In this method, dial gauge readings (e.g. compression) corresponding to the square root of time are plotted for the determination of time for completion of 90% consolidation (t_{90}) and then C_v is computed as:

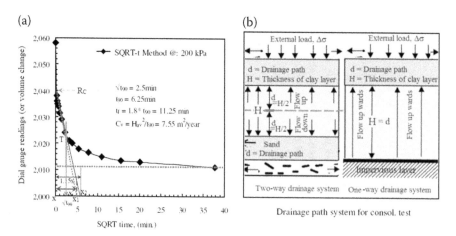

FIGURE 11.14 Typical plots: (a) Determination of t_{90} by Taylor's method; (b) Drainage path.

$$C_V = \left[\frac{T_V * H_D^2}{t_{90}} \right] (m^2/sec) \tag{11.10}$$

Where: T_V = Time factor = -0.9332 $\log_{10}(1.0.9)$-0.0851

H_D = Length of drainage path = $H_{av}/2$ for two-way drainage system (Figure 11.14b), and

H_{av} = Average thickness of the soil specimen under given stress increment.

This method is suitable for soils that give a consolidation curve with an initial straight-line portion. The procedure to determine the coefficient of consolidation by this method is given as follows:

1. Record the elapsed time readings for each stress increment for the time intervals as given in Table 11.4.
2. Draw a graphical plot between $\sqrt{t}$ reading (on X-axis) versus the compression readings (Y-axis) for a given stress increment to complete the "time-readings" consolidation curve, as shown in Figure 11.14a.
3. Now locate the straight-line portion on the initial part of the curve and extend it backward to cut the y-axis to locate the corrected zero dial gauge reading (R_C), as shown in Figure 11.14a.
4. Also, extend the initial straight-line portion forward to cut the x-axis at the point x_1 by a dotted line, as shown in Figure 11.14a.
5. From the R_C point on the y-axis, draw a straight line such that its abscissa at any point is 1.15 times that of the initial straight-line portion on a consolidation curve (e.g. $xx_2 = 1.15xx_1$). This line cuts the consolidation curve at point "T," as shown in Figure 11.14a.
6. Now draw a vertical line from point "T" onto the x-axis and record the time ($\sqrt{t_{90}}$) corresponding to 90% consolidation.
7. Determine t_{90} from ($\sqrt{t_{90}}$) and compute the coefficient of consolidation from Equation (11.10).

Similarly, for any desired stress increment, elapsed time readings can be recorded and the coefficient of consolidation can be determined. It may be noted that drainage path distance becomes half of the average soil specimen thickness for two-way drainage system. For a one-way or single drainage system, the drainage path distance is taken equal to the average thickness of soil specimen, as illustrated in Figure 11.14b.

11.6.3.2   Casagrande's Log Time Fitting Method

Casagrande (1936), based on the observation, suggested that the intersection of the straight-line portion of logt-U relation and asymptote of the lower portion gives the point corresponding to 100% degree of consolidation. In this method, a plot of the compression readings and logt is used and the time required for 5% consolidation (t_{50}) is determined (Figure 11.15a). Then the coefficient of consolidation is obtained as:

TABLE 11.4

Coefficient of Consolidation by √t-Method (Figure 11.14)

Initial height of soil specimen under current stress increment, H_o = 20 mm

Dial gauge least count, 1 Div. = 0.01 mm

Stress, σ' (kPa)	Time (min)	SQRT(t) √t (min)	Dial gauge readings Divns	ΔH (mm)	H (mm)	$H_{av} = (H_o + H_f)/2$ (mm)	$H_D = H_{av}/2$ (m)	t_{90} (min)	C_v m²/day
200	0	0	2058	0	20	19.765	0.009883	6.25	7.55
	0.083333	0.288675	2038	-0.2	19.8				
	0.166667	0.408248	2036	-0.02	19.78				
	0.25	0.5	2035	-0.01	19.77				
	0.5	0.707107	2034	-0.01	19.76				
	1	1	2031.5	-0.025	19.7-35				
	2	1.414214	2029	-0.025	19.71				
	5	2.236068	2024	-0.05	19.66				
	10	3.162278	2021	-0.03	19.63				
	15	3.872983	2020	-0.01	19.62				
	30	5.477226	2018	-0.02	19.6				
	60	7.745967	2016.5	-0.015	19.5-85				
	120	10.95445	2015	-0.015	19.57				
	240	15.49193	2013.5	-0.015	19.5-55				
	405	20.12461	2013	-0.005	19.55				
	1425	37.74917	2011	-0.02	19.53				

$$C_v = \left[\frac{T_{50} * H_D^2}{t_{50}} = \frac{0.197 * H_D^2}{t_{50}} \right] \ (m^2/s) \hspace{1cm} (11.11)$$

Where: $T_{50} = T_V$ = Time factor corresponding to 50% degree of consolidation.

$$= \frac{\pi}{4} * \left(\frac{50}{100} \right) = 0.19635 \approx 0.197$$

H_D = Length of drainage path = $H_{av}/2$ for a two-way drainage system (Figure 11.15b), and

H_{av} = Average thickness of the soil specimen under given stress increment.

It may be noted that this method also requires time-dial gauge readings in the secondary consolidation range beyond 100% degree of consolidation. The procedure to determine the coefficient of consolidation by this method is given as follows:

1. Record the elapsed time readings for each stress increment for the time intervals, as given in Table 11.5.
2. Draw a graphical plot between $\log_{10}$ (t) reading (on the x-axis) versus the compression readings (y-axis) for a given stress increment to complete the "time-readings" consolidation curve, as shown in Figure 11.15a.
3. Measure the vertical distance (z) on the curve between points corresponding to times t = 1 min. and t = 4 min., respectively, as shown in Figure 11.15a.
4. The corrected dial gauge reading (R_c) is obtained by drawing a horizontal line at a vertical distance (z) above the point corresponding to t = 1 min.
5. The consolidation curve logt-dial gauge readings or Volume change (ml) consists of two straight portions at each end with a curve joining the straight-line ends. Extend the two straight-line portions to intersect at point T.

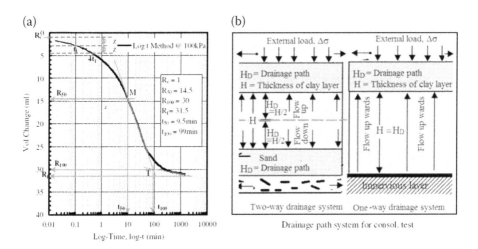

FIGURE 11.15 Typical plots: (a) Det. of t_{50} by Casagrande's method; (b) Drainage path.

TABLE 11.5

Determination of Coefficient of Consolidation by Log-t Method (Figure 11.15)

Initial height of soil specimen under current stress increment, H_o= 20 mm, Final ht., H_f = 19.665 mm, Dial gauge least count, 1 Div. = 0.01 mm

Stress, σ'	Time	Volume change, ΔV (ml)	Total vol. change	ΔH	H	H_{av} = $(H_o + H_f)/2$	H_D = $H_{av}/2$	T_{50}	C_v
(kPa)	(min)		(ml)	(mm)	(mm)	(mm)	(m)	(min)	m²/day
100	0	0	31.20			19.8325	0.00992	9.5	1.2
	0.0166667	1.7							
	0.0333333	2.1							
	0.05	2.3							
	0.0666667	2.45							
	0.0833333	2.65							
Data break as total data points are 5038									
	1445.0667	31.15							
	1445.65	31.15							
	1446.2333	31.15							
	1446.8167	31.15							
	1447.4	31.1							
	1447.9833	31.15							
	1448.5667	31.15							
	1449.15	31.15							
	1449.7333	31.15							
	1450.3167	31.05							

6. The abscissa (x-coordinate) of point T gives the time corresponding to 100% consolidation (t_{100}) on the log scale and the ordinate gives a dial gauge reading as R_{100} (Figure 11.15a).

7. The volume change or compression reading corresponding to 50% consolidation (R_{50}) is obtained from the relation:

$$R_C - R_{50} = \frac{1}{2}(R_C - R_{100}) \tag{11.12}$$

$$\Rightarrow R_{50} = R_C - \frac{1}{2}(R_C - R_{100}) \tag{11.13}$$

The abscissa (x-coordinate) of the point corresponding to R_{50} on the consolidation curve gives the time corresponding to 50% consolidation ($\log_{10}t_{50}$), which is used to determine "C_V" from Equation (11.11).

From the test results, it is observed that when the stress increment is low (low level stress < 150 kPa), the coefficient of consolidation also decreases, e.g. more time is required for draining out water from the soil specimen. Similarly, a test can be repeated for other stress increments and results be compared. However, the test results indicate that these two methods are suitable for fine grained soils and it is very difficult to catch up initial point of consolidation test in case of coarser soils such as fly ash or silt-dominant soils. Therefore, these methods are not recommended for silt or fine sand-dominated soils.

11.6.3.3   Rectangular Hyperbola Method (1987)

Sridharan et al. (1987) reported that Taylor's square root of time fitting method and Casagrande's log time fitting method are suitable for fine grained soils only and it is very difficult to catch – up initial point of consolidation test in case of coarser soils such as fly ash or silt dominant soils. Therefore, these methods are not recommended for silt or fine sand-dominated soils and proposed a new method known as "rectangular hyperbola method," which can be used for determination of "C_V" of coarser soils such as fly ashes and silt/fine sand-dominated soils. Sridharan et al. (1987) suggested to draw a graphical plot between ratio of t/δ (along y-axis) versus t (along x-axis), as shown in Figure 11.16a. From Figure 11.16a, the straight-line portion between 60% and 90% consolidation is located and "C_V" is obtained as:

$$C_V = \left[\frac{A \times m \times H_D^2}{c} = \frac{0.24 \times m \times H_D^2}{c} \right] \ (m^2/sec) \qquad (11.14)$$

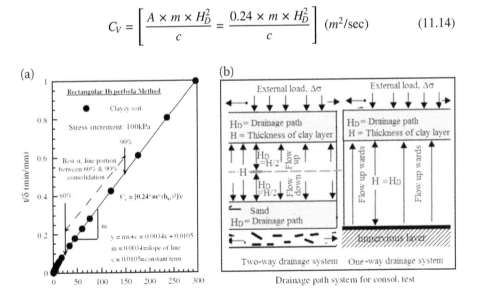

FIGURE 11.16   Typical plot for: (a) C_v by rectangular hyperbola method; (b) Drainage path.

TABLE 11.6

Coefficient of Consolidation Rectangular Hyperbola Method (Figure 11.16)

Initial height of soil specimen under current stress increment, $H_o = 20mm$, Final ht., $H_f = 19.70mm$

Dial gauge least count, 1 Div. = 0.01 mm $(C_v) = \frac{0.24 \times m \times d^2}{c}$ (m^2/sec)

Stress, σ' (kPa)	Time, t (min)	Volume change, $\Delta V = \delta V \cong \delta$ (ml = mm)	t/δ (min/mm)	ΔH mm	H mm	H_{av} mm	$H_D = H_{av}/2$	A	m	c m^2/sec	C_v m^2/day
100	0		0			19.85	0.00993	0.24	0.0034	0.0105	4.03
	0.0833		0.00087								
	0.1333		0.00136								
	0.1833		0.00167								
	0.25		0.00221								
	0.283		0.00242								
	0.35		0.00292								
		Data break, actual data points > 5,000									
	60		0.22945								
	90		0.32847								
	120		0.42553								
	147		0.51399								
	177		0.61034								
	207		0.70769								
	237		0.80612								
	267		0.89899								
	297		0.99664								

Where: m = Slope of straight-line portion of t/δ vs t plot

c = Vertical intercept of straight line in the units of m^2/sec or cm^2/sec depending on t, H

A = Constant equal to 0.24

H_D = Length of drainage path = $H_{av}/2$ for two-way drainage system (Figure 11.16b), and

H_{av} = Average thickness of the soil specimen under given stress increment.

The procedure to determine the coefficient of consolidation by this method is given as follows:

1. Record the elapsed time readings for each stress increment for the time in-tervals, as given in Table 11.6.
2. Draw a graphical plot between (t) readings (on x-axis) versus (t/δ) readings (y-axis) for a given stress increment to complete the "time-readings" con-solidation curve, as shown in Figure 11.16a.
3. From Figure 11.16a, the locate straight-line portion between 60 to 90% consolidation.
4. Extend the straight-line portion of (U = 60 – 90%) backward onto the y-axis.
5. The slope (m) and the intercept on the y-axis (c) of this extended straight line are determined.
6. The coefficient of consolidation is then determined using Equation (11.14).

It has been observed that the straight-line portion required for this method is readily obtained for most of the soils (Sridharan et al. 1999; Mir 2001). Hence, the method is found to be consistent and reliable for determination of the coefficient of con-solidation in most of the situations. Also, the method makes it easy to program the calculation of consolidation results including the coefficient of consolidation without necessarily plotting a graph of t versus (t/δ).

11.6.4   DETERMINATION OF COEFFICIENT OF PERMEABILITY (κ)

Once the coefficient of consolidation and coefficient of volume change are known from consolidation test, the coefficient of permeability can be computed indirectly from consolidation parameters as below:

$$k = \frac{C_v * a_v * \gamma_w}{1 + e_o} = C_v * m_v * \gamma_w$$

Where: γ_w = Unit weight of water

C_v = Coeff. of consolidation, and

e_o = initial void ratio.

$m_v = \left[\frac{a_v}{1+e_o}\right]$ = modulus of volume change or constrained modulus = 1/E

$$a_v = \frac{-\Delta e}{\Delta \sigma_v'} = \frac{(e_1 - e_2)}{(\sigma_{v2}' - \sigma_{v1}')} \quad (m^2/kN)$$

11.7 GENERAL COMMENTS

The 1-D consolidation test is a very laborious and time-consuming test. The technician/researcher/student should be well versed with setting up the consolidation apparatus and soil specimen preparation. The consolidation loads available in the laboratory are standards loads and soil specimen and the researchers must know to calculate desired loads and loading increment for a desired soil specimen size. The total load to be applied depends upon the soil type in the field to be consolidated under type of structure (such as embankment etc.).

Further, effect of friction between the inside consolidation ring and the soil specimen should be ascertained for both soil type and consolidation ring type. It has been observed that effect of friction is less in the floating type consolidation ring as compared to fixed-type consolidation ring. The variation of side friction for both types of consolidation rings is illustrated in Figure 11.17.

From Figure 11.17a, it is seen that in the fixed-ring consolidometer, soil moves downward relative to the consolidation ring and the frictional force tends to go upward to counterbalance the soil movement. Now if load P is applied on top of the

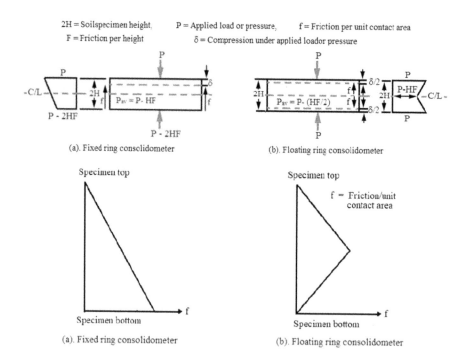

FIGURE 11.17 Variation of effect of friction between inside consolidation ring and the soil specimen for (a). Fixed-ring consolidometer; (b). Floating-ring consolidometer.

soil sample of 2H height in the consolidation ring, then the load at the base of the consolidation ring is (P – 2HF) and hence, the average load is (P- HF), where F is the frictional force per height.

Similarly, in the case of a floating-type consolidation ring, the soil sample moves towards the center from the top and bottom relative to the consolidometer ring. Therefore, the average load at the base of the consolidation ring is (P – HF/2). Thus, it is seen that the effects of side friction in reducing the average applied load to the soil specimen are smaller in the floating-ring consolidometer than in the fixed-ring consolidometer.

11.8   APPLICATIONS/ROLE OF CONSOLIDATION PARAMETERS IN SOIL ENGINEERING

The consolidation parameters (e.g. e_o, C_c, C_s, p_c, C_v, a_v, m_v, δc, etc.) are used in the settlement analysis of various structures built on compressible soils due to volume change. Consolidation settlement (δ_c) constitutes primary consolidation (due to removal of water) and secondary settlement (due to rearrangement of particles). Primary consolidation is the major component and it can be reasonably estimated by using consolidation parameters.

Total settlement (δ_t) is the magnitude of downward movement of a structure. Final settlement is the settlement value that is produced at the end or after the permanent downward displacement of the foundation due to consolidation, or the process in which soils decrease in volume due to loading.

11.8.1   Types of Settlement—Based on Mode of Occurrence, Various Types of Settlements

- Immediate settlement or elastic settlement (δ_e)
- Primary settlement or consolidation settlement (δ_c)
- Secondary settlement or progressive settlement (δ_s)

Total settlement is given by $\delta_t = \delta_e + \delta_c + \delta_s$ (Figure 11.18a).

- **Immediate consolidation settlement** (Figure 11.18b) occurs due to re-arrangement of soil particles in the soil mass under applied loading. This is generally very small and often ignored. According to elastic theory, for a homogeneous, isotropic, elastic medium, the immediate settlement is given by:

$$\delta_i = \delta_e = \left[\frac{q_{nf} * B}{E} * (1 - \mu^2)I_f \right] \tag{11.15a}$$

Where: $\delta e = \delta i$ = Elastic or immediate or distortion settlement
$\Delta \sigma = q_{nf}$ = P/A$_f$ = Net footing load

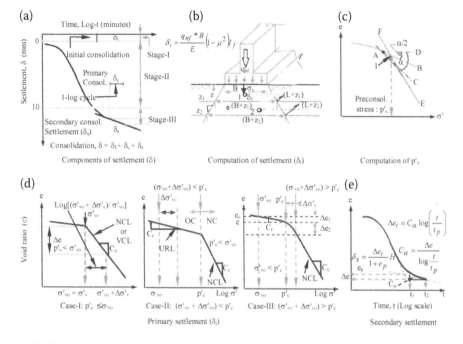

FIGURE 11.18 Computation of settlement from compressibility parameters.

B = Foundation width or diameter
E = Modulus of elasticity of soil
μ = Poisson's ratio, and
I_f = Influence factor for rigid footings.

- **Primary consolidation settlement** occurs due to expulsion of pore water from the soil pore voids in a saturated soil mass under applied loading. Based on the OCR value, soil is identified either as NC or OC and the proper equation for settlement analysis is chosen:

Soil is NC if OCR ≤ 1, and
Soil is OC If OCR > 1.
Where: OCR = $[p'_c/\sigma'_{vo}]$, p'_c = preconsolidation pressure and σ'_{vo} = present effective stress.

Case-I: For NC soils (OCR = 1 and $p'_c < \sigma'_{vo}$–Figure 11.18d): the primary consolidation settlement is computed as:

$$\delta_p = \left[\frac{C_c * H}{1 + e_o} \log_{10}\left(\frac{\sigma'_{vo} + \Delta\sigma'_v}{\sigma'_{vo}} \right) \right] \tag{11.15b}$$

Where: H = Thickness of soil stratum under consideration (m)

C_c = Comp. index from lab. oedometer tests
e_o = Initial void ratio from lab. oedometer tests
σ'_{vo} = Present vertical eff. overburden stress at the center of soil stratum under consideration (kPa)

$\Delta\sigma'$ = increase in loading due to external loading at the center of soil stratum under consideration (kPa), and

p'_c = Preconsolidation pressure from consolidation test (Figure 11.18c).

Case-II: If the soil is heavily over-consolidated (OCR > 1): we have to consider the case of recompression (OC zone) depending on the magnitude of p'_c, σ'_o and $\Delta\sigma'$.

   a. When $[(\sigma'_{vo} + \Delta\sigma'_v) < p'_c]$
     In this case, soil is heavily consolidated and consolidation occurs along the URL (unloading- reloading line). Figure 11.18d and the settlement is computed as:

$$\delta_p = \left[\frac{C_r * H}{1 + e_O} \log_{10}\left(\frac{\sigma'_{vo} + \Delta\sigma'_v}{\sigma'_{vo}} \right) \right] \tag{11.15c}$$

Where: $C_r = C_s$ = Slope of recompression or unloading curve. Other parameters are the same as in the case of NC
soils.
   b. When $[(\sigma'_{vo} + \Delta\sigma'_v) > p'_c \text{ and } \sigma'_{vo} < p'_c]$
     In this case, soil is lightly over consolidated and we have to consider two components of settlement: one along the URL (unloading-reloading line) and the other along the NCL (normally consolidated line) (Figure 11.18d). The equation to use in this case for settlement computation is:

$$\delta_p = \left[\frac{H}{1 + e_o} \left(C_r \log_{10} \frac{p'_c}{\sigma'_{vo}} + C_c \log_{10} \frac{\sigma'_{vo} + \Delta\sigma'_v}{p'_c} \right) \right] \tag{11.15d}$$

Where: All parameters have same meaning as in case I and in case of NC soils.

Secondary consolidation occurs due to plastic adjustment of the soil grains in a saturated soil mass after 100% primary consolidation (Figure 11.18e). This is very slow process and the settlement achieved is too small and often ignored. Secondary settlement is computed as:

$$\delta_s = \left[\frac{\Delta e_t}{(1 + e_p)} * H \right] \quad where \ \Delta e_t = C_\alpha \log_{10}\left(\frac{t}{t_p} \right) \tag{11.15e}$$

Where: H = Thickness of saturated soil stratum under consideration (m)

C_α = Slope of the last straight line portion (after primary consolidation)

e_p = Void ratio at the end of primary consolidation

t_p = Time at the end of primary consolidation

$C_\alpha \approx 0.04\ C_c$ for inorganic clays and silts

$\approx 0.05\ C_c$ for organic clays and silts

$\approx 0.075\ C_c$ for peats.

Typical allowable total settlements for foundation design are given in Table 11.7.

TABLE 11.7
Typical Values of Allowable Settlement for Foundation Design

Type of Structure	Typical Values of Allowable Settlement, δ_a (mm)
Office Buildings	≤25
Industrial Buildings	25–75
Bridges	50

11.9   SOURCES OF ERROR

The main sources of error may arise due to:

1. Effect of disturbance caused by trimming a given soil sample. For determination of engineering properties of in situ soils, it is mandatory that the soil sample should be collected as an undisturbed soil specimen and transported to a soil testing laboratory very carefully so that the in situ structure and moisture content remain intact and do not change due to sample disturbance. Likewise, reconstituted soil samples prepared in the laboratory should be trimmed very carefully to avoid or minimize the disturbance while preparing the soil sample. Sample disturbance may lead to inaccurate measurement of soil properties and it has been observed that soil disturbance is more prominent in small-sized samples compared to large-sized soil samples, as shown in Figure 11.19.
2. It may be ensured that the consolidation ring is filled with soil sample flush with the top surface so that the soil sample is completely confined laterally.
3. It may be ensured that the consolidation ring is made of non-corrosive material to avoid the galvanic effect on compressibility parameters.

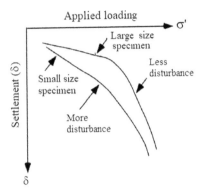

FIGURE 11.19 Effect of soil disturbance on preparation of soil specimen.

4. It should be ensured that neat and fully saturated porous stones are used to avoid any blockade of flow through porous stones during consolidation process.
5. It may be ensured that the consolidation rings are properly greased before filling the soil to avoid any frictional effect between specimen and consolidation ring side during the consolidation process.
6. It may be ensured that load increments are applied equally with a load increment of about 1. Inappropriate load increment factor. Higher load increment factor (>1) may not produce a qualitative consolidation curve for preconsolidation pressure.

11.10 PRECAUTIONS

1. Undisturbed soil samples from the field site should be transported carefully so that natural properties of the in situ remain intact.
2. Likewise, the reconstituted soil sample prepared in the laboratory should be trimmed carefully to avoid affect of sample disturbance.
3. A load increment factor of one (1) should be applied while changing the loads during the consolidation test.
4. Remolded soil samples should be prepared in the laboratory at OMC and 90% MDD (e.g. 0.9 ∗ MDD) because in the field, 100% degree of compaction is never achieved.
5. Make sure that the soil sample does not swell under the seating load such as in case of expansive clays. In such cases, apply the seating load higher than the standard seating load to avoid swelling.
6. Also, make sure that the soil sample does not compress under the seating load such as in the case of soft clays. In such cases, apply the seating load lower than the standard seating load.
7. Make sure that the filter papers and porous stones used are of adequate quality and fully saturated.
8. Make sure that the unloading is done by load increment of one-fourth of the present loading acting on the soil sample.
9. Determine the water content after completing the consolidation test for determination of final water content and final vopid ratio.

REFERENCES

ASTM D2435 / D2435M-11. (2011). "Standard Test Methods for One-Dimensional Consolidation Properties of Soils Using Incremental Loading." ASTM International, West Conshohocken, PA, www.astm.org.

ASTM D4546-14. (2014). "Standard Test Methods for One-Dimensional Swell or Collapse of Soils." ASTM International, West Conshohocken, PA, www.astm.org.

BS 1377. 1990. "Methods of Test for Soils for Civil Engineering Purposes-Part 5: Compressibility, Permeability and Durability Tests." British Standards, UK.

BS 1377. 1990. "Methods of Test for Soils for Civil Engineering Purposes-Part 6: Consolidation and Permeability Tests in Hydraulic Cells and with Pore Pressure Measurement." British Standards, UK.

Casagrande, A. 1936. "The Determination of the Preconsolidation Load and Its Practical Significance." Discussion D-34, Proceedings of the 1st International Conference on Soil Mechanics and Foundation Engineering, Vol. III, pp. 60–64, Cambridge, UK.

Cooling, L. F., and A. W. Skempton. 1941. "Some Experiments on the Consolidation of Clay." *J. Inst. Civ. Eng.* 16: 381–398.

IS: 2720 (Part 1). 1980. "Indian Standard Code for Preparation of Soil Samples." Bureau of Indian Standards, New Delhi.

IS: 2720 (Part 15). 1986. "Method of Test for Soils: Determination of Consolidation Properties." Bureau of Indian Standards, New Delhi.

IS: 2720 (Part 41). 1977/2002. "Method of Test for Soils: Determination of Swelling Pressure of Soils." Bureau of Indian Standards, New Delhi.

IS: 2720 (Part 40). 1977/2002. "Method of Test for Soils: Determination of Free Swell Index for Fine Grained Soils." Bureau of Indian Standards, New Delhi.

Ladd, C. C. 1971. "Settlement Analysis for Cohesive Soils." Research report R 64-17, Department of Civil Engineering, Massachusetts Institute of Technology, USA.

Mir, B. A. 2001. *The Effect of Fly Ash on the Engineering Properties of Black Cotton Soils.* M. E. thesis, Department of Civil Engineering, IISc, Bangalore.

Reddy, Krishna R. 2000. "Engineering Properties of Soils Based on Laboratory Testing." Department of Civil and Materials Engineering, University of Illinois at Chicago, USA.

Schmertmann, J. H. 1955. "The Undisturbed Consolidation Behaviour of Clay." *Trans. ASCE* 120: 1201–1233.

Sridharan, A., N. S. Murthy, and K. Prakash. 1987. "Rectangular Hyperbola Method of Consolidation Analysis." *Geotechnique* 37(3): 355–368.

Sridharan, A., H. B. Nagaraj, and N. Srinivas. 1999. "Rapid Method of Consolidation Testing." *Canadian Geotechnical Journal* 36: 392–400.

Taylor, D. W. 1942. "Research on Consolidation of Clays." MIT, Dept. Civil and Sanit. Engrg. No. 82.

Terzaghi, K. 1923. "The Calculation of the Permeability Factor of the Clay from the Course of the Hydrodynamic Stress Phenomena." Academy of Sciences in Vienna. Mathematics and Science Class. Meeting Reports. Department II a. Vol. 132, No. 3/4, 125–138.

12 Unconfined Compression Strength of Soils

References: IS: 2720 (Part 1); 2720 (Part 10); ASTM D2166 / D2166M; ASTM D1587; BS 1377-Part 7

12.1 OBJECTIVES

The main objective of conducting unconfined compression strength (UCS) test is to ascertain the short-term stability of various structures on saturated cohesive soils.

A "UCS" test is a special type of the triaxial compression test in which a soil sample is directly sheared without lateral confinement (e.g. $\sigma_3 = 0$). This test is applicable to undisturbed, slightly disturbed (remolded), and undisturbed cohesive soils.

Another advantage of the UCT is that the failure surface will tend to develop in the weakest portion of the clay sample.

12.2 DEFINITIONS AND THEORY

The unconfined compression strength (UCS) test is an effortless and expeditious test to establish the shear strength law for saturated cohesive soils, e.g. $\tau = c_u (\because \phi = 0)$. In this test, maximum unconfined compression strength $(q_u)_{max}$ of cohesive soils is determined with no lateral support and undrained shear strength (c_u) is taken equal to half of the unconfined compression strength (e.g. $c_u = 1/2 * q_u)_{max}$). The shear strength parameters obtained by UCS tests are based on total stress approach since no pore water pressure is measured during this test. The "UCS" test is generally conducted to ascertain the short-term stability of various structures such as railway embankments, pavement subgrades, slopes etc. which are subjected to rapid loading on saturated cohesive soils. The "UCS" test is suitable for saturated clayey soils, which can be collected as an undisturbed soil sample in a thin-walled core cutter or SPT or UCS seamless tubes. The undrained cohesion $(s_u$ or $c_u)$ of a soil sample in a most critical condition is taken as one-half of the maximum unconfined compressive stress at failure (e.g. s_u or $c_u = q_u/2$), as illustrated in Figure 12.1 under $(\phi_u = 0)$ condition (e.g. saturated soils, S = 1).

However, with elapsed time, the excess pore water pressure (u) vanishes and the effective stress $(\sigma' \approx \sigma$ as $u_w \rightarrow 0)$ is increased. Therefore, the shear strength parameters are obtained based on effective stress approach and the law of shear strength expressed as: $s = \tau_f = c' + \sigma' \tan \phi'$ must be used (Reddy 2000). The clays

DOI: 10.1201/9781003200260-12

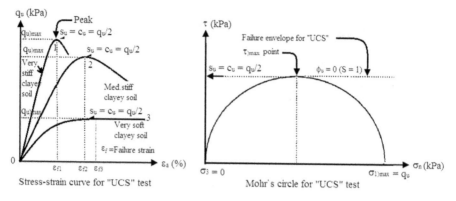

FIGURE 12.1 Unconfined compression test and test parameters.

have very low permeability and when tested in a strain controlled "UCS" test under rapidly applied loads, the excess pore water pressure generated do not have enough time to drain out from the clay sample. Thus, there is no change in volume and the water content and the void ration remain unchanged. However, pore water pressure is not measured during the "UCS" test; therefore, strength parameters obtained are based on total stress approach. An indication of typical values of shear strength is given by the following classification of clays based on consistency (Table 12.1).

TABLE 12.1
Unconfined Compression Strength Values

Sr. No.	Soil Consistency	q_u (kPa)	q_u (t/m^2)
1	Very Soft	< 25	< 2.5
2	Soft	25–50	2.5–5.00
3	Medium	50–100	5.00–10.0
4	Stiff	100–200	10. 0–20.0
5	Hard	200–400	20.0–40.0
6	Very Hard	> 400	> 40.0

12.2.1 PRINCIPLE OF "UCS" TEST

The basic principle of "UCS" test is that the strength parameters of a cohesive soil sample are measured only under major principal stress and the minor principle or confining stress is zero ($\sigma_1 > 0$ & $\sigma_3 = 0$), as shown in Figure 12.2. Therefore, the undrained shear strength for "UCS" test is expressed as:

$$\tau_f = s_u = c_u = \left[\frac{q_{u)max}}{2} = \frac{\sigma_{1)max}}{2} \right] (\because \sigma_3 = 0) \qquad (12.1)$$

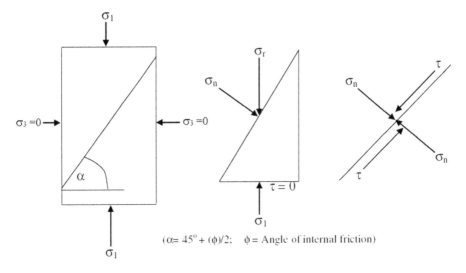

FIGURE 12.2   Principle of unconfined compression test.

The test is applicable to slightly disturbed (remolded) and undisturbed cohesive soils. Cohesionless soils (sands and gravels) cannot be subjected to this kind of test because they do not form unsupported prism and cylindrical sample. The undrained cohesion of a clayey soil sample is computed by using the following expression:

$$\sigma_1 = [\sigma_3 \tan^2 \alpha + 2c_u \tan \alpha] \tag{a}$$

Where: σ_1 = Major principal stress at failure (total stress approach)
$\quad \sigma_3$ = Minor Principal Stress at failure
$\quad \alpha$ = Failure angle with major principal plane = $45° + \phi/2$
$\quad \phi$ = Angle of internal friction

However, in unconfined compression test, $\sigma_3 = 0$ (as there is no lateral confinement), and $\sigma_1 = q_{u)max}$ (unconfined compressive stress). Therefore, equation [a] can be modified as:

$$q_{u)max} = [2c_u \tan(45^o + \phi'/2)] \tag{b}$$

Or

$$c_u = \left[\frac{q_{u)max}}{2 \tan(45^o + \phi'/2)} \right] \tag{c}$$

If the soil sample is fully saturated (e.g. $S = 1$, and hence $\phi = 0$ concept), equation [c] can be modified as:

$$c_u = s_u = \left[\frac{q_{u)max}}{2} \right] \qquad \text{(d)}$$

Also, based on total stress approach, the law of shear strength for soils is generally expressed as (e.g. Mohr-Coulomb's equation):

$$s = \tau_f = [c + \sigma_{n)f} \tan \phi] \qquad (12.2)$$

Where: τ_f = Shear resistance at
$\sigma_{n)f}$ = Normal stress at failure

$$For\, saturated\, soils,\ \phi = 0,\ \Rightarrow s = [\tau_f = c] \qquad \text{(e)}$$

Thus, in the case of the "UCS" test, shear strength of soil is (by comparing equations [d] and [e]):

$$s = \tau_f = c_u = \frac{q_{u)max}}{2}$$

12.3   METHOD OF TESTING

The undrained cohesion of cohesive soils on an in situ specimen and remolded (disturbed) specimen is determined in laboratory using a compression machine (a loading frame consists of two metal plates, either hand operated or machine driven) by means of strain-controlled mechanism.

12.3.1   PRE-REQUISITE FOR UNCONFINED COMPRESSION STRENGTH (UCS) TEST

Since strength is an engineering property and some of the physical and index tests such as soil grading, specific gravity, consistency limit tests and compaction tests are pre-requisite for conducting this test. The "UCS" test is conducted either on undisturbed (in situ) saturated clayey soil specimen or remolded or reconstituted soil specimen as per requirement.

12.3.2   SOIL TESTING MATERIAL

The undisturbed soil specimen at natural moisture content is collected directly from the field in a "UCS" tube, extracted and trimmed in the laboratory for testing. The soil specimen is assumed to be virtually saturated (in in-situ condition).

12.3.3   SIZE AND PREPARATION OF SOIL SPECIMEN

The maximum size of soil particle in the "UCS" test specimen should not exceed "D/10"mm for a specimen of 30 to 72 mm diameter and "D/6"mm for a specimen of diameter greater than 72 mm. In all cases, the sample diameter (D) should be larger

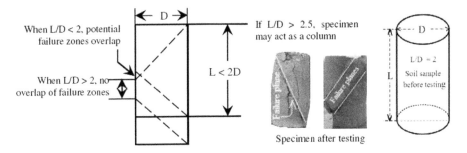

FIGURE 12.3   L/D ratios for any soil compression test (UCS, Triaxial tests, etc.).

than 30 mm (ASTM D2166). It may be noted that larger particles present in the test specimen tend to reduce "s_u." The "UCS" consists of trimming a soil specimen of some diameter (D) to an adequate length (L) ratio, L/D. The L/D ratio of the test specimen should be large enough to avoid interference of potential 45° failure planes (Figure 12.3) and small enough not to obtain a "column" failure. The **ASTM** specifies the L/D ratio to satisfy these criteria is:

$$2 < \frac{L}{D} < 2.5 \tag{12.3}$$

The L/D ratio shall be not less than 2. The unconfined compression test, often called a "U" test, is a test at natural water content (e.g. for in-situ test specimens). It is currently employed more than other shear tests on cohesive soils. This is a special type of triaxial test without lateral confinement.

For "UCS" test on clayey soils, soil specimen passing through a 425 μm IS sieve may be used. For silt-dominated soils, a soil specimen passing through a 2 mm IS sieve may be used. If remolded soil specimen based on OMC & MDD (compaction test) is to be prepared, then the weight of soil mass may be calculated as per "UCS" tube size chosen. If a reconstituted soil specimen is to be used, then about 10 kg soil may be taken for preparation of slurry and consolidated on laboratory floor up to the desired water content or stress/consistency of soil mass. However, it may be noted that soil slurry be prepared at water content of about 1.5 * LL to 2 * LL (LL-liquid limit of soil mass) thoroughly mixed and de-aired before consolidation to desired consistency (in situ state of soil). For soil specimen preparation, refer to Section 10.4 (Test No. 10).

12.4   TESTING EQUIPMENT AND ACCESSORIES

The following equipment and accessories are required for the "UCS" test (Figure 12.4):

1. Unconfined compression testing machine including dial gauge and load cell or proving ring
2. UCS tubes, 76 mm diameter, 200 mm long
3. 0.425 mm IS sieve

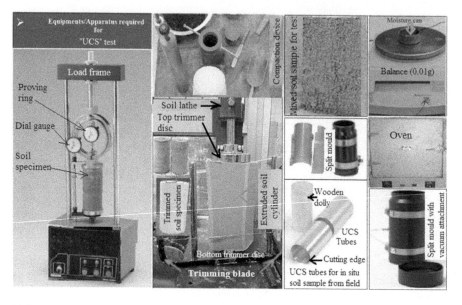

FIGURE 12.4 Experimental setup and accessories for "UCS" test.

4. A tray for mixing soil sample
5. Weighing balance accuracy 0.01 g
6. Split mold of internal diameter of 38 mm and 76 mm long
7. Vernier calipers
8. Soil lathe and accessories for trimming undisturbed soil specimen
9. Static compression machine and accessories if remolded soil specimen is to be prepared
10. Moisture content cans
11. Stopwatch
12. Distilled water
13. Squeeze bottle
14. Oil or grease
15. Oven
16. Porcelain evaporating dish
17. Sample extruder
18. Desiccator
19. Scale (0.5 mm least count)

12.5 TESTING PROGRAM

1. Check and prepare the apparatus (in this test, power-operated compression machine is used).
2. Clean the "UCS" tubes or Shelby tubes and apply a thin layer of grease inside for collection of in-situ samples.
3. Clean the ground surface in the field at the desired location and drive the "UCS" tubes into the ground vertically downward by a wooden mallet.

4. Excavate the soil surrounding the "UCS" tubes carefully and seal them in airtight plastic bags so that there is no moisture loss from the soil sample.

5. Extrude the soil sample from the "UCS" tube sampler or Shelby tube sampler in the soil testing laboratory and cut a soil specimen of standard initial size of 38 mm diameter (D_o) by 76 mm (L_o) long (e.g. L/D = 2) in a split mold. If the sampler is of higher diameter, then the test specimen may be trimmed by using a soil lathe or a trimmer carefully so that the diameter is 38 mm.
   Note: Make sure that the soil specimen is extruded out from the in the same direction it was collected in the field so that there is less resistance encountered by the soil specimen while extruding it out.

6. Find the volume of split mold from its known dimensions, V (cm^3).

7. Take weight of soil specimen after extruding it to the nearest to 0.01 g, W_s (g).

8. If a remolded sample is to be prepared in the laboratory, then firstly, find MDD and OMC for the soil sample. Calculate the dry weight of the soil sample based on split mold dimensions [diameter-D_o = 38 mm, length-L_o = 76 mm, cross-sec. area-A = $\pi D^2/4$ = 11.342 cm^2, and volume-V = A $*$ L = 86.2 cm^3 of split mold] and compaction characteristics (e.g. using OMC and MDD, refer Section 10.4 Test No. 10 for details). For remolded specimen, dry weight of soil sample is taken as: W_s = 0.9MDD $*$ V (g) and water content is taken as: w = (OMC/100) $*$ W_s (ml). The compacted soil specimen is prepared by mixing dry soil with desired water content and compacted in the split mold by static compression (Figure 12.5).

9. For remolded soil specimen prepared in step 6, above, take weight of compacted moist soil nearest to 0.01 g W_s (g).

10. Determine water content either from trimmings in step 3 or left-out soil sample in step 6.

11. Calculate the deformation (δL) corresponding to 20% strain ($\varepsilon_a)_f$ as:

$$dL = (e_a)_f * L_o = 0.20 * 76 = 15.2 \, mm$$

Where L_o = Initial specimen length (as measured in step 3 or step 6).
Note: Determination of δL for assumed strain of 20% will help in determining the number of dial gauge divisions per minute to fix the desired strain rate for shearing the soil sample.

12. Now place the soil specimen in the compression machine carefully and set the load cell or proving ring and dial gauge readings to zero.

13. Apply compressive load at desired strain rate. A rate of strain of approximately of 1.5 to 2% per minute is frequently employed for the strain controlled test. In case of stress controlled test, an application of one-tenth to one-fifteenth of the estimated strength every 0.5 to 1 minute is commonly employed.

14. Take readings approximately every 0.5 mm (for every 50 divisions deformation with least count of 0.01 mm) of vertical deflection until the proving ring pointer moves in opposite direction, e.g. soil specimen has failed. However, it is recommended to record at least 6 to 8 readings beyond failure point.

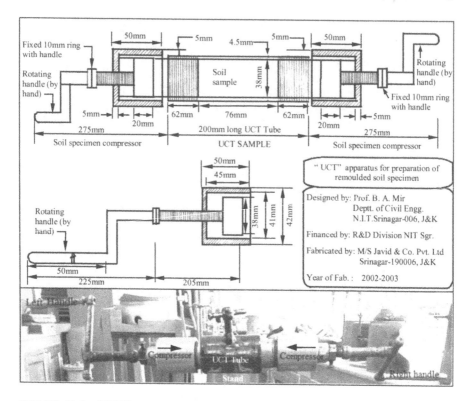

FIGURE 12.5 "UCS" test apparatus for preparation of remolded soil specimen (NIT Sgr).

Note: It may be noted that a well-defined peak or failure point will be observed in case of OC clays (e.g. stiff clays). When the sample fails, the load will decrease rapidly beyond the peak point. It is recommended to record at least 6 to 8 readings or until the load becomes constant. This will help in producing the critical state of the soil and hence the ultimate or residual strength.

However, in the case of soft clays, the stress-strain curve will be non-linear and no well-defined peak will be produced. Therefore, in this case, the "barreling" failure will be produced as shown in Figure 12.6. In such a case, the test may be stopped after achieving 20% axial strain as failure strain.

15. At the end of test, dismantle the specimen assembly from the compression testing machine and remove the failed soil specimen carefully.
16. Identify the nature of the failure plane and measure the angle (α) between the cracks and the horizontal line, as shown in Figure 12.6. The angle between the horizontal and the failure plane should actually be measured at the time the failure plane is formed. Then find the value of $\phi_u = (\alpha - 45°) * 2$.
17. Take a representative sample of the failed specimen for determination of the final moisture content after the test as per standard codal procedure (see Test No. 1).
18. Repeat the test for different samples if any.

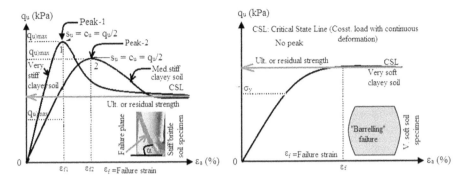

FIGURE 12.6 Typical stress-strain curves and failure patterns for stiff and soft clayey soils.

12.6 OBSERVATION DATA SHEET AND ANALYSIS

12.6.1 DETERMINATION OF WATER CONTENT AND DRY UNIT WEIGHT

Test observations, data analysis, and calculations for determination of water content and dry unit weight are given in Table 12.2.

12.6.2 DETERMINATION OF UNDRAINED SHEAR STRENGTH BY "UC" TEST

Test observations, data analysis, and calculations for determination of maximum unconfined compressive stress "UCS" and undrained shear strength (c_u or s_u) are given in Table 12.3.

Calculations are:

1. For each test (refer to Table 12.3), compute change in vertical length (δL) in (Col. 2):

$$\delta L = Strain\ Divns.\ (Col.1) * Dial\ gauge\ L.\ C.$$

2. Now compute axial strain in (Col. 3):

$$\varepsilon_a = \frac{\delta L}{L_o} * 100\ (\%)\ [\text{Where: } L_o = \text{ Initial length } of\ the\ soil\ specimen]$$

3. Compute applied load on the soil specimen, P in (Col. 5):
   P = Proving ring divns. * L.C. of Proving ring/divn (kg)
4. Compute corrected area, A_c, in (Col. 6):

$$A_c = \frac{A_o}{1 + (\varepsilon_a/100)}\ (cm^2)\ [A_o = Initial\ x - sec.\ area\ of\ the\ soil\ specimen]$$

TABLE 12.2

Determination of Water Content and Dry Unit Weight

Initial length of the soil specimen, L_o = 76 mm, Initial Diameter of soil sample, D_o = 38 mm, Initial xec. sec. area of soil sample, A_o = 11.342 cm^2, Initial volume of soil specimen, Vo = 86.2 cm^3, 1 cm^3 = 10^{-6}m^{3}Sp. gravity, G = 2.63, Unit weight of water, γ_w = 10 kN/m^3, ρ_w = 1 g/cm^3, Least count of dial gauge, LC = 0.01 mm

Sl. No.	Observations and Calculations	Trial (Soil core) No.		
		1	2	3
Observation				
1	Split Mold No.			
2	Internal diameter, D_o (cm)			
3	Internal height, H_o (cm)			
4	Volume of cutter, V_o (cm^3) = $\pi/4 * (D^2 * H)$			
Calculations: Bulk unit weight, γ_b (kN/m^3)				
5	Weight of wet soil, W_s (gm)			
6	Bulk Unit Weight, $\gamma_b = W_s/V_o$ (g/cm^3)			
Calculations: Water content, w (%)				
7	Porcelain Dish No.			
8	Weight of empty Porcelain Dish (W_1)			
9	Weight of Porcelain Dish + wet soil (W_2)			
10	Weight of Porcelain Dish + dry soil (W_3)			
11	Weight of water $W_w = W_2 - W_3$			
14	Weight of dry soil solids, $W_D = W_3 - W_1$			
15	Water content (fraction), w = $(W_w)/(W_D)$			
Calculations: Field dry unit weight, γ_d				
16	Field dry unit weight, $\gamma_d = \gamma_b/(1 + w/100)$ (kN/m^3)			
18	Average dry unit weight, γ_d (kN/m^3)			
19	Void ratio, $e = (G * \gamma_w/\gamma_d) - 1$			
20	Degree of saturation, $S = w * G/e$			
21	Porosity, $n = e/(1 + e)$			

Area correction is applied because during compression, the diameter of the soil sample changes and hence the cross-sectional area also changes. However, it may be noted that the the volume of the specimen is assumed constant through the test.

5. Compute unconfined compressive stress, q_u in (Col. 7): $q_u = P/A_c$ (kPa)
6. Now compute undrained shear strength, c_u in (Col. 8): $c_u = q_u/2$ if specimen is fully saturated

Or, $c_u = q_u/(2\tan\alpha)$ if soil is partially saturated.
7. Plot the graph of axial strain, ε_a in percent (Col. 3) versus stress, q_u (Col. 7), as shown in Figure 12.7.
8. Determine the peak stress from the stress-strain plot, which is taken as the maximum unconfined compression strength, $q_{u)max}$ of the soil specimen.

TABLE 12.3

Determination Undrained Shear Strength by "U" Test (Specimen-1)

Stress-Strain Characteristics of "U" TEST

Strain L.C. (mm/dvn) = 0.01	PR L.C. (kg/dvn) = 0.17	D_o (mm) = 38	A_o cm^2) = 11.34
SITE: Flood Channel Srinagar	$1 cm^3 = 10^{-6} m^3$	L_o (mm) = 76	V_o (cc) = 86.19
Sample type: Disturbed/Remolded		BH: I	Depth (m): 1.5

Strain Divns	Change in length, δL (mm)	Strain (%)	PR Divns	Load, P (kg)	A_c (cm^2)	q_u (kPa)	c_u (kPa)	c_u (t/m^2)
0	0	0.00	0	0	11.34	0.00	0.00	0.00
50	0.5	0.66	5	0.85	11.42	7.45	3.72	0.37
100	1	1.32	12	2.04	11.49	17.75	8.88	0.89
150	1.5	1.97	17	2.89	11.57	24.98	12.49	1.25
200	2	2.63	25	4.25	11.65	36.49	18.24	1.82
250	2.5	3.29	29	4.93	11.73	42.04	21.02	2.10
300	3	3.95	38	6.46	11.81	54.71	27.36	2.74
350	3.5	4.61	47	7.99	11.89	67.21	33.60	3.36
400	4	5.26	53	9.01	11.97	75.26	37.63	3.76
450	4.5	5.92	56	9.52	12.05	78.97	39.49	3.95
500	5	6.58	60	10.2	12.14	84.02	42.01	4.20
550	5.5	7.24	64	10.88	12.23	88.99	44.50	4.45
600	6	7.89	69	11.73	12.31	95.26	47.63	4.76
650	6.5	8.55	71	12.07	12.40	97.32	48.66	4.87
700	7	9.21	73	12.41	12.49	99.35	49.67	4.97
750	7.5	9.87	76	12.92	12.58	102.68	51.34	5.13

(Continued)

TABLE 12.3 (Continued)

Stress-Strain Characteristics of "U" TEST

Strain L.C. (mm/dvn) = 0.01
SITE: Flood Channel Srinagar
Sample type: Disturbed/Remolded

PR L.C. (kg/dvn) = 0.17
$1cm^3 = 10^{-6} m^3$

D_o (mm) = 38
L_o (mm) = 76
BH: I

A_o cm^2 = 11.34
V_o (cc) = 86.19
Depth (m): 1.5

Strain Divns	Change in length, δL (mm)	Strain (%)	PR Divns	Load, P (kg)	A_c (cm^2)	q_u (kPa)	c_u (kPa)	c_u (t/m^2)
800	8	10.53	79	13.43	12.68	105.95	52.98	5.30
850	8.5	11.18	81	13.77	12.77	107.84	53.92	5.39
900	9	11.84	84	14.28	12.86	111.00	55.50	5.55
950	9.5	12.50	85	14.45	12.96	111.49	55.74	5.57
1,000	10	13.16	87	14.79	13.06	113.25	56.63	5.66
1,050	10.5	13.82	89	15.13	13.16	114.98	57.49	5.75
1,100	11	14.47	90	15.3	13.26	115.38	57.69	5.77
1,150	11.5	15.13	90	15.3	13.36	114.49	57.25	5.72
1,200	12	15.79	92	15.64	13.47	116.13	58.07	5.81
1,250	12.5	16.45	93	15.81	13.57	116.48	58.24	5.82
1,300	13	17.11	94	15.98	13.68	116.80	58.40	5.84
1,350	13.5	17.76	95	16.15	13.79	117.11	58.55	5.86
1,400	14	18.42	96	16.32	13.90	117.39	58.70	5.87
1,450	14.5	19.08	97	16.49	14.02	117.66	58.83	5.88
1,500	15	19.74	98	16.66	14.13	117.91	58.95	5.90
1,550	15.5	20.39	99	16.83	14.25	118.13	59.07	5.91
1,600	16	21.05	100	17	14.37	118.34	59.17	5.92
1,650	16.5	21.71	101	17.17	14.49	118.53	59.26	5.93

1,700	17	22.37	102	17.34	14.61	118.69	59.35	5.93
1,750	17.5	23.03	103	17.51	14.73	118.84	59.42	5.94
1,800	18	23.68	104	17.68	14.86	118.97	59.49	5.95
1,850	18.5	24.34	105	17.85	14.99	119.08	59.54	5.95
1,900	19	25.00	106	18.02	15.12	119.17	59.58	5.96
1,950	19.5	25.66	107	18.19	15.26	119.24	59.62	5.96
2,000	20	26.32	108	18.36	15.39	119.29	59.64	5.96
2,050	20.5	26.97	108	18.36	15.53	118.22	59.11	5.91
2,100	21	27.63	109	18.53	15.67	118.24	59.12	5.91
2,150	21.5	28.29	109	18.53	15.82	117.17	58.58	5.86
2,200	22	28.95	110	18.7	15.96	117.16	58.58	5.86
2,250	22.5	29.61	111	18.87	16.11	117.13	58.56	5.86
						$\varepsilon_f(\%)$	$q_{u)max}(kPa)$	$c_{u)max}(t/m^2)$
						26.32	119.29	11.93

Figure 12.7 (Soil specimen-1)

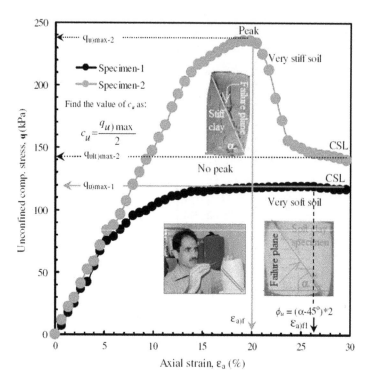

FIGURE 12.7 Stress-strain curves for clayey soils (specimens 1 and 2)

　　　Note: It may be noted that "q_u" is taken as peak compressive stress for the
soil specimen. If 20% strain occurs before the peak stress, then the stress
corresponding to 20% strain should be taken as q_u. For tests continued to
20% strain without reduction of axial load, the unconfined compressive
strength as rule shall be taken as the compressive stress at 15% strain (such
as for soil specimen-1: very soft clayey soil does not exhibit any peak and
axial load remained constant with increasing deformation). For soft clays, q_u
is defined as $\sigma 1$ at a strain level of 15%. For stiff clays, q_u is defined as the
peak of the ($q_u - \varepsilon_a$) curve (e.g. peak point value from stress-strain curve).

9. Draw Mohr's circle for $\sigma 3 = 0$ and $\sigma 1 = q_{u)max}$.

10. Compute the sensitivity of the soil sample for the known values of undisturbed compression strength (q_u) and remolded compression strength
($q_{u)R}$) as follows:

$$S_T = \frac{q_{u)undisturbed}}{q_{u)remoulded}} \qquad (12.4)$$

Similarly, the test data for soil sample 2 is summarized in Table 12.4.

TABLE 12.4
Determination Undrained Shear Strength by "U" Test (Specimen-2)

Stress-Strain Characteristics of "U" TEST

Strain L.C. (mm/dvn) = 0.01

SITE: Flood Channel Srinagar

Sample type: Disturbed/Remolded

PR L.C. (kg/dvn) = 0.17

$1\ cm^3 = 10^{-6}\ m^3$

D_o (mm) = 38

L_o (mm) = 76

BH: I

$A_o\ cm^2 = 11.34$

V_o (cm³) = 86.19

Depth (m): 1.5

Strain Divns	Change in length, δL (mm)	Strain (%)	PR Divns	Load, P (kg)	A_c (cm²)	q_u (kPa)	c_u (kPa)	c_u (t/m²)
0	0	0.00	0	0	11.34	0.00	0.00	0.00
50	0.5	0.66	8	1.36	11.42	11.91	5.96	0.60
100	1	1.32	15	2.55	11.49	22.19	11.09	1.11
150	1.5	1.97	20	3.4	11.57	29.39	14.69	1.47
200	2	2.63	28	4.76	11.65	40.87	20.43	2.04
250	2.5	3.29	36	6.12	11.73	52.19	26.09	2.61
300	3	3.95	41	6.97	11.81	59.03	29.52	2.95
350	3.5	4.61	50	8.5	11.89	71.50	35.75	3.57
400	4	5.26	59	10.03	11.97	83.78	41.89	4.19
450	4.5	5.92	62	10.54	12.05	87.43	43.72	4.37
500	5	6.58	69	11.73	12.14	96.62	48.31	4.83
550	5.5	7.24	77	13.09	12.23	107.07	53.53	5.35
600	6	7.89	87	14.79	12.31	120.11	60.06	6.01
650	6.5	8.55	95	16.15	12.40	130.22	65.11	6.51
700	7	9.21	103	17.51	12.49	140.17	70.09	7.01
750	7.5	9.87	109	18.53	12.58	147.26	73.63	7.36

(Continued)

TABLE 12.4 (Continued)

Stress-Strain Characteristics of "U" TEST

Strain L.C. (mm/dvn) = 0.01

SITE: Flood Channel Srinagar

Sample type: Disturbed/Remolded

PR L.C. (kg/dvn) = 0.17

$1 \text{ cm}^3 = 10^{-6} \text{ m}^3$

$A_o \text{ cm}^2 = 11.34$

$V_o \text{ (cm}^3) = 86.19$

Depth (m): 1.5

D_o (mm) = 38

L_o (mm) = 76

BH: I

Strain Divns	Change in length, δL (mm)	Strain (%)	PR Divns	Load, P (kg)	A_c (cm²)	q_u (kPa)	c_u (kPa)	c_u (t/m²)
800	8	10.53	118	20.06	12.68	158.26	79.13	7.91
850	8.5	11.18	126	21.42	12.77	167.75	83.87	8.39
900	9	11.84	138	23.46	12.86	182.36	91.18	9.12
950	9.5	12.50	147	24.99	12.96	192.80	96.40	9.64
1,000	10	13.16	156	26.52	13.06	203.07	101.54	10.15
1,050	10.5	13.82	163	27.71	13.16	210.58	105.29	10.53
1,100	11	14.47	169	28.73	13.26	216.66	108.33	10.83
1,150	11.5	15.13	172	29.24	13.36	218.81	109.40	10.94
1,200	12	15.79	177	30.09	13.47	223.42	111.71	11.17
1,250	12.5	16.45	181	30.77	13.57	226.69	113.34	11.33
1,300	13	17.11	185	31.45	13.68	229.87	114.94	11.49
1,350	13.5	17.76	189	32.13	13.79	232.98	116.49	11.65
1,400	14	18.42	192	32.64	13.90	234.79	117.39	11.74
1,450	14.5	19.08	194	32.98	14.02	235.32	117.66	11.77
1,500	15	19.74	196	33.32	14.13	235.81	117.91	11.79
1,550	15.5	20.39	196	33.32	14.25	233.88	116.94	11.69
1,600	16	21.05	190	32.3	14.37	224.84	112.42	11.24
1,650	16.5	21.71	181	30.77	14.49	212.41	106.20	10.62

1,700	17	22.37	163	27.71	14.61	189.68	94.84	9.48
1,750	17.5	23.03	149	25.33	14.73	171.92	85.96	8.60
1,800	18	23.68	136	23.12	14.86	155.58	77.79	7.78
1,850	18.5	24.34	134	22.78	14.99	151.97	75.98	7.60
1,900	19	25.00	133	22.61	15.12	149.52	74.76	7.48
1,950	19.5	25.66	132	22.44	15.26	147.10	73.55	7.35
2,000	20	26.32	132	22.44	15.39	145.79	72.90	7.29
2,050	20.5	26.97	132	22.44	15.53	144.49	72.25	7.22
2,100	21	27.63	133	22.61	15.67	144.28	72.14	7.21
2,150	21.5	28.29	133	22.61	15.82	142.96	71.48	7.15
2,200	22	28.95	133	22.61	15.96	141.65	70.83	7.08
2,250	22.5	29.61	133	22.61	16.11	140.34	70.17	7.02
Figure 12.7 **(Soil specimen-2)**						**ε_f(%)**	**$q_{u)max}$(kPa)**	**$c_{u)max}$(t/m^2)**
						19.74	235.81	23.58

TABLE 12.5

Based on the Sensitivity, Clays can be Classified

Sensitivity	Soil Description	Sensitivity	Soil Description
1–2	Slightly sensitive	8–16	Slightly quick
2–4	Medium sensitivity	16–32	Medium quick
4–8	Sensitive	32–64	Very quick
8–16	Very sensitive	>64	Extra quick

Based on the sensitivity, clays can be classified as per Table 12.5.

12.7 GENERAL COMMENTS

The undrained shear strength obtained from the ":UCS" test is a representative value; therefore, it is recommended to determine the average value of "qu" by conducting two to three tests at each site. It may be noted that the shear strength of remolded clayey is always smaller or lower side as compared to the shear strength of undisturbed clayey soil at the same water content, which can be quantified by its sensitivity.

Note: It should not be inferred that the strength of a soil after remolding is necessarily less than that of the undisturbed soil. A soil which is remolded and then consolidated may have greater strength. At a pressure greater than the maximum pre-compression pressure or preconsolidation pressure, a remolded specimen would exist at a lower void ratio than it would if undisturbed. In other words, remolding prior to consolidation to a large pressure results in a more dense soil. The strength increases from this increased denseness may offset the structural strength of the soil, with the result that the remolded soil has a greater strength than the undisturbed soil.

The unconfined test may give misleading results on heterogeneous soils because the lack of lateral support may be too severe a boundary condition. A check in the triaxial machine is desirable for such soils.

The ratio of compressive to shear strength equal to two (e.g. $s_u = q_u/2$ or $q_u/s_u = 2$) is based on a combination of theoretical and empirical considerations, as explained in Figure 12.7.

Mohr;s circle is shown in Figure 12.8 for an unconfined compression test. Since no pore water pressure is measured in the "UCS" test; therefore, strength parameters are expressed/obtained in terms of total stress approach. From Figure 12.8, the shear strength AB is equal to:

$$AB = AC * Cos\phi$$

Or

$$\tau = AC * Cos\phi = \frac{\sigma_{1)max}}{2} = \frac{q_u}{2} \qquad (12.5)$$

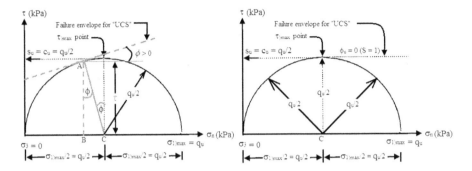

FIGURE 12.8 Mohr's circle for unconfined compression test.

Where: ϕ = Frictional angle.

The frictional angle, ϕ, is normally about $30°$; therefore, from Equation (12.4), we have:

$$s = \tau_f = 0.43 * q_u \approx 0.5 * q_u = \frac{q_u}{2}$$

The results mentioned above indicate that there is something inherent in the unconfined test which gives a value of compressive strength lower than that given by other types of tests. The reason for the lower strength is probably due to the severe condition of no lateral support to the specimen. In view of the foregoing, the value of "s_u" is commonly taken 0.5 times the compressive strength as measured by the unconfined compression test, e.g.:

$$s_u = c_u = \tau_f = \frac{q_u}{2}$$

12.8   APPLICATIONS/ROLE OF "UCS" IN SOIL ENGINEERING

In soil engineering practice, the "UCS" test is generally adopted to ascertain the short-term stability of various structures built on saturated cohesive soils under dynamic loading where no volume change takes place due to expulsion of water. Since the clays have very low permeability, therefore, the undrained cohesion is evaluated quickly by the "UCS" test in undrained conditions.

12.9   SOURCES OF ERROR

The main sources of error in the "UCS" test are:

1. If the proper type of soil is not tested. It may be noted that the "UCS" test is suitable for undisturbed NC clays or slightly OC undisturbed clays, which can be tested without lateral confinement.

2. If the test specimen is not representative of soil sample. It may be noted that only representative soil sample may be tested. It may be ensured that re-presentative soil sample is taken from same undisturbed soil mass and it has not been disturbed in the field. Thus, the sampling process may affect the shear strength and any disturbance while collecting undisturbed soil sample may reduce the shear strength, as shown in Figure 12.9.

3. The effect of end restraints may affect the undrained shear strength. Therefore, it may be ensured that the test soil specimen should not be too short to avoid effect of end restraints. Also, the test soil specimen should not be too long to avoid its buckling. Therefore, to avoid effect of end restraints, the L/D ratio should be maintained equal to 2 or greater than 2 (Figure 12.10).

4. Loss of initial water content has direct bearing on the strength of clayey soils. Therefore, it may be ensured that there is no moisture loss during transporting of undisturbed soil sample from the field to the lab or during trimming the soil sample in the laboratory.

5. Rate of loading too fast has a direct bearing on the strength of clayey soils. Fast strain rate is applied when dissipation of pore water pressure is not allowed under rapid loading. However, if the strain rate is too fast during the test, it may overestimate the shear strength.

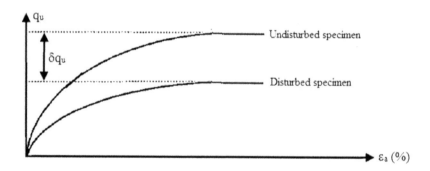

FIGURE 12.9   Effects of sampling disturbance of stress-strain behavior.

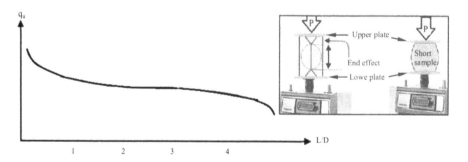

FIGURE 12.10   Effects of end restraints on unconfined comp. strength of soil.

6. If the soil specimen is disturbed while trimming, it will not result in desired dimensions of the specimen and hence, the test parameters will not be correct.

12.10 PRECAUTIONS

1. The "UCS" tube sampler or Shelby tube sampler may be applied with a thin layer of grease before collection of in situ soil samples to avoid affect of side friction between sampler and the soil sample.
2. The in situ soil samples collected in the field should be properly sealed in airtight bags to avoid moisture loss.
3. The in situ soil samples should be transported carefully to avoid any disturbance to soil samples.
4. The compression testing machine, dial gauge, and proving ring should be calibrated carefully before start of test so that actual loads are applied to the soil sample during the test.
5. Soil samples may be prepared in the laboratory carefully to prevent moisture loss and disturbance to their natural conditions.
6. Remolded soil specimens should be prepared in the laboratory at desired MDD and OMC by static compression in a specified mold. The extra soil if any above or below the mold may be removed carefully.
7. Make sure that the soil specimen is extruded out from the in the same direction it was collected in the field so that there is less resistance encountered by the soil specimen while extruding it out.
8. The loading of the soil specimen must be applied at a constant strain rate. With $L/D = 2$, the following speeds are usually appropriate for a motorized unit.

Specimen diameter (mm)	
Approximate platen speed (mm/min)	
38	1.5
50	2
75	3
100	4

At the end of the test, determine the final water content of the soil sample.

REFERENCES

ASTM D2166/D2166M. 2013. "Standard Test Method for Unconfined Compressive Strength of Cohesive Soil." ASTM International, West Conshohocken, PA, www.astm.org.
ASTM D1587. (2013). "Practice for Thin-Walled Tube Sampling of Soils for Geotechnical Purposes." ASTM International, West Conshohocken, PA, www.astm.org.
BS 1377 (Part 7). 1990-clause 8. "Shear Strength Tests (Total Stress): The Unconsolidated

Undrained Triaxial Compression Test, Without Pore Water Pressure Measurement." British Standards, UK.

IS: 2720 (Part 1). 1980. "Indian Standard Code for Preparation of Soil Samples." Bureau of Indian Standards, New Delhi.

IS: 2720 (Part 10). 1973. "Method of Test for Soils: Determination of Shear Strength Parameter by Unconfined Compression Test." Bureau of Indian Standards, New Delhi.

Reddy, Krishna R. 2000. "Engineering Properties of Soils Based on Laboratory Testing." Department of Civil and Materials Engineering University of Illinois at Chicago, USA.

Skempton, A.W., and V. A. Sowa. 1963. "The Behaviour of Saturated Clays During Sampling and Testing." *Geotechnique* 14(4): 269–290.

Skempton, A. W., and P. La Rochelle. 1965. "The Bradwell Silp: A Short-Term Failure in London Clay." *Geotechnique* 15(3): 221–242.

13 Vane Shear Test for Cohesive Soils

References: IS: 2720 (Part 1); 2720 (Part 30); IS 4434 (1978); BS 1377-Part 7 ASTM D 4648; ASTM D 2166

13.1 OBJECTIVE

The main objectives of conducting the vane Shear Test for finding undrained strength are:

1. Vane Shear Test is conducted for characterization of saturated clays of soft to medium consistency, highly sensitive or very soft clays, which are highly susceptible to sampling disturbance.
2. Vane Shear Test is suitable for very soft soils when their undisturbed samples cannot be collected in the field or cannot be prepared in the laboratory due to sampling disturbance.
3. Vane Shear device is portable and the test can also be conducted in the field if so desired.
4. The main objective of "VST" is that it can be adopted only for highly sensitive and normally consolidated clays to determine undrained shear strength.

13.2 DEFINITIONS AND THEORY

Shear strength of sensitive or soft clay deposits is difficult to obtain accurately in the laboratory by conventional "UCS" or triaxial tests as getting undisturbed samples is very difficult because of sampling disturbance. Therefore, the vane shear test is an alternative test in which undrained shear strength of too sensitive or soft clays can be determined. This test is suitable for characterization of saturated clays of soft to medium consistency, highly sensitive or very soft clays without the sample being disturbed by sample preparation or soils which are fissured or highly susceptible to sampling disturbance. The schematic diagram of the VST apparatus is shown in Figure 13.1a. The laboratory vane shear is 10 mm in diameter, 10 mm in height, and 1 mm thick, while the field vanes have diameters ranging from 50 mm to 150 mm. In the laboratory vane shear test, a properly trimmed and undisturbed soft clayey soil sample is placed in a cup and the shear vane is inserted into the specimen up to the desired depth and rotated in the sample by applying torque. It may be noted that the torque is gradually applied to the upper end of the torque rod until the soil fails in shear due to the rotation of the vanes. It is assumed that the undrained shear strength (c_u) is constant

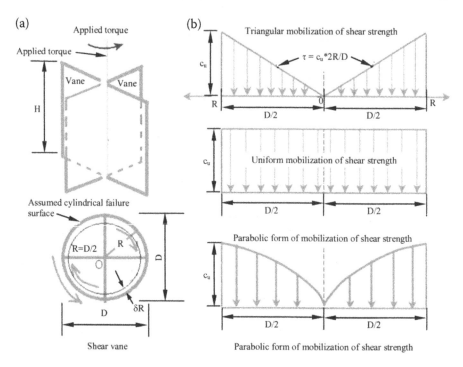

FIGURE 13.1 (a). Vane shear apparatus and calculation of torque due to shear stress on cylinder ends; (b). Variations of shear strength mobilization.

throughout the sheared soil sample. The applied torque is measured by a torsion spring of specified stiffness by recording the angle of twist (θ). When soil is stressed to its shear strength, the vanes will rotate (@ 0.1°/sec or 1° per minute or with a rate of 1 revolution per second) in the soil. The resistance to applied torque in the soil sample is mobilized throughout the vertical and horizontal faces of the soil sample of diameter "D" and the diameter of vane. Since the soil fails along a cylindrical surface, the shearing resistance can be calculated from the vane dimensions and the applied torque. It may be noted that the undrained strength varies as zero at the center and maximum value at the outer surface (e.g. $R = D/2$), as shown in Figure 13.1. Also, there could various types of variation of mobilization of shear strength from the center of the torque to the outer end as shown in Figure 13.1(b). Assuming that distribution of shear resistance is linearly increasing with increasing radius of the soil sample, then the shear stress can be expressed as:

$$\tau = \left[c_u \frac{2R}{D} \right] \binom{R \to 0 \ at \ centre}{R \to D/2 \ at \ outer \ surface} \tag{13.1}$$

The total shearing resistance is computed by taking moments about the axis of the torque shaft, and with reference to Figure 13.1a, the torque T is given by:

$$\left[(\pi DH)*c_u*(D/2) + 2\int_0^{D/2} c_u*R*(2\pi R\delta R) \right]$$
$$T_{\max} = (Cylinderical\ face + Two\ ends - top\ \&\ bottom) \qquad (13.2)$$

Where: R = Elemental ring radius $+D/2$

δR = Thickness of elemental ring, and

τ = Shear stress contribution to torque about "O"

= $(2\pi R * \delta R) * c_u * R$.

Thus, solving Equation (13.2), the torque (T) is related to the undrained strength (c_u) on soft clays and the resistance to rotation consists of as under:

Case 1: When the top of the vanes is below the soil surface as shown in Figure 13.2a:

1. Resisting moment on the cylindrical surface: $\left[c_u * \pi DH * \dfrac{D}{2} \right] = M_s$
2. Resisting moment at the top end of the shear vane:

$$\left[c_u * \left(\tfrac{\pi}{4}D^2 \right) * \frac{2}{3}\frac{D}{2} \right] = M_{eT}$$

3. Resisting moment at the bottom end of the shear vane:

$$\left[c_u * \left(\tfrac{\pi}{4}D^2 \right) * \frac{2}{3}\frac{D}{2} \right] = M_{eB}$$

Where: πDH = Surface area of soil specimen and $D/2$ is moment arm $\pi DH * c_u$ = Shearing force at the cylindrical surface $\pi DH * c_u * D/2$ = Resisting moment of the shear force along the cylindrical face of the soil cylinder, and Factor = $2/3 = \alpha$ = is taken for uniform stress distribution and so on.

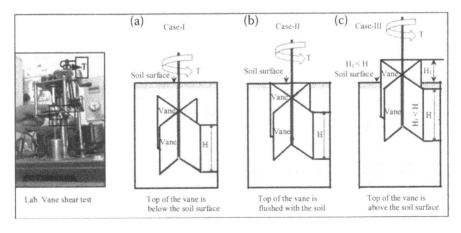

FIGURE 13.2 Position of vanes for determination of torque at failure in the soil specimen.

Therefore, total torque at failure is calculated as under (Figure 13.1):

$$T_{max} = \left[c_u * \left(\frac{\pi}{4}D^2 \right) * \frac{2}{3}\frac{D}{2} \right] + \left[c_u * \pi DH * \frac{D}{2} \right] + \left[c_u * \left(\frac{\pi}{4}D^2 \right) * \frac{2}{3}\frac{D}{2} \right]$$

$$= \pi D^2 c_u \left[\frac{D}{6} + \frac{H}{2} \right] \tag{13.3}$$

Equation (13.3) is valid only:

- For uniform stress distribution shear resistance throughout soil specimen (e.g. cylindrical face and top and bottom ends)
- The top surface or end of the vane is below the soil surface (Figure 13.2a).
- Undrained shear strength (c_u) is constant on the cylindrical sheared surface and both ends of the sheared soil sample.

Thus, the torque, T, at failure can be expressed as:

$$T = \pi c_u \left[\alpha \frac{D^3}{4} + \frac{D^2 H}{2} \right] \tag{13.4}$$

Or

$$c_u = \frac{T}{\pi \left[\alpha \frac{D^3}{4} + \frac{D^2 H}{2} \right]} \tag{13.5}$$

Where: $\alpha = 1/2$ for triangular mobilization of undrained shear strength
$\alpha = 2/3$ for uniform mobilization of undrained shear strength, and
$\alpha = 3/5$ for parabolic mobilization of undrained shear strength.

Equation (13.4) is referred to as *Calding's equation,* for the calculation of shear resistance at the top and bottom ends of the shear vane. However, several types of distribution of shear strength mobilization at the ends of the soil cylinder as follows (Figure 13.1b):

1. *Triangular distribution:* In this case, the shear strength mobilization is maximum (c_u) at cylindrical face of the soil sample (e.g. R = D/2) and decreases linearly to zero at the centeer (e.g. R = 0).
2. *Uniform distribution:* In this case, the shear strength mobilization is constant (e.g. c_u) from cylindrical face of the soil sample to the center of the soil sample.
3. *Parabolic distribution:* In this case, the shear strength mobilization is maximum (c_u) at cylindrical face of the soil sample (e.g. R = D/2) and decreases parabolically to zero at the center (e.g. R = 0).

Case 2: When the top of the vanes is flush with the soil surface as shown in Figure 13.2b:

Equation (13.3) can be modified if the top of the vane is flush with the soil surface. Then, the torque at failure is calculated as under:

$$T_{max} = \pi D^2 c_u \left[\frac{D}{12} + \frac{H}{2} \right] \qquad (13.6)$$

In this case, the top of the vane is flush with the soil surface in the cup and only the bottom end shears.

Case 3: When the top of the vanes is above the soil surface as shown in Figure 13.2c:

Equation (13.6) can be further modified if the depth of the vane inside the soil specimen in the cup is H_1 ($H_1 < H$), then the torque at failure is calculated as under:

$$T_{max} = \pi D^2 c_u \left[\frac{D}{12} + \frac{H_1}{2} \right] \qquad (13.7)$$

Hence, for known values of T_{max}, D, H, H_1, c_u can be computed.

However, it may be noted that the maximum torque (T_{max}) applied to cause shear failure in the soil sample should be equal to the total resisting moments due to shear force along the cylindrical face of the soil sample (M_s) and the two ends ($M_{eT} + M_{eB}$), as shown in Figure 13.3a. Figure 13.3b shows the assumed shear stress distribution. Thus, the maximum torque is:

$$T_{max} = [M_s + (M_{eT} + M_{eB})] \qquad (13.8)$$

Sometimes, VST is conducted to find undrained cohesion separately for top, bottom, and sides of the soil cylinder. In this case, undrained cohesion of the clay in the vertical direction (c_{uv}) on the sides of the cylindrical specimen and in the horizontal direction (c_{uh}) along the ends of cylindrical specimen is calculated separately as below (Figure 13.4):

$$T_{max} = c_{uv} \frac{\pi D^2}{2} H + c_{uh} \frac{\pi D^3}{6} \qquad (13.9)$$

Since the applied torque is measured by a torsion spring of specified stiffness by recording the angle of twist (θ), therefore, it can be also be determined from the spring constant and the angle of twist by the following correlation:

(a)   (b)

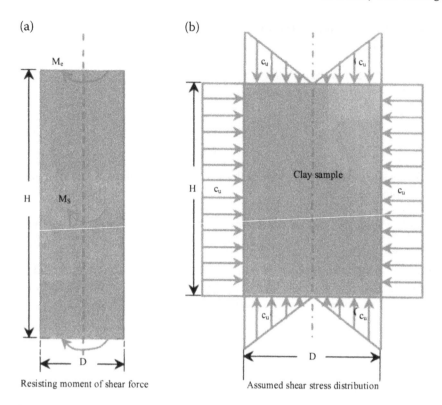

Resisting moment of shear force   Assumed shear stress distribution

FIGURE 13.3   Resisting moment of shear force and assumed shear stress distribution in a VST.

$$T_{max} = \left[\frac{(\theta_{initial} - \theta_{final}) * \pi * k}{180} \right] = \left[\frac{\Delta\theta}{180} * k \right] \qquad (13.10)$$

Where: $\theta_{initial}$ = Initial angle of twist (pointer)
θ_{final} = Final angle of twist (pointer), and
k = Spring constant.

When the maximum torque (T_{max}) is obtained, the test is further continued by rotating the vanes to measure the residual torque (T_r) so that the soil sensitivity is also measured as:

$$S_T = \left[\frac{T_{max}}{T_r} \right]$$

The soil sensitivity quantifies the loss of undrained cohesion due to sampling disturbance. Sensitivity of the soil can be determined if the test is repeated after turning the sample several times and allowing the soil to remold. The test is repeated on the remolded soil specimen and the shear strength in the remolded state is determined. Then, the sensitivity is defined as:

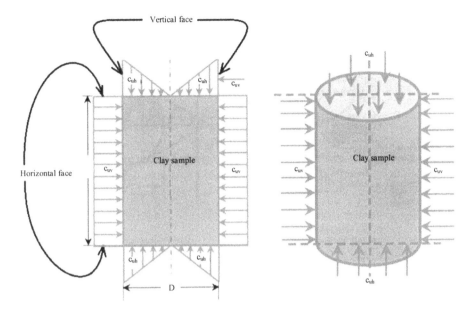

FIGURE 13.4 Assumed shear stress distribution along horizontal and vertical faces of a soil specimen in a VST.

$$S_T = \left(\frac{c_u)_{undisturbed\ soil\ specimen}}{c_u)_{disturbed\ soil\ specimen}} \right) \qquad (13.11)$$

Based on the value of sensitivity, natural soft clay deposits are grouped into four categories, as given in Table 13.1.

Quick clays are highly flocculent, which acquire liquid consistency on re-molding. Sensitivity is related to liquidity index (LI), because loss of strength is associated with greatest flocculation when the water content of soil is close to liquid limit (LL). Soils deposited in a marine environment tend to be highly sensitive. Some quick clays can have sensitivity even greater than 100. Over-consolidated clays are found to be insensitive.

TABLE 13.1
Clay classification based on sensitivity

Sensitivity, S_T	Clay Classification	Sensitivity, S_T	Clay Classification
< 1.00	In-sensitive clay	4.00–8.00	Sensitive clay
1.00–2.00	Little sensitive clay	8.00–16.00	Extra sensitive clay
2.00–4.00	Moderately sensitive clay	>16.00	Quick clay

13.3 METHOD OF TESTING

The undrained cohesion of clays either on in situ (undisturbed specimen) or re-molded (disturbed) specimen is determined in a laboratory using a vane shear apparatus, either hand operated or machine driven.

13.3.1 PRE-REQUISITE FOR VST

Since the VST is suitable for highly sensitive and very soft clays without a sample being disturbed, the "UCS" test should be conducted either on an undisturbed (in situ) saturated clayey soil specimen or remolded or reconstituted soil specimen as per requirement.

13.3.2 SOIL TESTING MATERIAL

The undisturbed soil specimen at natural moisture content is collected directly from the field in a Shelby tube and extracted and trimmed to the desired specimen size in the laboratory for testing. The soil specimen is assumed to be virtually saturated (in in situ condition), normally consolidated inorganic clay, of a consistency ranging from soft to medium stiff. The procedure also applies to remolded and reconstituted inorganic clays prepared in the laboratory. For a remolded soil specimen, a soil paste of desired consistency (soft to medium soft) is carefully mixed and filled into the vane shear cup so as to avoid any air bubbles entrapped in the cup.

For reconstituted soil specimen, soil slurry in adequate quantity should be prepared at a water content of about $1.5 * LL$ to $2 * LL$ (LL-liquid limit of soil mass), thoroughly mixed and de-aired before consolidation to desired consistency (in situ state of soil) on the laboratory floor. For soil specimen preparation, refer to Section 9.4 (Test No. 9).

13.4 TESTING EQUIPMENT AND MATERIAL

The following equipment and accessories are required for "VST" (Figure 13.5):

1. Fine-grained soil passing 0.425 mm IS sieve
2. Undisturbed soil specimen from field
3. Vane shear testing machine including set of calibrated springs (usually 4) of different stiffness, to allow for a range of soil strengths
4. 38 diameter sampling tube close to the edge
5. Weighing balance accuracy 0.01 g
6. Spatulas, trimming knives, steel rule
7. Drying oven and moisture cans for obtaining water content
8. Soil lathe and accessories (cutting tools and a straightedge) for undisturbed field specimen trimming
9. A tray for mixing soil
10. Stopwatch
11. Distilled water
12. Squeeze bottle

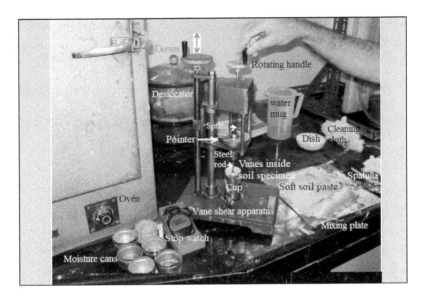

FIGURE 13.5 Vane shear apparatus and other required accessories for lab. VST.

13.5 TESTING PROGRAM

1. Prepare the soil specimen as per requirements, e.g. either undisturbed or remolded soil specimen for vane shear test.
2. Check and prepare laboratory vane shear apparatus for testing (in this test power operated vane shear device is used).
3. In the case of undisturbed soil specimen collected from field in sampler tube, clamp the sample tube securely and keep its axis vertical and with the end to be rested uppermost.
4. Remove any end cap, wax seal, or packing material and trim the sample above the tube so that its upper end is flat and perpendicular to the tube axis.
5. Clamp the tube in position, select the spring (note k-value), and adjust the pointer carefully.
6. Measure the initial angle of twist (initial reading of pointer before rotating) and insert the vane into the soil specimen as per desired Case 1, Case 2, or Case 3, as explained in Section 13.2.
7. Check the initial reading of the twist or pointer ($\theta_{initial}$) and rotate the vanes as per required rate of rotation until the soil specimen fails.
   Note: It may be noted that during the VST, it is assumed that applied torque is effective only up to the cylindrical surface and the two ends of the soil sample. However, this is not true in practice and some more area outside this boundary of the soil sample will be affected by the applied torque due to rotation of vanes. Therefore, the above assumption may result in erratic results as the applied torque is not fully resisted by the soil sample and some part of it is also used to mobilize some area outside the cylindrical surface of the soil sample under test.

8. Record the final angle of twist (θ_{final}) and compute the undrained shear strength (c_u) using Equation (13.10).
9. Remold the soil specimen and repeat steps 5 to 8 to obtain a remolded shear strength (c_{uR}).
10. Remove the vanes from the soil specimen and clean the VST apparatus for the next test trial.
11. Measure the water content and density as per codal procedure (Refer to Lab. Test 1 & 2).
12. Repeat the above steps on a fresh sample at least once.

13.6   OBSERVATION DATA SHEET AND ANALYSIS

Test observations, data analysis, and calculations for determination of undrained shear strength are given in Table 13.2.

TABLE 13.2
Determination of undrained shear strength by VST

Initial length of the soil specimen in the cup, L_o = 76 mm, Initial Diameter of soil sample, D_o = 38 mm, Initial xec. sec. area of soil sample, A_o = 11.342 cm^2, Initial volume of soil specimen, V_o = 86.2 cm^3, 1 cm^3 = 10^{-6}m^3 Sp. gravity, G = 2.63, Unit weight of water, γ_w = 10 kN/m^3, ρ_w = 1 g/cm^3, Spring constant, k = (N/cm)

Sl. No.	Observations and Calculations	Position No.		
		1	2	3
1	Height of vane, H (cm)			
2	Diameter of vane, D (cm)			
3	Wet weight of soil specimen, W (g)			
4	Spring constant, k (N/cm)			
5	Initial angle, $\theta_{initial}$			
6	Final angle, θ_{final}			
7	Torque, T (kN-m) by Equation (13.9)			
8	Undrained shear strength, cu (kN/m^2) by Equation (13.4)			
9	Water content, w (%)			
10	Bulk unit weight, γ_b = W/V$_o$ (kN/m^3)			
11	Dry unit weight, γ_d = γ_b/[1 + (w/100)] (kN/m^3)			
12	Undrained shear strength, c_{uR} (kN/m^2)			
13	Sensitivity, S_T = c_u/c_{uR}			
14	Soil classification			

13.7   GENERAL COMMENTS

- The VST is suitable for normally consolidated clayey soils and highly sensitive clays. The VST can be conducted either in the field or in the laboratory.

The thickness of the vanes has direct bearing on undrained cohesion of clays. Higher vane thickness may reduce undrained cohesion due sample remolding. The VST is not suitable for very stiff clays. Many researchers have reported that VST results (e.g. c_u) are not reliable for foundation design in case of highly plastic soils. Therefore, Bjerrum (1972) suggested a correction based on the degree of plasticity of soils as given below:

$$c_{u)design} = [\beta * c_{u)VST}] \tag{13.12}$$

Where: β = Correction factor = $1.7 - 0.54 \log (PI)$
PI = Plasticity index.

13.7.1 OTHER ALLIED METHODS FOR DETERMINING UNDRAINED SHEAR STRENGTH

Similar to the vane shear apparatus, there are other allied laboratory devices like the Tor-vane shear device and cone penetrometers (Figure 13.6), which can be used to determine the undrained shear strength of clays.

- **Tor-Vane Shear Test**

A Tor-vane shear test device is a hand-operated device fitted with rigid fins and a calibrated spring, as shown in Figure 13.6(a). This device can be used either in the field or in the laboratory for determination of undrained cohesion of clays. In this method, the rigid fins are inserted into the soil sample and rotated/twisted under gradually applied torque until the soil sample fails. The shearing resistance to applied torque near the surface of soil sample can be recorded through a calibrated proving ring.

(a)   (b)

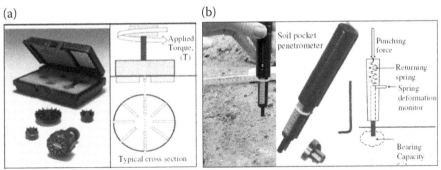

Tor-vane shear device   Pocket penetrometer device

FIGURE 13.6 Allied laboratory devices for determination of the undrained shear strength of clays: (a). Tor-vane shear device, and (b). Pocket penetrometer test device.

• **Pocket Penetrometer**

A Pocket Penetrometer is a hand-operated punching probe (Figure 13.6b) that is used to determine the undrained cohesion of clays. In this method, the probe is pushed into the soil mass until soil fails and the "UCS" (q_u) is recorded by a calibrated spring as penetration force. This failure punching force gives an indicative value of *bearing capacity of the soils*. Unlike the Tor-vane shear device, this punching probe can be used both in the field and in the laboratory.

13.8   APPLICATIONS/ROLE OF "VST" IN SOIL ENGINEERING

The main application of "VST" is the determination of undrained strength for short-term analysis and design of various structures on clays as this method is very quick and cost effective.

The vane shear test is suitable for saturated clays of soft to medium consistency, fissured or highly plastic clays susceptible to sampling disturbance.

13.9   SOURCES OF ERROR

The main sources of error in the "VST" are:

• The thickness of the vanes has direct bearing on undrained cohesion of clays. Higher vane thickness may reduce undrained cohesion due sample remolding.
• The VST is not suitable for very stiff clays.
• It may be noted that during the VST, it is assumed that applied torque is effective only up to the cylindrical surface and the two ends of the soil sample. However, this is not true in practice and some more area outside this boundary of the soil sample will be affected by the applied torque due to rotation of vanes. Therefore, the above assumption may result in erratic results as the applied torque is not fully resisted by the soil sample and some part of it is also used to mobilize some area outside the cylindrical surface of the soil sample under the test.
• Shear strength of clays is not constant throughout the cylindrical surface and on top and bottom ends. This may result in highly erratic results.

13.10   PRECAUTIONS

1. "VST" is only suitable for highly sensitive and very soft clays susceptible to sampling disturbance.
2. Use a well-calibrated and desired stiffness spring.
3. The soil specimen should always be pushed in the sampling tube or the mold along the same direction in which it enters the main tube in the field.
4. Two ends of the soil specimen should be perpendicular to the long axis of the specimen.
5. The loading of the soil specimen should be at a constant rate.

REFERENCES

ASTM D 4648. "(2016). Test Method for Laboratory Miniature Vane Shear Test for Saturated Fine-Grained Clayey Soil." ASTM International, West Conshohocken, PA, www.astm.org.

ASTM D2166/D2166M. 2013. "Standard Test Method for Unconfined Compressive Strength of Cohesive Soil." ASTM International, West Conshohocken, PA, www.astm.org.

Bjerrum, L., and N. E. Simons. 1960. "Comparison of Shear Strength Characteristics of Normally Consolidated Clays." *Proceedings of Research Conference on Shear Strength of Cohesive Soils*, ASCS, 711–726.

Bjerrum, L. 1972. "Embankments on Soft Ground."ASCE Conference on Performance of Earth and Earth-Supported Structures. Purdue University, 2, 1–54.

BS 1377 (Part 7). 1990-clause 8. "Shear Strength Tests (Total Stress): The Unconsolidated Undrained Triaxial Compression Test, Without Pore Water Pressure Measurement." British Standards, UK.

IS: 2720 (Part 1). 1980. "Indian Standard Code for Preparation of Soil Samples." Bureau of Indian Standards, New Delhi.

IS: 2720 (Part 10). 1973. "Method of Test for Soils: Determination of Shear Strength Parameter by Unconfined Compression Test." Bureau of Indian Standards, New Delhi.

IS: 2720 (Part 10). 1973. "Method of Test for Soils: Laboratory Vane Shear Test." Bureau of Indian Standards, New Delhi.

IS 4434. 1978. "Method of Test for Soils: Code of Practice for In-Situ Vane Shear Test for Soils." Bureau of Indian Standards, New Delhi.

Skempton, A. W. 1957. "Discussion on 'The Planning and Design of New Hong Kong Airport.'" *Proceedings, Institution of Civil Engineers* 7(3): 305–307.

Skempton, A.W., and V. A. Sowa. 1963. "The Behaviour of Saturated Clays During Sampling and Testing." *Geotechnique* 14(4): 269–290.

Skempton, A. W., and P. La Rochelle. 1965. "The Bradwell Silp: A Short-Term Failure in London Clay." *Geotechnique* 15(3): 221–242.

14 Direct Shear Test (DST) for Soils

References: IS: 2720 (Part 1); 2720 (Part 13); ASTM D3080; ASTM D 5321; ASTM D6528; BS 1377-Part 7

14.1 OBJECTIVES

The main objectives of conducting the DST are:

- The DST is a simple and quick test compared to the triaxial test.
- In this test, quick drainage of pore water in the soil sample is usually easy because of thin depth of the soil sample (e.g. 20 mm only)
- DST is generally suitable for coarse grained soils. However, fine grained soils can also be tested.

14.2 DEFINITIONS AND THEORY

The direct shear test (DST) is a method of soil testing to determine the shear strength parameters (c&ϕ) of soils under drained, undrained, or consolidated-undrained conditions on predetermined failure surfaces (Reddy 2000). The strength of a soil system is the ultimate internal resistance of the soil system to balance the applied loads to avoid the failure within the soil system. According to Mohr (1900), the failure within the soil system is not caused by the extreme stress of one type e.g. normal stress or shear stress, but by a critical combination of both. The failure occurs when for the attained normal stress, the associated shear stress penetrates the failure envelope (Figure 14.1a) and the law of shear strength for mixed soils or "c-ϕ" soils is expressed as:

$$\tau_f = [c' + (\sigma_n - u)\tan\phi'] = [c' + \sigma'_n \tan\phi'] \qquad (14.1)$$

Where: c' = Effective cohesion
  σ_n = Total normal stress
  u = Pore water pressure
  σ'_n = Effective normal stress
  ϕ' = Effective friction angle, and
  τ_f = Shear stress at failure (but not maximum shear stress) associated with attained normal stress, which introduce failure in the soil system.

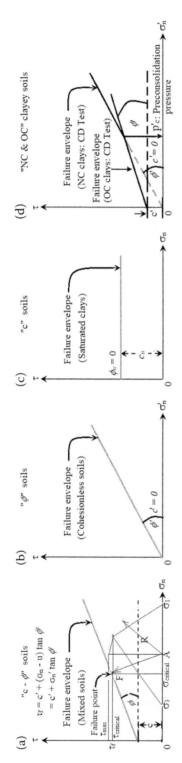

FIGURE 14.1 Failure envelopes and law of shear strength for granular and cohesive soils.

The shear strength characteristics are summarized as follows:

1. The shear strength of granular (e.g. cohesionless) soils is practically or effectively independent of time because of their free draining characteristics (e.g. highly pervious soils). The failure envelope for cohesionless soils, also known as "ϕ" soils, is shown in Figure 14.1b and the law of shear strength is expressed as:

$$\tau_f = [\sigma'_n \tan \phi'] \tag{14.2}$$

The shear strength of granular soils (such as sands) is generally taken as drained shear strength based on effective stress approach as there is complete dissipation of pore pressure under applied loading.

2. The shear strength of saturated cohesive soils (generally clays) is taken solely as apparent cohesion defined as undrained strength, c_u ($\phi_u = 0$ concept), under rapid loading because of their low permeability. The failure envelope for saturated cohesive soils, also known as "c" soils, is shown in Figure 14.1c and the law of shear strength is expressed as:

$$t_f = [c_u] = [s_u] \tag{14.3}$$

In case saturated clays subjected to rapid loading (such running trains and vehicles, etc.), there is practically no chance of dissipation of pore water pressure because of their low permeability (k $<$ 10^{-9}m/s). Therefore, shear strength is exclusively measured in terms of undrained cohesion based on total stress approach.

3. Drained shear strength or long-term residual strength of either normally consolidated or over-consolidated clays is solely dependent on time because of their very low permeability or poor draining characteristics that require a slow rate of loading so that there is complete dissipation of pore water pressure.

In the case of normally consolidated clays under the consolidated drained (CD) test (long-duration test due to low permeability), the apparent cohesion ceases and the shear strength of dry cohesive soils or NC clays is ignored (e.g. c'$\approx$ 0). This is due to the fact that during the "CD" test under a very slow rate of shearing, the pore water pressure dissipates very slowly by squeezing the clay particles, which become rough enough that their strength is solely taken in terms of frictional resistance (or angle of shearing resistance) between the clay particles as ϕ' (c' = 0). However, OC clays exhibit both cohesion and frictional characteristics. The failure envelopes for NC and OC clays soil are shown in Figure 14.1d and the law of shear strength is expressed as:

$$\tau_f = [\sigma'_n \tan \phi'] \quad for \ NC \ clays \tag{14.4a}$$

$$\tau_f = [c' + \sigma'_n \tan \phi'] \quad for \ OC \ clays \tag{14.4b}$$

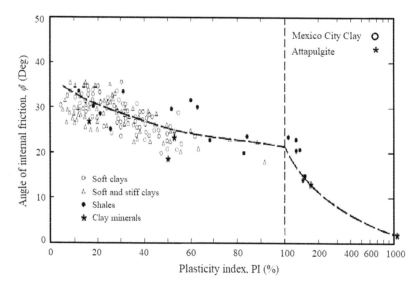

FIGURE 14.2   Variation of angle of shearing resistance (ϕ') with increasing clay plasticity (after Terzaghi et al. 1996).

It may be noted that the index properties (or consistency limits) are indicative of engineering properties. Therefore, empirical relationships have been developed for clays between the angle of shearing resistance (ϕ') and plasticity index (PI), as shown in Figure 14.2. From Figure 14.2, it is observed that the angle of shearing resistance decreases slightly with increasing plasticity characteristics (e.g. PI) of clays. This is understandable because when clay mineral content increases, the plasticity increases and the angle of shearing resistance decreases. Further, it can be seen from Figure 14.2 that drained frictional angle values can be $\pm 10°$ in variance to the dotted line.

Similarly, empirical relationships have been developed for granular soils (e.g. sands and gravels) between the angle of shearing resistance (ϕ'), dry unit weight (γ_d), and as relative density (R_D), shown in Figure 8.2 (Test no. 8). From Figure 8.2, it may be concluded that shear strength is impacted by various influencing factor such as relative density, particle size distribution, shape of particles, void ratio, effective stress state, clay characteristics, etc. Angular-shaped particles have a higher angle of internal friction value compared to sub-rounded or rounded particles. Thus, stress history, shape, and soil grading of coarse grained soils have direct bearing on the shear strength of coarse grained soils.

Keeping in view the vital role of angle of shearing resistance (ϕ') in the stability analysis of slopes, landslides, and foundations, it is highly recommended that the angle of shearing resistance (ϕ') should be determined in the soil testing laboratory, preferably by consolidated-drained (CD) triaxial tests or consolidated-undrained (CU) triaxial tests with pore water pressure measurement (u_w) so as to get drained or effective strength parameters (e.g. ϕ'). However, a triaxial test is a time consuming and expensive test; therefore, the direct shear test (CD) is recommended for short-term analysis or end of construction for various structures.

14.2.1 Drained and Undrained Conditions Soil Tests

In soil engineering practice and from stability analysis point of view, it is highly desired to understand the basic mechanism of drained and undrained shear strength of soils. Since the soil mass is a very complex and heterogeneous triparticulate system, which comprises soil solids, liquid fluid (water), and gases (air) in the void pores of the soil mass. It may be noted that the soil is a universally available material used as construction material or as foundation medium. However, it is the water and air that make it problematic material for construction purposes. When water and air enter into soil void pores, it becomes unsaturated soil, generally known as a three-phase soil system. If it is presumed that only air occupies all the void pores in the soil mass, then it is fully dry and unsaturated soil, known as a two-phase soil system. However, this is a rare case. When only water occupies all the void pores in the soil mass, it becomes fully saturated soil, and referred to as a two-phase soil system. The water present in the soil voids pores (e.g. the space between soil solid particles) is called pore water, and the pressure exerted on water in in situ state of soil mass (but not by externally applied loading) is known as hydrostatic pore water pressure (u_w) and the condition is known as hydrostatic condition in the field or in situ hydrostatic condition (e.g. saturated soil mass under hydrostatic pore water pressure is in equilibrium). When the saturated soil mass is subjected to externally applied loads, it is resisted by water (because water is incompressible), which is generally known as excess pore water pressure (u_e) in addition to hydrostatic pore water pressure already in the saturated soil mass. If the pore water pressure (u_e) is allowed to drain out or dissipate out of soil mass under applied loading, then it is known as a drained condition. This may happen if sufficient time is given to allow pore water pressure to dissipate under applied loading until it vanishes (e.g. $u_e = 0$) and hydrostatic condition is regained. Thus, the drained condition is established when the external load is applied very slowly such that pore water pressure dissipates from the soil completely and the soil mass reaches a state of equilibrium. Hence, such tests conducted in the soil testing laboratory are generally known as drained tests or consolidated drained (CD) tests.

However, if rapid loading (such as running trains or vehicles) is applied on a saturated soil mass such that the excess pore water pressure developed has no chance to dissipate out due to low permeability of soils (e.g. clays), the condition is known as undrained condition. Thus, undrained condition is established when the external load is applied rapidly (e.g. dynamic loads) such that pore water pressure does not dissipate from the soil mass and the soil mass reaches a state of equilibrium with respect to pore water pressure response due to the changes in rapid loading. Hence, such tests conducted in the soil testing laboratory are generally known as undrained tests or unconsolidated undrained (UU) tests or sometimes consolidated undrained (CU) tests if the soil mass is allowed to consolidate before subjecting to rapid loading. Such tests can be simulated in the soil testing laboratory by conducting either direct shear test or triaxial tests.

Since the water content plays a vital role for fine grained soils (particularly clays of low permeability), it is recommended to the understand the concept of drained and undrained conditions of a saturated soil mass for stability analysis of various problems for sustainable development without soil failure.

14.2.2 SHEAR STRENGTH IN UNDRAINED OR DRAINED CONDITION?

The Father of Soil Mechanics "Karl von Terzaghi" has said that "unfortunately, soils are made by nature and not by man, and the products of nature are always complex." Thus, soils are very complex and heterogeneous in nature and do not posses constant shear strength parameters unlike man-made materials such as steel or concrete. Therefore, it is mandatory to test each soil sample every time whenever collected from the desired location in the field. Depending on the type of structure and site conditions, shear strength evaluation in undrained or drained or in both conditions if necessary. However, it should be kept in mind the critical situation to select a drain or undrain condition depending upon the type of structure and loading condition. For example, in case of slope stability problems, drained strength is always less than undrained strength, which is considered as the most critical condition. But in the case of stability of an embankment, undrained strength is always less than the drained strength and hence considered a most critical condition.

Once the type of structure and type of loading is known, soils tests are conducted in the laboratory under drained or undrained conditions as may be required. The Mohr-Columb envelope is plotted for both drained and undrained conditions for determination of shear strength parameters as shown in Figure 14.3.

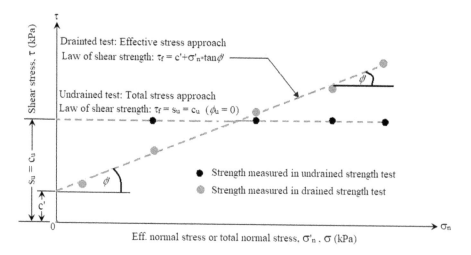

FIGURE 14.3 Typical failure criteria for drained and undrained shear strength for saturated clay.

14.2.3 STRAIN-CONTROLLED TESTS

Soil samples are conducted in the laboratory either by strain-controlled or stress-controlled direct shear test apparatus. Depending on the soil type (e.g. sands, clays or mixed soils), strain-controlled tests are conducted under a constant strain rate or constant shear displacement in the direct shear apparatus in which the lower part of the direct shear box is moved relative to the top half of the box until the soil sample fails. The shear force is measured by a calibrated horizontal proving ring and horizontal displacement is measured by a linear variable displacement transducer (LVTD). The shear stress is computed for each test under a specified normal stress and the Mohr-Coloumb envelope is plotted between shear stress and normal stress, as illustrated in Figure 14.3. However, in the case of a stress-controlled test, the tests are conducted under constant rate of loading or force until soil sample fails. It may be noted that in case of strain-controlled test, peak strength, and ultimate strength can be observed, whereas only peak strength can be observed in a stress-controlled test. Furthermore, a strain-controlled test can be easily conducted compared to a stress-controlled test where it is difficult to maintain a constant rate of loading during the test. However, the stress-controlled test has an advantage over the strain-controlled test as it simulates the field conditions close to the actual in situ state of soil in the field.

14.2.4 PRINCIPLE OF THE DIRECT SHEAR TEST

The basic principle of the direct shear or box shear test is that the soil sample is sheared through a predetermined shear failure plane under specified normal stress (e.g. 50 kPa or 100 kPa or 15 kPa) as per standard codal procedures. The changes in shear stress (τ), horizontal displacement (Δ_H), vertical deformation (δ_V), and the volume change ($\Delta V = \delta_V$*specimen surface area) corresponding to each specified normal stress is measured at a constant normal stress (σ_n). The direct shear apparatus, test accessories, and the principle of the direct shear test is illustrated in Figure 14.4, and the law of shear strength for mixed soils or "c-ϕ" soils is expressed as:

$$\begin{aligned} \tau_f &= [c + \sigma_n \tan \phi'] \rightarrow (Total\ Stress\ Approach),\ and \\ &= [c' + \sigma'_n \tan \phi'] \rightarrow (Effective\ Stress\ Approach) \end{aligned} \tag{14.5}$$

Based on DST results, failure envelope, and the Mohr's circle can be plotted as shown in Figure 14.4(c). From Figure 14.4(c), it is seen that the failure envelope penetrated the Mohr's circle at point "F," which is known as the failure point with its coordinates as normal stress ($\sigma_n = \sigma_y$) and shear stress ($\tau_f = \tau_{xy}$). The coordinates of the failure point can be obtained by drawing the Mohr's circle as illustrated in Figure 14.4(c):

- Draw line "FO" normal to the failure envelope (F_1FF_2) through point "F" or any desired point on the failure envelope.

- The line "OF" represents the radius of the Mohr's circle, which is equal to $[(\sigma_1 - \sigma_3)/2]$ and point "O" is the center of the Mohr's circle on normal stress axis.

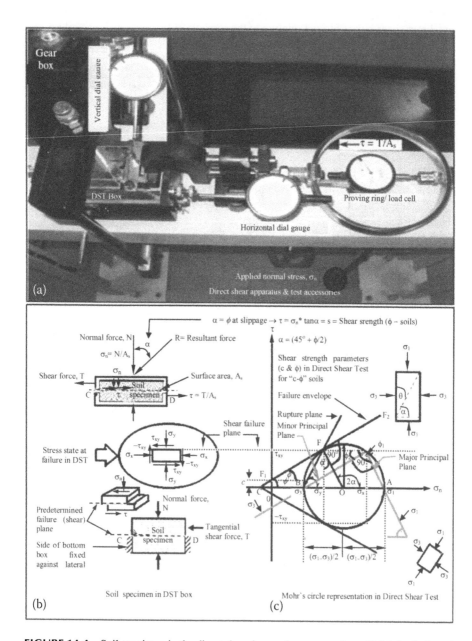

FIGURE 14.4   Soil specimen in the direct shear box and representation of Mohr's circle and stress state at failure.

- σ_1 and σ_3 are major and minor principal stresses, which can be obtained from the geometry of the Mohr's circle.
- In DST, the failure plane is predetermined on the horizontal plane. Therefore, the origin of planes at point "P" (known as pole) is located by drawing a horizontal line through failure point "F," as shown in Figure 14.4(c).
- The point "B" represents the minor principal stress and point "A" represents the major principal stress on the abscissa-normal stress axis.
- Now join the point "P" with points "B" and "A" on the normal stress axis.
- The line "PB" represents the direction of the minor principal stress and the line "PA" represents the direction of the major principal stress.
- The coordinates of failure point "F" can be now obtained from Mohr's circle by drawing the vertical line down to σ-axis (σ_y) and horizontal line to τ-axis (τ_{xy}) from point F.
- The rupture plane can also be drawn from point B through failure point F, which is inclined at angle "α."
- From the geometry of the Mohr's circle, and the ΔFOC as $2\alpha = (90° + \phi)$, which yields $\alpha = (45° + \phi/2)$.

Similarly, for each test under a given normal stress σ, τ versus Δ_H, and ΔV versus Δ_H are plotted in Figure 14.5. The shear stress at failure is taken/measured corresponding to the peak point and the ultimate or residual strength is taken when the shear stress-shear strain curve attains critical state, as shown in Figure 14.5(a).

Soil may contract or dilate during shearing, which mostly depends on its initial density, as seen in Figure 14.65(b). From Figure 14.5(b), it is also seen that OC soils (dense sands and stiff clays) dilate, whereas NC soils (loose sands and soft clays) contract. This is understandable because the soil particles are densely packed, which readjust and realign themselves during shearing, as illustrated in Figure 14.5(c). Similarly, the critical void ratio is also attained at a large horizontal deformation, as shown in Figure 14.5(d).

14.3 METHOD OF TESTING

The shear strength of soils either in situ (undisturbed specimen) or remolded (disturbed) specimen is determined in the laboratory using the direct shear test apparatus by means of a strain-controlled mechanism.

14.3.1 PRE-REQUISITE FOR DIRECT SHEAR TEST (DST)

Since strength is an engineering property and some of the physical and index tests such as soil grading, specific gravity, consistency limit tests, compaction, and relative density tests are a pre-requisite for conducting the direct shear test (DST). The "DST" test is conducted on natural materials and compacted materials (e.g. c-ϕ soils, ϕ-soils, and c-soils). Relative density test is a pre-requisite for sandy soils, whereas the compaction test is pre-requisite for c-ϕ soils and c-soils if the DST test is to be conducted on remolded or disturbed soil specimen. The DST test can also be conducted on undisturbed, saturated clayey soils or reconstituted soil specimen for determination of

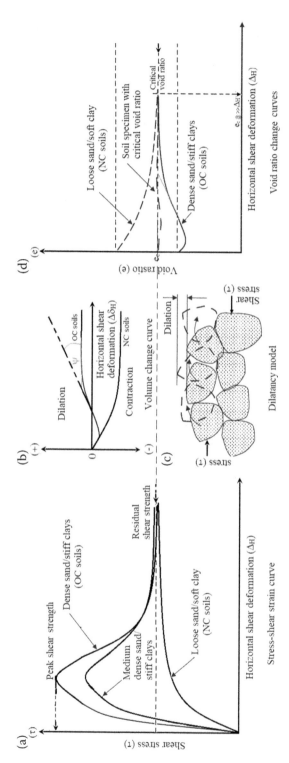

FIGURE 14.5 Direct shear test result: (a). Stress–shear strain curve; (b). Volume change curve; (c). Dilatancy model; and (d). Void ration change curves.

undrained shear strength as per requirement. However, neither undisturbed soil specimen can be collected in the field nor compaction test can be conducted in case of cohesionless (sandy soils). Therefore, relative density test is pre-requisite for preparation of remolded specimen for sandy soils for conducting the direct shear test.

14.3.2 SOIL TESTING MATERIAL

The undisturbed soil specimen of saturated clayey soils at natural moisture content is collected directly from the field in core cutters and extracted and trimmed in the laboratory for testing. However, the soil sample collected should be normally consolidated with soft to medium stiff consistency.

For a remolded clayey soil specimen, the soil specimen passing through the 425 μm IS sieve may be used. For silt-dominated soils, the soil specimen passing through the 2 mm IS sieve may be used. However, for preparation of remolded soil specimen at OMC and 0.9*MDD, standard Proctor compaction test is a prerequisite for all remolded soil samples.

If a reconstituted soil specimen is used, then consistency limit tests (LL & PL) are conducted on the same soil for preparation of soil slurry (it may be noted that soil slurry is prepared at water content of about 1.5*LL to 2*LL and thoroughly mixed and de-aired before consolidation to desired consistency). For soil specimen preparation, refer to Section 10.4 (Test No. 10).

For a remolded sandy soil specimen, a soil specimen passing through a 4.75 mm IS sieve may be used. However, for a remolded sandy soil specimen, a relative density test is conducted for the same soil for determination of maximum and minimum void ratio. For the relative density test on sandy soils, refer to Test No. 10.

14.3.3 SIZE OF SPECIMEN

The maximum size of the soil particle in the "DST" test specimen should not exceed "H/10" mm where "H" is the height of the soil specimen.

14.4 TESTING EQUIPMENT AND ACCESSORIES

For determination of shear strength of soil samples, the following equipment and accessories are required (Figure 14.6):

1. Direct shear testing machine including porous stones, grid plates (two plain and two perforated), base plate, loading cap, dial gauges, proving ring, gear box, funnel, controlled drive unit, weight hanger, lever arm with counter balance and loading yoke (Figure 14.6)
2. Air-dried sandy soils passing 4.75 mm IS sieve
3. Undisturbed soil specimen from field
4. Air-dried fine-grained soil passing 2mm (silty soils) & 0.425mm (clayey soils) IS sieves
5. Core cutters for collection of undisturbed soil samples (in case of clayey soils) from field
6. Weighing balance accuracy 0.01 g

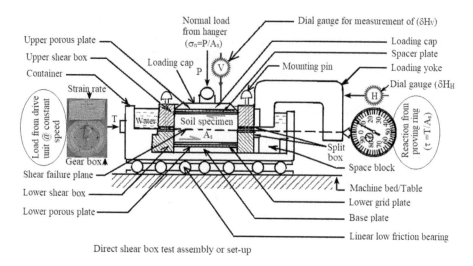

Direct shear box test assembly or set-up

FIGURE 14.6   Direct shear testing machine setup and other accessories.

7. Moisture content cans
8. Squeeze bottle
9. Soil lathe for trimming undisturbed soil specimen to desired size
10. Static compaction machine if remolded soil specimen is to be prepared in the lab
11. Glass Perspex sheet for mixing soil
12. Relative density apparatus and required accessories
13. Sample extruder
14. Oven
15. Desiccator
16. Oil or grease
17. Two fixing screws and two spacing screws
18. Porcelain evaporating dish
19. De-aired water supply
20. Calibrated weight
21. Stopwatch

14.5   TESTING PROGRAM

Based on testing requirements and soil type available, the type of direct shear test can be chosen and the soil specimen can be prepared accordingly.

14.5.1   Sandy Soil or Cohesionless Soil

For sandy soils, generally a consolidated drained direct shear test is chosen.

1. Check and prepare the DST machine along with all necessary accessories (Figure 14.6). Be sure that the DST box is carefully assembled and fixed in

position. Do not forget to saturate the porous stones before the test if damp or saturated sand is used.

2. Clean the DST box and measure its actual *dimensions* carefully. Generally, a standard DST box or mold of size 6 cm*6 cm*2 cm is often used, which gives an initial volume, V_o = 72 cm^3 (1 cm^3 = 10^{-6} m^3).

3. Take adequate quantity of air-dried sand and determine maximum and minimum void ratios (e_{max} and e_{min}) in the laboratory (Refer to Lab. Test No. 10).

   Note: Generally for normal sands, the range for max^m and min^m void ratio is: e_{max} = 0.87 – 0.94, e_{min} = 0.29 – 0.34.

   Further, the sandy soil taken for DST should be air-dried and not oven-dried. If the soil is oven-dried, then it should be allowed to come into moisture equilibrium with the atmosphere. This is because the oven drying may alter the angle of internal friction. Also, the soils in nature would hardly exist in an oven-dried state.

4. Assume the relative density of sand (e.g. the compactness of the sandy soil) for determination of its void ration in natural state:

$$R_D = \left[\frac{e_{max} - e_o}{e_{max} - e_{min}} \right] (\%) \qquad (14.6)$$

   Assuming R_D = 0.6 (R_D = 36–65%: in medium stiff compactness in the field), maximum void ration in loosest state, e_{max} = 0.90, and minimum void ratio in densest state, e_{min} = 0.30, then the void ration in in situ condition (e_o) from Equation (14.6) is 0. 54.

5. Calculate the dry unit weight of sand for e_o = 0.54, G = 2.65, and γ_w = 10 kN/m^3 given by:

$$\gamma_d = \left[\frac{G * \gamma_w}{1 + e_o} \right] \qquad (14.7)$$

6. For computed dry unit weight (γ_d), calculate the weight of dry sand: $W_D = \gamma_d * V_o$ (g).

   Take care that the consistency of units is maintained (1 g = 10^{-5} kN). Thus, the calculated weight of dry sand is to be compacted in the DST box to about 5 mm from the top, make the surface of the soil level by using a leveler, and measure the height or thickness of the soil specimen as shown in Figure 14.7.

   Note: A more accurate measurement of the soil volume can be obtained if this measurement (H_o) is made after the normal load has been applied. However, this procedure is very difficult and only a small difference is involved.

7. Place the upper porous stone and loading cap on top of the soil specimen carefully (use perforated grid plates for CU and CD Tests instead of plain

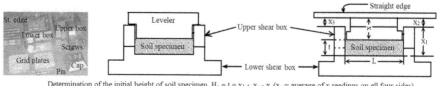

Determination of the initial height of soil specimen, $H_o = t = x_1 + x_a - x$ (x_a = average of x readings on all four sides)

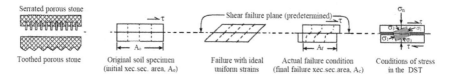

FIGURE 14.7   Preparation of remolded soil sample and conditions of stresses for DST.

grid plates). The conditions of stresses in the DST box are illustrated in Figure 14.7.

8. Attach one dial gauge with the container to record the horizontal shear movement and other on the loading cap on the loading cap to record the vertical movement.

9. Attach the proving ring with the loading yoke for load measurement during shearing. Contact can be observed by a slight movement of dial gauges and the proving ring if correctly attached. Record initial readings.

10. Saturate the soil sample if required so for sufficient time before the start of the test.

11. Mount the loading yoke on the ball placed on the loading cap.

12. Apply the desired normal load (e.g. normal stress, σ_n).
    Note: For the DST, three samples are prepared at the same density and tested under three different normal loads of 50 kPa, 100 kPa, and 150 kPa.

13. Separate the two parts of the DST box by advancing the spacing screws in the upper half of the shear box (e.g. remove the vertical lock screws or mounting pins and raise the upper half of the shear box by turning the spacing screws). The separation between two parts of the DST box should be slightly larger (by eye) than the largest soil particles in the soil specimen.
    Note: For most soils, a spacing of approximately 1 mm between (or the size of maximum soil particle) the two halves of the DST box is satisfactory.

14. Adjust the dial gauges and the proving ring to read zero and record the initial dial gauge (displacement) and proving readings (load) before the start of the test before shearing the soil specimen.

15. Before proceeding with the test, remove the mounting pins and the spacing screws before the start of the test.

16. Set the **desired strain rate** and record the proving ring reading (e.g. applied horizontal load) and dial gauges (for horizontal and vertical displacements) at every 50 divisions for a strain-controlled test.
    Note: A rate of shearing displacement of approximately 1.2 mm/minute is often used. For undrained tests, the rate of strain is 1 to 1.5 mm/minute in

clays and 1.5 to 2.5 mm/minute in sand. For drained tests, the rate of strain is 0.0005 to 0.02 mm/minute in clays and 0.2 to 1 mm/minute in sands. It may be noted that a rapid shear of a saturated soil may throw stresses into the pore water, thereby causing a decrease in the strength of a loose soil or an increase in the strength of a dense soil. *In the stress-controlled test*, increments of shear force can be added either at regular intervals of time or after displacement has ceased under the existing force or applied load. A load increment equal to some percentage e.g. 10% of the estimated shear strength can be used.

17. Continue the test until the specimen fails. Take about 6 to 10 readings even after specimen fails. Or continue the test to a horizontal displacement of approximately 15 to 20% of the length of the soil specimen or unless a constant shear force is obtained.

18. When the soil specimen fails, remove the soil from the DST box and repeat the test on identical soil specimen (e.g. with same dry weight, unit weight, etc.) under increasing normal stress of 50 kPa, 100 kPa, and 150 kPa, respectively.

19. Calculate the horizontal, vertical displacements, void ratio, and shear stress for each test.

20. Determine the final water content of the soil specimens as per codal procedures.

21. The observation of test data and calculations are given in Tables 14.1 and 14.2.

22. The direct shear test response and the sample of typical shear stress-shear strain curves for the DST are shown in Figure 14.8(a).

23. Plot the Mohr-Coulomb failure envelope to obtain shear strength parameters (c *and* ϕ) for desired test conditions (e.g. undrained and drained as the case may be) and evaluate the major and minor principle stress for the chosen test conditions.

14.5.2 COHESIVE SOIL OR CLAYEY SOIL

DST can be conducted either on undisturbed saturated clayey soils or remolded/compacted cohesive soils as per requirement. Since clayey soils or clays exhibit low permeability; therefore, either "UU" or "CU" test is preferred on these soils. Recommended procedure is given as below:

1. Check and prepare the DST machine along with all necessary accessories (Figure 14.8). Be sure that DST box is carefully assembled and fixed in position.

2. Check and record actual dimensions of the DST box. Generally, a standard DST box or mold of size 6 cm*6 cm*2 cm is often used, which gives an initial volume, $V_o = 72$ cm^3 (1 cm$^3 = 10^{-6}$ m^3).

3. For undisturbed clayey soil specimen, collect an in situ soil sample from the field in a core cutter and transport the same to the soil lab carefully without any disturbance.

TABLE 14.1

Determination of water content and dry unit weight of soil specimen

Sl. No.	Observations and Calculations	Trial (Soil core) No.		
		1	**2**	**3**
Before test: Calculation of Water content, w and γ_d (kN/m^3)				
1	DST box no.	**B1**	**B1**	**B1**
2	Initial height, H_o (cm)			
3	Initial length, L_o (cm)			
4	Initial xec. sec. area, $A_o = L_o*L_o$ (cm^2)			
5	Initial volume, $V_o = A_o *H_o$ (cm^3) [1 cm^3= 10^{-6} m^3]			
6	Weight of empty DST box, W_1 (g)			
7	Weight of DST box + wet soil, W_2 (g)			
8	Weight of soil used in the test, $W_s = W_2-W_1$ (g)			
9	Unit weight of soil, $\gamma = (W_s/V_o) *10$ (kN/m^3) [$\gamma = \rho*g$]			
10	Moisture can no.	**C1**	**C2**	**C3**
11	Wt. of empty moisture can, W_{C1} (g)			
12	Wt. of moisture can + dry soil, W_{S1} (g)			
13	Weight of water $W_w= W_{S1} - W_{C1}$ (g)			
14	Weight of dry soil solids, $W_D = W_{S1} - W_{C1}$ (g)			
15	Water content (fraction), $w = (W_w)/(W_D)$			
16	Dry unit weight of soil, $\gamma_d= (\gamma/1+w) *10$ (kN/m^3) [$\gamma=\rho*g$]			
17	Initial void ratio, $e = (G_*\gamma_w/\gamma_d) - 1$			
18	Degree of saturation, $S = w_*G/e$			
During test				
1	Weight of loading yoke + normal load			
2	L.C. of proving ring for shearing load			
3	Normal stress applied, σ_n (kPa)			
4	Shear stress at failure, τ_f = (No. of shear loading Divns. at failure*LC)/A_o (kPa) (Table 14.2)			
5	Plot Mohr-Coulomb failure envelope between σ_n and τ_f			
6	Plot shear stress-shear strain curve between τ and ε_s (Figure 14.9b)			
7	Maximum shear stress at failure from shear stress-shear strain plot (Figure 14.9b)			
8	Shear strain at failure, ε_s (%) from shear stress-shear strain plot (Figure 14.9b)			
9	Shear strength parameters from Mohr-Coulomb failure envelope (c and ϕ) (Figure 14.12)			
10	Plot Mohr's circle and find Principal stresses (σ_1 and σ_3) (Figure 14.12)			
11	Angle of rupture plane, α from Mohr's circle (Figure 14.12)			
After test: Calculations: Calculation of water content, w and shear stresses at failure				
1	Wt. of empty moisture can, W_{C1} (g)			
2	Wt. of moisture can + dry soil, W_{S1} (g)			

(Continued)

TABLE 14.1 (*Continued*)

Sl. No.	Observations and Calculations	Trial (Soil core) No.		
		1	2	3
3	Wt. of moisture can + wet soil, W_{S2} (g)			
4	Weight of water $W_w = W_{S2} - W_{S1}$ (g)			
5	Weight of dry soil solids, $W_D = W_{S1} - W_{C1}$ (g)			
6	Water content (fraction), $w = (W_w)/(W_D)$			

4. Prepare at least three soil samples of standard size 6 cm*6 cm*2 cm by trimming the undisturbed soil sample collected from the field carefully such that the soil specimen are of the same size to fit into the DST box.
5. Prepare the DST box, e.g. assemble the two halves of the DST box by the fixing screws or mount pins and attach the base plate to the lower part of the DST box.
6. Place a saturated porous stone (in case of saturated soil) in the lower part of the box.
7. For an undrained test, place the grid plate on the porous stone, keeping the serrations of the grid at right angles to the direction of shear. Also, for consolidated undrained (CU) and consolidated drained (CD) tests, use perforated grid plates in place of plain plates.
8. Weigh the DST box with the base plate, porous stone, and grid plate (W_1 g)
9. Place the soil specimen prepared in step 4 in the shear box carefully.
10. Weigh the DST box with the soil specimen (W_2 g). The difference between step 10 and step 8 will give the actual weight of soil specimen (Ws = W_2 - W_1, g) taken for the test.
11. Place the upper grid plate, porous stone, and loading cap in order on the soil specimen (Figure 14.6).
12. Place the DST or shear box inside the container fixed with linear low friction bearings (Figure 14.6) and mount it on the loading frame.
13. Mount the loading yoke on the ball placed on the loading cap.
14. Attach one dial gauge with the container to record the horizontal shear movement and other on the loading cap on the loading cap to record the vertical movement.
15. Attach the proving ring with the loading yoke for load measurement during shearing. Contact can be observed by a slight movement of dial gauges and proving ring if correctly attached. Record initial readings.
16. Saturate the soil sample if so desired for sufficient time before the start of the test.
17. Separate the two halves of the DST box by removing the vertical lock screws or mounting pins and raise the upper half of the shear box by turning

TABLE 14.2
Determination of shear strength parameters by "DST test"

Least count of shear dial gauge, L.C. (mm/divn) = 0.01, Least count of normal dial gauge: L.C. (mm/divn) = 0.0254
Least count of proving ring, L.C. (kg/dvn) = 0.272, L_o (mm) = 60, H_o (mm) = 20, A_o cm²) = 36, V_o (cm³) = 72 (1 cm³ = 10^{-6}m³)
SITE: Flood Channel Srinagar $\quad W_d$ = 120 g, γ_d = 16.6kN/m³, G = 2.65, e_o = 0.59 $\quad$ 1 (kg/cm²) = 100 kPa = 10 t/m²
Sample type: Disturbed/Remolded $\qquad$ BH: 1 $\quad$ Sample No. 2, Depth: 1.5m, $\Delta e = \Delta H (1+ e_o)/H_o$, $e = e_o - \Delta e$

Elapsed Time		Shear displacement	Normal displacement				Shear force,			Corrected area, A_c (cm²)	Shear stress, $\tau = T/A_c$ (kg/cm²)	Undrained strength (in case of sat. clays), $c_u = \tau$ (kg/cm²)
Divns	ΔH_H (mm)	Shear strain, $\varepsilon_s = \Delta H_H/L_o$ (%)	Divns	ΔH_V (mm)	Δe	e	Divns.	T (kg)	T (kg)			
0	0.00	0.00	0	0.0000	0.00	0.59	0	0.00	0.00	36.00	0.00	
10	0.10	0.17	-0.5	-0.0127	0.00	0.59	5	1.36	1.36	35.94	3.78	
20	0.20	0.33	-1	-0.0254	0.00	0.59	12	3.26	3.26	35.88	9.10	
30	0.30	0.50	-2	-0.0508	0.00	0.59	17	4.62	4.62	35.82	12.91	
40	0.40	0.67	-2	-0.0508	0.00	0.59	25	6.80	6.80	35.76	19.02	
50	0.50	0.83	-2	-0.0508	0.00	0.59	29	7.89	7.89	35.70	22.10	
75	0.75	1.25	-2.5	-0.0635	-0.01	0.60	38	10.34	10.34	35.55	29.07	
100	1.00	1.67	-3.5	-0.0889	-0.01	0.60	47	12.78	12.78	35.40	36.11	
150	1.50	2.50	-5.5	-0.1397	-0.01	0.60	56	15.23	15.23	35.10	43.40	
200	2.00	3.33	-6.5	-0.1651	-0.01	0.60	63.00	17.14	17.14	34.80	49.24	
250	2.50	4.17	-8.5	-0.2159	-0.02	0.61	69	18.77	18.77	34.50	54.40	
300	3.00	5.00	-10.5	-0.2667	-0.02	0.61	73	19.86	19.86	34.20	58.06	
350	3.50	5.83	-12	-0.3048	-0.02	0.61	79	21.49	21.49	33.90	63.39	
400	4.00	6.67	-12	-0.3048	-0.02	0.61	84	22.85	22.85	33.60	68.00	
450	4.50	7.50	-12	-0.3048	-0.02	0.61	87	23.66	23.66	33.30	71.06	
500	5.00	8.33	-12	-0.3048	-0.02	0.61	91	24.75	24.75	33.00	75.01	
550	5.50	9.17	-12	-0.3048	-0.02	0.61	94	25.57	25.57	32.70	78.19	
600	6.00	10.00	-12	-0.3048	-0.02	0.61	95	25.84	25.84	32.40	79.75	
650	6.50	10.83	-13	-0.3302	-0.03	0.62	97	26.38	26.38	32.10	82.19	

										$\tau_{j,max}$(kPa)
700	7.00	11.67	-13	-0.3302	-0.03	0.62	99	26.93	31.80	84.68
750	7.50	12.50	-13	-0.3302	-0.03	0.62	100	27.20	31.50	86.35
800	8.00	13.33	-13	-0.3302	-0.03	0.62	102	27.74	31.20	88.92
850	8.50	14.17	-13	-0.3302	-0.03	0.62	103	28.02	30.90	90.67
900	9.00	15.00	-13	-0.3302	-0.03	0.62	103	28.02	30.60	91.56
950	9.50	15.83	-13	-0.3302	-0.03	0.62	102.00	27.74	30.30	91.56
1,000	10.00	16.67	-13	-0.3302	-0.03	0.62	101.00	27.47	30.00	91.57
1,050	10.50	17.50	-13	-0.3302	-0.03	0.62	100.00	27.20	29.70	91.58
1,100	11.00	18.33	-13	-0.3302	-0.03	0.62	99.00	26.93	29.40	91.59
1,150	11.50	19.17	-13	-0.3302	-0.03	0.62	98.00	26.66	29.10	91.60
1,200	12.00	20.00	-13	-0.3302	-0.03	0.62	96.50	26.25	28.80	91.14

$\tau_{j,max}$(kPa) = 92

$c_{(u)max}$(kPa)

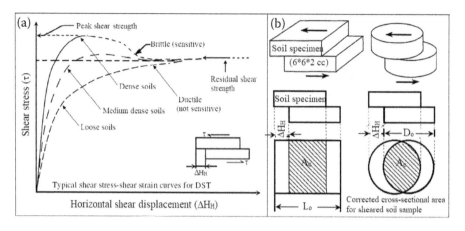

FIGURE 14.8 Shear stress-shear strain curve and Mohr's circle representation for DST.

the spacing screws. The separation between two parts of the DST box should be slightly larger (by eye) than the largest soil particles in the soil specimen.

18. Apply the desired normal load. For consolidated undrained (CU) and consolidated drained (CD) tests, allow the soil to consolidate fully under applied normal load. However, this step is avoided for unconsolidated undrained test.

    Note: For DST, three samples are prepared at the same density and tested under three different normal loads of 50 kPa, 100 kPa, and 150 kPa.

19. Adjust the dial gauges and the proving ring to read zero and record the initial dial gauge (displacement) and proving readings (load) before the start of the test before shearing the soil specimen.

20. Before proceeding with the test, remove the mounting pins and the spacing screws before start of the test.

21. Set the ***desired strain rate*** and record the proving ring reading (e.g. applied horizontal load) and dial gauges (for horizontal and vertical displacements) at every 50 divisions for a strain-controlled test.

22. Continue the test until the specimen fails. Take about 6 to 10 readings even after the specimen fails. Or, continue the test to a horizontal displacement of approximately 15 to 20% of the length of the soil specimen or unless a constant shear force is obtained.

23. When the soil specimen fails, remove the soil from the DST box and repeat the test on identical soil specimen (e.g. with same dry weight, unit weight, etc.) under increasing normal stress of 50 kPa, 100 kPa, and 150 kPa, respectively.

24. Calculate the horizontal, vertical displacements, void ratio, and shear stress for each test.

25. Determine the final water content of the soil specimens as per codal procedures.

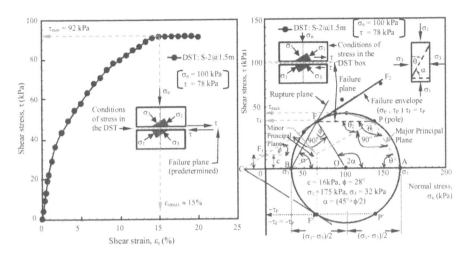

FIGURE 14.9 Shear stress-shear strain curve and Mohr's circle representation for S-2 in DST.

26. The observation of test data and calculations are given in Tables 14.1 and 14.2.
27. Plot the Mohr-Coulomb failure envelope to obtain shear strength parameters (*c and ϕ*) for desired test conditions (e.g. undrained and drained as the case may be) as shown in Figure 14.9 for DST for soil specimen 2. The test data is given for soil specimen 2 under normal stress of 100 kPa in Table 14.2.

14.6 OBSERVATION DATA SHEET AND ANALYSIS

14.6.1 DETERMINATION OF WATER CONTENT AND DRY UNIT WEIGHT

Test observations, data analysis, and calculations for determination of water content and dry unit weight are given in Table 14.1.

14.6.2 DETERMINATION UNDRAINED SHEAR STRENGTH BY "DST" TEST

Test observations, data analysis, and calculations for determination of undrained shear strength (c_u or s_u) are given in a data sheet presented in tabular form as below (Table 14.2).
Calculations are:

1. **DST box details:**
   - Wt. of the DST box with base plate, porous stone, and grid plate: W_1 (g)
   - Wt. of the DST box with base plate, porous stone, grid plate, and soil: W_2 (g)
   - Weight of dry soil: $W_d = W_2 - W_1$
   - Shear box dimensions = 6 cm × 6 cm × 2 cm

- Thickness of the soil sample: H_o = 2 cm
- Cross-sec. area of the DST box/soil specimen: A_o = 6 × 6 = 36 cm^2 = 36 × 10^{-4} m^2
- Volume of the soil: V_o = A_o × H_o = 72 cm^3= 72 × 10^{-6} m^3
- Weight of the soil sample, W_d = W_2 − W_1 = 120 g = 12 × 10^{-4} kN
- Dry unit weight of soil sample, γ_d = W_d/V_o = 12 × 10^{-4} kN /72 × 10^{-6} m^3 = 16.6 kN/m^3
- G_S = 2.65, γ_w = 10 kN/m^3
- Initial Void ratio, e_o = $G_S\gamma_w/\gamma_d$ − 1 = 0.59

2. **Data calculations** (Table 14.2)
   - For each set of shear displacement readings, calculate change in horizontal length of soil specimen: strain (col. 3):
   - ΔH_H (mm) = Shear strain Divns. (col. 2)* Least count of shear dial gauge
   - For each set of shear displacement readings, calculate shear strain (col. 4):

$$\varepsilon_s = \frac{\Delta H_H}{L_o}*100 \ (\%) \ \text{[Where: } L_o = \text{Initial length } of \ the \ soil \ specimen\text{]}$$

   - For each set of normal displacement readings, change in vertical height or thickness of soil specimen: strain (col. 6):
   - ΔH_V(mm) = Normal strain Divns. (col. 5)* Least count of normal dial gauge
   - For each set of normal displacement readings, calculate change in void ratio:

$$\Delta e = \Delta H_V (1 + e_o)/H_o$$

Where: e_o = Initial void ratio = $[(G*\gamma_w/\gamma_d)-1]$, γ_d = W_d/V_o & H_o = Initial sample height

   - For each set of normal displacement readings, calculate the void ratio: e = e_o − Δe
   - Calculate applied shear load on the soil specimen (col. 5):
     T = Proving ring divns. (col. 9)* L.C. of Proving ring/divn (kg)
   - For each set of readings, calculate corrected area, A_c (col. 11).

The corrected area A_c of the sheared specimen for the square box of length L_o, is calculated as (refer to Figure 14.8b): $A_c = L_o*(L_o − \Delta H_H)$
and for the cylindrical box of internal diameter D_o, corrected area A_c is:

$$A_c = \frac{D_o^2}{2}\left(\theta − \frac{\Delta H_H}{D_o} \sin \theta\right)$$

Where: $\theta = \cos^{-1}\left(\frac{\Delta H_H}{D_o}\right)$ in radians Or

$$A_c = A_o \left[\frac{2}{\pi} \left(\cos^{-1} \left(\frac{\Delta H_H}{D_o} \right) - \frac{\Delta H_H}{D_o} \left(1 - \left(\frac{\Delta H_H}{D_o} \right)^2 \right) \right) \right] \qquad (14.8)$$

It may be noted that the area correction in Equation (14.8) has negligible effect on the final calculated shear strength parameters, and therefore, it is generally ignored. However, the error on stresses due to non-contact area during the test in a square DST box of initial xec. sec. area, $A_o = (L_o * L_o)$ is:

$$\frac{\Delta A}{A_o} = \left[\frac{A_o - A_c}{A_o} \right] = \left[\frac{\Delta H_H}{L_o} \right]$$

And for a cylindrical shear box $\left(A_o = \frac{\pi D_o^2}{4} \right)$ is:

$$\frac{\Delta A}{A_o} = \left[\frac{A_o - A_c}{A_o} \right] = \left[1 - \frac{2\theta}{\pi} + \frac{2\Delta H_H}{\pi D_o} \sin \theta \right]$$

For the standard DST box size of 60 mm diameter with horizontal displacement (ΔH_H) of 10 mm, the error due to non-contact area on normal and shear stresses may be as high as about 15–20%.

- For each set of readings, calculate shear stress (col. 12): $\tau = T/A_c$
- For each set of readings, calculate c_u (col. 13): $c_u = \tau$ if specimen is fully saturated

(only applicable to saturated clays or cohesive soils)

- Plot the graph of shear strain, ε_s in percent (col. 4) versus stress, τ (col. 12), as shown in Figure 14.9. Test data for soil sample 2 tested under normal stress of 100 kPa is given in Table 14.2.
- Determine the peak shear stress and shear strain from this graph.
- Plot the Mohr-Coulomb failure envelope for "DST" for three sets of normal stresses of 50, 100, and 150 kPa against their shear stress values and compute shear strength parameters (c & ϕ), as shown in Figure 14.9. Test data for soil sample 2 tested under normal stress of 100 kPa is given in Table 14.3.
- Plot Mohr's circle on the Mohr-Coulomb failure envelope and compute principal stresses, as shown in Figure 14.9. The procedure for plotting Mohr's circle is given in Section 14.2.4 (Figure 14.4c).

14.7   RESULTS AND DISCUSSIONS

Undrained tests are easily conducted by the DST without pore water pressure measurement; however, it is bit difficult to conduct drained test. The value of angle

TABLE 14.3

Test data for plotting of Mohr-Coulomb failure envelope

Proving ring details			τ_f = No. divns* Value of 1 divn. stress (kg/cm²)		
Proving No.	PR-2-K	Test No.	Applied normal stress, σ_n (kPa)	No. of shear divns. at failure	Shear stress at failure, τ_f (kPa)
Load capacity of proving ring (kg)	200	1	50	54	40.82
No. of divisions in proving ring	736	2	100	103	77.85
L. C.-1 divn (kg/divn) [1 kg = 0.01kN]	0.272	3	150	127	95.99
Surface area of DST box (cm²)	36	"c" from failure envelope (kPa)			16
Value of 1 divn. (kg/cm²) [kg/cm²=100 kPa]	0.0076	"ϕ" from failure envelope (Deg)			28
Plot Mohr's circle from failure envelope (Figure 14.9)		Major principal stress, σ_1 from Mohr`s circle (kPa)			175
		Minor principal stress, σ_3 from Mohr`s circle (kPa)			32

of frictional resistance for angular particles will be higher than rounded or sub-rounded particles.

14.8   GENERAL COMMENTS

Since the shear failure plane is predetermined in the DST, therefore, this test is not as reliable as compared to the triaxail test. In the triaxial test, the soil sample fails along the weakest plane during shearing. This test is easy inexpensive and quick test as compared to the triaxial test. Further, the DST is suitable for sandy soils only. However, in case of cohesive soils being tested in a DST, it may be noted that:

- Excessive strain rate or rate of loading should be avoided; otherwise, excessive values of shearing strength may be obtained. This is due to viscous resistance of the liquid filling the pores of the soil. This factor has only a smaller influence on the results if the test is slow.

FIGURE 14.10   Schematic diagram for progressive failure in DST.

- *A sample of cohesive soil should completely consolidate before the test;* otherwise, a part of the normal load *N* applied to the sample will be carried by the moisture, and the inter-granular pressure will be less than the normal stress applied. Hence, according to Coulomb's formula, the shearing strength of the soil will be less than its true value.
- It has been found by investigators at Harvard and the Massachusetts Institute of Technology that the *previous history of the sample,* particularly its voids ratio at the beginning of the test, has a considerable influence on the results. There is *progressive failure* in the box, as shown in Figure 14.10. Different sections of the box are under different strains, so that the box does not represent a uniform unit.

14.9   APPLICATIONS/ROLE OF "DST" IN SOIL ENGINEERING

The DST is a very easy and quick method of soil testing for determination of shear strength parameters. This test finds application for lightweight structures. This test is conducted to get the residual shear strength as well as peak shear strength and shear strain characteristics of the soils.

14.10   SOURCES OF ERROR

Following are possible errors that would cause inaccurate determinations of strength characteristics:

1. Main source of error in the DST is that the shear failure plane is pre-determined. Therefore, the soil sample may not fail along the weakest failure plane.
2. There is a non-uniformity of shear stress distribution over the surface of the soil sample tested (e.g. stresses are maximum at the edges and minimum at the center) (Wroth 1987).
3. Highly permeable soils cannot be tested in an undrained condition in this test.
4. Since the shear box is of small size, therefore, the lateral restraint by the side walls of the shear box will have a direct bearing on the shear strength of soils in this test.
5. Since the lower half of the shear box moves laterally relative to the upper half, the non-contact area will have direct bearing on the test results as the surface area deceases along the predetermined shear failure plane.

14.11   PRECAUTIONS

1. Make sure that the two halves of the shear box are separated from each other by at least the size of maximum particle size in the soil sample.
2. Make sure that the pins or screws are removed before starting the shearing soil sample, else there will be no movement of lower half box relative to upper half.

3. Make sure that the upper half of the shear box is in contact with the load cell or proving ring, else it will not produce a qualitative shear stress-shear strain plot due to softening at the beginning of the test.

4. Make sure that all the requisite initial readings of dial gauge, proving ring, and their least counts are recorded.

5. Make sure that the normal stress applied is uniform throughout the test.

6. Make sure that perforated grid planes in place of plain grid plates and fully saturated porous stones and filter papers are used for drained tests.

REFERENCES

ASTM D6528-07. 2007. "Standard Test Method for Consolidated Undrained Direct Simple Shear Testing of Cohesive Soils." *Annual Book of ASTM Standards*, Vol. 04.08. ASTM International, West Conshohocken, Philadelphia, PA, www.astm.org.

ASTM D3080 / D3080M-11. 2011. "Standard Test Method *for Direct Shear Test* of Soils Under Consolidated Drained Conditions." ASTM International, West Conshohocken, PA, www.astm.org.

BS 1377 (Part 7). 1990-clause 8. "Shear Strength Tests (Total Stress): The Unconsoldiated Undrained Triaxial Compression Test, Without Pore Water Pressure Measurement." British Standards, UK.

Hvorslev, M. J. 1949. "Subsurface Exploration and Sampling of Soils for Civil Engineering Purposes." U.S. Army Corps of Engineers, U.S. Waterways Experiment Station, Vicksburg, MS.

IS: 2720 (Part 39). 1977. "Method of Test for Soils: Determination of Shear Strength Parameter by Direct Shear Test." *Bureau of Indian standards*, New Delhi.

Mohr, O.1900. "Which Circumstances Determine the Elastic Limit and the Breakage of a Material?" *Time of Ver Deut Ing* 44:1524–1530.

Reddy, Krishna R. 2000. "Engineering Properties of Soils Based on Laboratory Testing." Department of Civil and Materials Engineering University of Illinois at Chicago, USA.

Terzaghi K., R. B. Peck, and G. Mesri. 1996. *Soil Mechanics in Engineering Practice*. 3rd ed. New York: John Wiley & Sons.

Taylor D. W. 1939. "A Comparison of Results of Direct Shear and Cylindrical Compression Tests." Proc. ASTM.

Wroth, C. P. 1987. "The Behaviour of Normally Consolidated Clay as Observed in Undrained Direct Shear Tests." *Géotechnique* 37(1): 37–43.

15 Shear Strength of Soils by Triaxial Test

References: IS: 2720 (Part 11); 2720 (Part 12); ASTM D2850; ASTM WK3821; ASTM D4767; BS 1377-Part 7; BS 1377: Part 8 (Clause-7); BS 1377: Part 8 (1990-Clause 8)

15.1 OBJECTIVES

- The main objective of conducting the triaxial test on soils is to measure soil shear parameters (s_u or c' and ϕ'), which are used for critical assessment of stability of various civil engineering structures under short-term and long-term stability analysis under controlled drainage and loading applications.
- Another main objective of conducting the triaxail test on soils is because of better control of initial stresses and loading stress paths than unconfined compression and direct shear tests.
- A triaxial test is highly preferred over the direct shear test because the soil sample fails along the weakest plane rather than on a predetermined failure plane in the case of a direct shear test.

15.2 DEFINITIONS AND THEORY

A triaxial test is a novel technique for testing soil samples subjected to orthogonal stresses existing in the field under controlled drainage conditions. These orthogonal stresses are generally known as principal stresses (e.g. σ_1, σ_2, σ_3, where: $\sigma_2 = \sigma_3$). Since the DST has numerous limitations, therefore, A. Casagrande conceived the need for triaxial test for reliable measurement of soil strength parameters. A triaxial test is a most suitable and widely used soil test in the geotechnical laboratory for evaluation of shear parameters (c, ϕ & c', ϕ') under controlled drainage and loading conditions. These parameters are required in the design of various civil engineering structures under short-term and long-term stability analysis. The triaxial tests are mostly conducted on high-quality undisturbed soil samples and reconstituted soil samples under very high confined pressures or as desired based on a research problem. These soil samples can be tested in isotropic, anisotropic, and K_o conditions as desired by the researchers. A typical experimental setup for a triaxial testing is illustrated in Figure 15.1. In this test, a soil sample is saturated (known as saturation stage) and then consolidated (known as consolidation stage) under desired cell or confined pressure (generally known as minor principal stress, $\sigma_3 = \sigma_c = \sigma_r$ in case of the triaxial compression test). Once the consolidation stage is completed, the soil sample is sheared (known as shearing stage) under specified strain rate and constant cell pressure under drained or undrained conditions as required by the researcher.

DOI: 10.1201/9781003200260-15

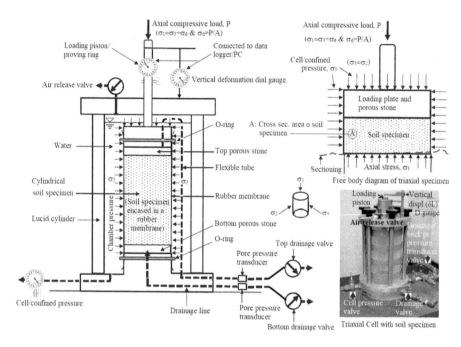

FIGURE 15.1 A typical experimental setup for triaxial test.

From Figure 15.1, it can be seen that if the drainage valve is kept open and the pore water pressure is allowed to drain during shearing stage, the test is known as a consolidated drained (CD) test. Thus, in this test, effective shear strength parameters (c' & ϕ') are measured in an effective stress approach. However, it may be noted that a drained test can take a long time to complete depending upon the type of soil (cohesive or cohesionless) being tested (e.g. clays may take about a week to drain completely).

Similarly, if the drainage valve (Figure 15.1) is kept closed and pore water pressure is not allowed to drain out and it is measured by a calibrated pore pressure transducer under applied vertical deviator stress during the shearing stage, then the test is known as consolidated undrained (CU) test. In this type of test, pore water pressure is measured and total stresses and effective stress are not equal as in case of drained test (e.g. in a drained test, total stress is equal to effective stress because no pore water pressure is measured). This test may be completed within a short period of time (1–2 hours for clays and 30 minutes for sands) depending upon soil sample type being tested. Thus, in this test, both total shear strength parameters (c and ϕ) in total stress approach and effective shear strength parameters (c' and ϕ') in an effective stress approach are measured. Further, if a soil sample is not consolidated, but directly sheared after saturation stage without pore pressure measurement, then the test is known as unconsolidated undrained (UU) test. This test is also known as rapid test, which can be completed within 15–30 minutes of time. Thus, in this test, only total shear strength parameters (c and ϕ) in total stress approach.

Therefore, to summarize, it is seen that a soil sample can be tested under different drainage and loading conditions as per research program as described below:

a. Consolidated drained (CD) test or slow (S) test
b. Consolidated undrained (CU) test or consolidated quick (Q_c or C-Q_U) test with/without pore water pressure measurement
c. Unconsolidated undrained (UU) or unconsolidated quick (Q_u) test without pore water measurement

The above test combinations are illustrated in Figure 15.2.

It may be noted that if the above triaxial tests (CD or CU) are isotropically consolidated before shearing, then these tests are symbolically represented as:

a. Isotropically consolidated drained (CID) test
b. Isotropically consolidated undrained (CIU) test with/without pore water pressure measurement

It may be noted that the above triaxial tests have been conducted by increasing the axial stress (σ_1) keeping confined stress (σ_3) constant. In these tests, the soil sample is always compressed in vertical direction which results in decrease in length and increase in lateral dimension (e.g. diameter). Therefore, these tests are also known as triaxial compression (TC) tests. However, if the axial stress is decreased and the confined stress is kept constant, then this test is known as triaxial extension (TE) test. In a triaxial extension test, a soil sample is compressed by lateral stress thereby increasing its length decreasing its diameter. Similarly, by increasing and decreasing axial and lateral stress, a soil sample can be tested in different conditions such as triaxial compression (TC), triaxial extension (TE), lateral compression

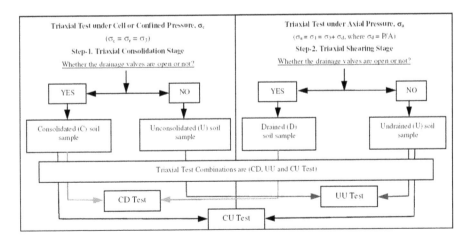

FIGURE 15.2 Schematic diagram of various triaxial test combinations under consolidation and shear stages.

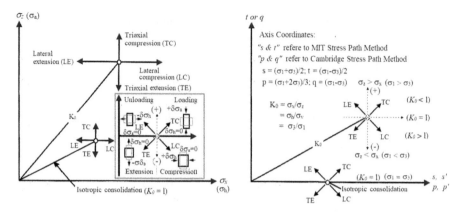

FIGURE 15.3   Types of triaxial tests and stress paths in "$\sigma_x - \sigma_z$" and "$s - t$"//"p-q" planes.

(LC), and lateral extension (LE), as illustrated in Figure 15.3. The stress paths for these triaxial tests are also shown in Figure 15.3.

Similarly, the variation of stresses for K_o-consolidated triaxial tests from different construction of different types of structures is shown in Figure 15.4.

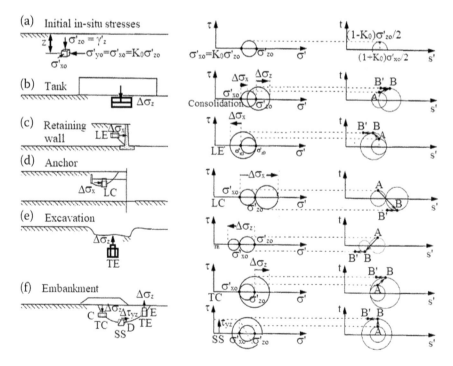

FIGURE 15.4   Variation of stresses for K_o-consolidated triaxial tests from different construction of different types of structures.

The parameter "K_o" is defined as the ratio of horizontal stress to the vertical stress because the horizontal stress cannot be determined directly in the field as the vertical stress. Therefore, by assuming the "K_o" parameter equal to one (1), the effect of overburden stresses on the soil mass can be simulated by the isotropic triaxial compression tests in the field. In an isotropic consolidated triaxial compression (e.g. TC), the vertical stress ($\sigma_1 = \sigma_z$) is increased under constant horizontal or confined or radial stress ($\sigma_h = \sigma_x = \sigma_c = \sigma_r$) such that:

$$\sigma_x = \sigma_y = \sigma_c, \quad and \quad \tau_{xy} = \tau_{yz} = \tau_{yz} = 0 \tag{15.1}$$

Similarly, for known values of pore water pressure, the effective stresses are (Figure 15.1):

$$\sigma'_x = \sigma'_y = \sigma'_z = \sigma_c \quad and \quad \tau_{xy} = \tau_{yz} = \tau_{zx} = 0 \tag{15.2}$$

Further, the "K_o" helps in the simulation of overburden stresses on the soil mass in the field under the following loading conditions:

$$\sigma'_x = \sigma'_y, \quad \tau_{xy} = \tau_{yz} = \tau_{zx} = 0, \quad and \quad \varepsilon_x = \varepsilon_y = 0 \tag{15.3}$$

However, it may be noted that different axial and radial stresses can also be applied in a "K_o" test (e.g. $\sigma'_x \neq \sigma'_z$) during a triaxial test on the soil sample. The various types of triaxail consolidation tests (e.g. "K_o," isotropic or anisotropic) under different loading (e.g. change in axial and radial stresses during shearing) and drainage (e.g drained or undrained) conditions are summarized in Table 15.1.

15.2.1 WHY CONDUCT A TRIAXIAL TEST?

Since the natural soil is tri-particulate, discontinuous, and heterogeneous materials, therefore, it is necessary to use most reliable testing methodology to measure accurately the mechanical properties of a discontinuous material. It worth mentioning here that the soil material is used both as a construction material and as a foundation medium for the construction of various engineering structures such as earthen dams, railway or road embankments, air fields, slopes, retaining walls, buildings, etc. for which the reliable measurement of the shear strength parameters is mandatory. However, it may be noted that in case of large engineering structures such as embankments or dams constructed on or in cohesive soils, the construction time is too long and pore water pressure may dissipate very slowly under this gradually applied loading with time. Thus, in such situations, the undrained shear strength parameters ($c_u > 0$, $\phi = 0$) can be determined based on total stress approach by unconsolidated-undrained (UU) triaxial compression tests, which are generally referred to as "Short-Term Stability Analysis" of large structures on saturated cohesive soils. In the "UU" triaxial compression test, undrained cohesion is determined on total stress approach as the pore water pressure can not be measured in this test. In case of very soft or highly sensitive clays such marine clays, the

TABLE 15.1

Triaxial Tests Under Different Loading and Drainage Conditions

Sr. No.	Types of Triaxial Test	Type of Consolidation	Drainage Conditions	Axial/Vertical Stress	Horizontal/Lateral Stress	Full Test Name
1	CD, CID, or CIDC	Isotropic	Drained	Increasing	Constant	Isotropically consolidated drained triaxial compression (TC)
2	CDE or CIDE	Isotropic		Decreasing		Isotropically Consolidated drained. triaxial extension (TE)
3	CK_0DC	K_0		Increasing		K_0 consolidated drained triaxial compression
4	$C K_0DE$	K_0		Decreasing		K_0 consolidated drained triaxial extension
5	CDLC or CIDLC	Isotropic		Constant	Increasing	Isotropically consolidated drained lateral triaxial compression (LC)
6	COLE or CIDLE	Isotropic			Decreasing	Isotropically consolidated drained lateral triaxial extension (LE)
7	$C K_0DLC$	K_0			Increasing	K_0 consolidated drained lateral triaxial compression
8	$C K_0DLE$	K_0			Decreasing	K_0 consolidated drained lateral triaxial extension
9	CU, CIU, or CIUC	Isotropic	Undrained	Increasing	Constant	Isotropically consolidated undrained triaxial compression
10	CK_0U or $C K_0UC$	K_0		Increasing	Constant	K_0 consolidated undrained triaxial compression
11	CUE or CIUE	Isotropic		Decreasing	Constant	Isotropically consolidated undrained triaxial extension
12	$C K_0UE$	K_0		Decreasing	Constant	K_0 consolidated undrained triaxial extension
13	CULC or CIULC	Isotropic		Constant	Increasing	Isotropically consolidated undrained lateral triaxial compression
14	CULE or CIULE	Isotropic			Decreasing	Isotropically consolidated undrained lateral triaxial extension
15	CK_0ULC	K_0			Increasing	K_0 consolidated undrained lateral triaxial compression
16	CK_0ULE	K_0			Decreasing	K_0 consolidated undrained lateral triaxial extension
17	UU or UUC	None		Increasing	Constant	Unconsolidated undrained triaxial compression
18	UUE	None		Decreasing	Constant	Unconsolidated undrained triaxial extension

undrained cohesion (c_u) is determined either by field or laboratory vane shear test (VST). However, it may be ensured that soil samples are fully saturated before "UU" or "VST" to avoid inaccurate measurement of undrained strength. The saturated cohesive soils can be identified based on the consistency of soils, which varies over a wide range of classes of consistencies such as very soft ($c_u \approx < 12.5$ kPa), soft ($c_u \approx 12.5$–25 kPa), medium stiff ($c_u \approx 25$–50 kPa), stiff ($c_u \approx 50$–100 kPa), and very stiff ($c_u \approx > 100$ kPakPa).

In addition to short-term stability analysis based on total stress approach, long-term stability can also be carried out based on effective stress approach by using either consolidated-undrained (CU) or consolidated-drained (CD) triaxial compression tests to assess the value of most critical factor of safety. It may be noted that the "CU" test is conducted on saturated soils with pore water pressure measurement for assessing the effective shear parameters (e.g. $c' > 0$, $\phi' > 0$). However, in the case of the "CD" test, no pore water pressure is measured and the shear parameters are same as effective shear parameters (e.g. $c = c' > 0$, $\phi = \phi' > 0$). In the case of granular soils such as sands, and normally consolidated clays, effective shear strength is measured in terms of internal shearing angle ($\phi' > 0$) and cohesion is zero (e.g. $c \approx 0$). In the case of over-consolidated clays, effective shear strength is measured in terms of internal shearing angle ($\phi' > 0$) and cohesion (e.g. $c' > 0$). But before conducting the desired type of triaxial tests, it may be ensured that the field conditions are carefully ascertained and well calibrated equipment are used with high degree of precision to simulate the field conditions by model tests in the laboratory. Also, it is mandatory to assess the discrepancy if any between field and laboratory test results so as to minimize the inaccurate measurements of soil properties.

15.2.2 BASIC PRINCIPLE OF A TRIAXIAL COMPRESSION TEST?

Since all the engineering structures are constructed on the ground or in the ground or beneath the ground surface, the soil is going to compress mainly in vertical direction compared to horizontal or radial direction. Therefore, the basic concept in a triaxial compression test is to compress the soil sample in vertical and horizontal directions with different magnitude of loading in a triaxial cell (e.g. $\sigma_v > \sigma_h$ or $\sigma_1 > \sigma_3$). Generally, a cylindrical soil sample is consolidated in the triaxial compression test under desired constant lateral or radial or horizontal (e.g. σ_h or σ_3 or σ_r and $\sigma_h = \sigma_v$) followed by shearing by increasing deviator stress (e.g. $\sigma_d = P/A$) and constant horizontal stress (e.g. $\Delta\sigma_3 = 0$). However, it should be ensured that the cylindrical soil sample has a L/D ratio equal to 2 (e.g. L/D = 2). A typical experimental setup of a triaxial compression test with a cylindrical soil sample inside a triaxial cell to be test is shown in Figure 15.5.

It may be noted that based on research requirements or a specified field problem (loading conditions as shown in Figure 15.3), triaxial extension (TE) or triaxial lateral compression (LC) may also be conducted in the laboratory. However, there are very rare cases in actual soil engineering practice. The stress state during triaxial compression of a soil sample is shown in Figure 15.6.

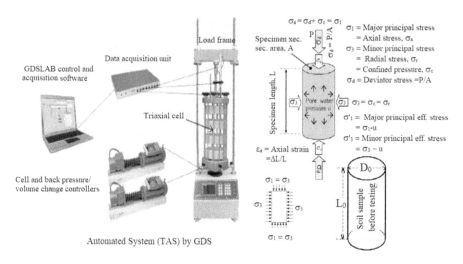

FIGURE 15.5 Typical experimental setup of triaxial compression test with cylindrical soil sample inside triaxial cell.

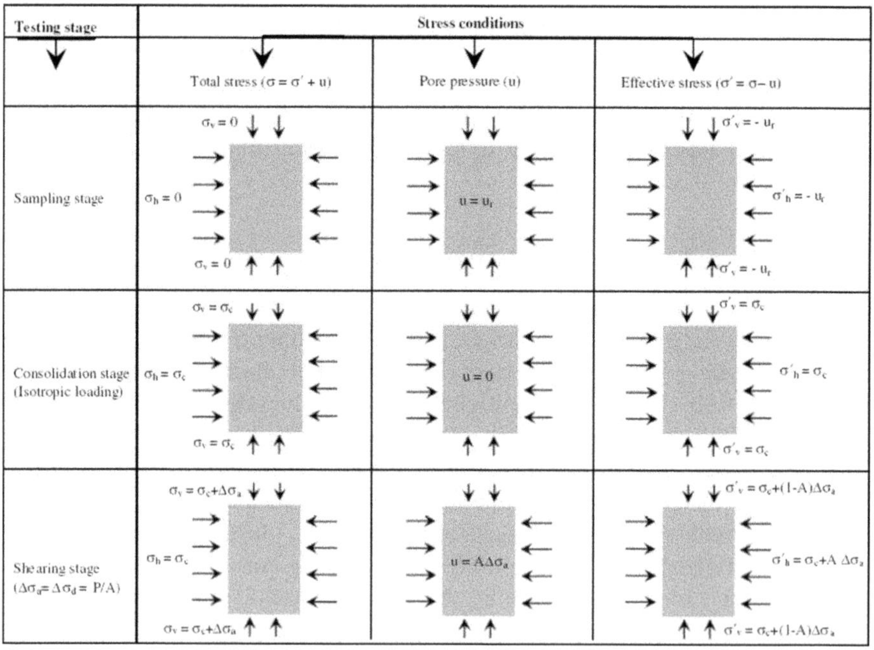

FIGURE 15.6 Stress state during triaxial compression of a soil sample.

15.2.3 DRAINAGE BOUNDARY CONDITIONS IN TRIAXIAL COMPRESSION TESTS

Primarily, there are two types of drainage boundary conditions allowed in consolidation and shear stages of triaxial tests (as illustrated in Figure 15.2):

1. **Drainage not allowed:** In this case, drainage is not allowed either in consolidation stage (known as Unconsolidated-U) or in shearing stage (known as Undrained). Hence, the test is known as the UU Test or Quickest test or Q_U-Test and may take about 10–15 minutes to complete the test. Therefore, this test is most suitable for clays soils (highly impervious soils where pore water pressure takes very long time to dissipate). This test generates total shear strength parameters (s_u and ϕ_u = 0 for saturated clays). However, if arrangements are made to measure pore water pressure during the shearing stage, effective shear strength parameters can be obtained.

2. **Drainage allowed and not allowed:** In this case, drainage is allowed in consolidation stage (known as consolidated-C) and not allowed in shearing stage (known as undrained-U). Hence, the test is known as a CU Test or Quick Test with pore water pressure measurement in the shearing stage. Hence, this test generates total and effective shear strength parameters (c, ϕ; c', ϕ').

3. **Drainage allowed:** In this case, drainage could be allowed either vertically (one-way drainage from either bottom end or top end or two-way drainage from top and bottom ends as desired) or radially (e.g. side drain, either inward or outward, which has the shortest drainage path compared to the vertical drainage path). Thus, if drainage is allowed during the consolidation stage, the volume change (ΔV) can be recorded at a constant back pressure (e.g. increased pore water pressure if applied during the saturation stage), and the test is called "Consolidated (C)," and drainage allowed during shearing stage, the test is called "Drained (D)." Hence the test is known as consolidated-drained (CD) compression test and generates effective shear strength parameters (c'_d & ϕ'_d). However, in the case of NC clays, c'_d = 0.

The various drainage boundary conditions corresponding to a desired triaxial compression test type with rate of loading (e.g. strain rate) and resulting shear parameters are given in Table 15.2.

15.2.4 Skempton's Pore Pressure Parameters

The concept of pore pressure parameters (B & A) was conceived by Skempton (1954), which play a very important role in the understanding of triaxial tests and in describing the soil behavior. The B-Parameter helps in assessing whether 100% degree of saturation (e.g. S = 1) has been achieved during the saturation stage of the triaxial test. It may be noted that the triaxial consolidation test can not be performed if degree of saturation is not achieved about 0.98–1.0. The B-parameter is defined as the ratio of incremental pore water pressure to the corresponding incremental cell pressure during the saturation stage, e.g. B = $[\Delta u_w/\Delta \sigma_c]$.

Similarly, A-Parameter helps in assessing the stress history of the soil sample in the shearing stage and it is related to the over-consolidation ratio (OCR), pore water pressure, and the increasing deviator stress. Therefore, it is very essential to know about these parameters for better understanding of triaxial testing of soils.

TABLE 15.2
Summary of Drainage Boundary Conditions in Triaxial Compression Tests

Sr. No.	Test Type	Drainage Conditions		Rate of Axial Strain (%/min.) during Shearing Stage	Strength Parameters Obtained
		Consolidation Stage (σ_c)	Shear Stage (σ_d)		
1	Quick-undrained–no excess pore pressure measurement (Q_U)	No drainage	No drainage	Typically fastest, reaching failure criterion in 5–10 minutes (0.3 – .0%/min)	Total stress (c_u)
2	Unconsolidated–undrained with pore pressure measurement (UU)	No drainage	No drainage	Strain rate slow enough to allow pore pressure equalization and measurement	Effective stress (c' and ϕ')
3	Consolidated quick–undrained (C–Q_U)	Full drainage usually consolidated under in-situ effective stress (σ'_c)	No drainage	Strain rate as for Q_U test. Usually, three consolidated different cell pressures	Total stress (c_u)
4	Consolidated - undrained with pore pressure measurement (CU)	Full drainage, three specimens usually consolidated under different effective cell pressures	No drainage	Strain rate slow enough to allow pore pressure equalization and measurement (0.1 – 0.5%/min)	Effective stress (c' and ϕ')
5	Consolidated–drained with volume change measurement (CD)	Full drainage as for CU test–record ΔV and maintain constant back pressure	Drainage allowed	Strain rate must be slow enough to prevent pore pressure buildup (0.01 – 0.05%/min)	Effective stress (c_d and ϕ_d)

15.2.4.1 Pore Pressure Parameter: B-factor

The B-value, mentioned above (saturation stage), is not a soil property but an experimentally useful parameter, whose theoretical derivation was first put forward by Bishop and Eldin (1950):

$$B = \left[\frac{\Delta u}{\Delta \sigma_c} \right] \qquad (15.4)$$

$\Rightarrow \Delta u = B * \Delta \sigma_c$, when B = 1 saturated soil, $\Delta u = \Delta \sigma_c$

The derivation of the above Equation (15.4) by Bishop and Eldin (1950) involves the following assumptions:

- The solid material(s) forming the soil particles is elastic and isotropic and the soil pores are interconnected.
- The solid material behaves as an isotropically elastic material under a change in effective stress.
- The pore water pressure distribution in the void pores of the soil matrix is linearly compressible.

Therefore, a soil sample is necessarily saturated before next stage of triaxial compression test. If it takes lot of time to saturate the soil sample under cell pressure, then back pressure (e.g. increased positive pore water pressure) can be applied through bottom drainage by maintaining a minimum difference of about 10 kPa between cell pressure and the pore water pressure. In such a case, check the value of applied cell pressure, let us assume it is 50 kPa and the corresponding pore water pressure read through pore pressure transducer is 20 kPa (e.g. B = 20/50 = 0.4). Now increase the pore water pressure value to 40 or 45 kPa and open the bottom drainage valve. The water will enter into the soil sample and the pore water pressure in the soil sample will start increasing until it becomes constant at about the cell pressure value and record it (let us assume it as u_{w1} = 40 kPa). Now close the drainage valve and check both cell pressure (50 kPa) and increased pore water (u_{w1} = 40 kPa) are constant for sufficient time. Increase the cell pressure by 50 kPa (e.g. $\Delta \sigma_c$ = 50 kPa) to make a total cell pressure 100 kPa and watch the increase in pore water pressure under applied increment of cell pressure until it becomes constant (let us assume it is u_{w2} = 85 kPa). Now determine the incremental increase in pore water pressure is $\Delta u_w = (u_{w2} - u_{w1})$ = 85 – 40 = 45 kPa. Then, the B-parameter is computed as B = $\Delta u_w / \Delta \sigma_c$ = 45/50 = 0.9, which is still far below the value of 0.98 to 1.0. Therefore, apply one more increment of back pressure and repeat above procedure till B = 0.98 – 1.0. Once the saturation stage is complete, close all drainage valves and increase the cell pressure to the desired value under which the soil sample is to be consolidated.

Thus, it is seen that the degree of saturation is dependent on the B-Parameter. If B-parameter is not equal to 1, the soil sample cannot be declared a saturated soil sample and fit for the next stage of triaxial test. The process of saturation of a soil sample cell pressure is shown in Figure 15.7(a) and in this process, typically, $B \geq 0.98$ is used to confirm full specimen saturation. However, if the B-value is too

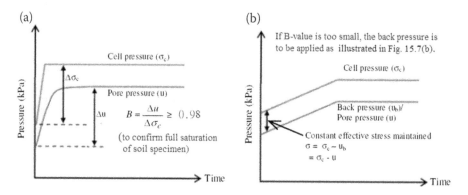

FIGURE 15.7 Saturation of a soil sample by applying: (a). Cell pressure, and (b). Cell pressure and Back pressure.

small, the back pressure is to be applied as explained above and the process of saturation by applying back pressure is illustrated in Figure 15.7(b).

15.2.4.2 PORE PRESSURE PARAMETER: A-FACTOR

Unlike the B-parameter, the A-Parameter also helps in assessing the stress history of the soil sample in the shearing stage. It may be noted that the A-Parameter is not a constant and it is related to over-consolidation ratio (OCR), pore water pressure, and the increasing deviator stress. The value of pore water pressure parameter "A" (defined as the change in pore water pressure for unit change in deviator stress) reported at failure is symbolically represented as:

A_f: for saturated soils and
$\bar{A}$: for partially saturated soils.

In undrained tests, the A-parameter is expressed as:

$$\Delta u = [\Delta u_1 - \Delta u_2] = [B\,1 * \Delta\sigma_3 + B * A * (\Delta\sigma_1 - \Delta\sigma_3)]$$
$$= B[\Delta\sigma_3 + A * (\Delta\sigma_1 - \Delta\sigma_3)] \tag{15.5}$$

Where: Δu_1 = Final pore water pressure at the end of consolidation stage under (σ_3), and
Δu_2 = Pore water pressure generated during shearing stage under increasing deviator stress ($\Delta\sigma_d = \Delta_1 - \Delta_3$).
In the "UU" test, the pore water pressure developed in the soil specimen is:

$$u = [u_c + u_d] \tag{15.6}$$

Where: $u_c = B * \sigma_3 = B * \sigma_c$ (from saturation stage)
u_d = Pore water pressure due to deviator stress during shearing stage
Likewise, pore water pressure developed under the applied deviator stress ($\Delta\sigma_d$) is:

$$u_d = [A * \Delta\sigma_d] \qquad (15.7)$$

Where:

$$\Delta\sigma_d = [\sigma_1 - \sigma_3] \qquad (15.8)$$

Combining Equation (15.6 to 15.8), we get:

$$u = [u_c + u_d] = [B * \sigma_3 + \overline{A} * (\sigma_1 - \sigma_3)] \qquad (15.9)$$

Where: $\overline{A}$ is defined as the change in pore water pressure for unit change in deviator stress for saturated soils.

For saturated soils, B = 1, and during shearing stage σ_3 is constant and A = A_f (pore water pressure parameter at failure), we have from Equation (15.9):

$$u = \left[A_f * (\sigma_1 - \sigma_3) \right]$$

Or

$$A_f = \left[\frac{u_e - u_o}{\sigma_d - \sigma_{d)o}} \right] = \left[\frac{\Delta u}{\Delta\sigma_d} \right] \qquad (15.10)$$

Where: u_o = Initial pore water at the start of shearing stage

u_e = Excess pore water pressure generated during shearing stage under increasing deviator stress

$\Delta\sigma_{d)o}$ = Deviator stress just at the start of shearing stage (usually it is zero as first reading), and

$\Delta\sigma$ = Deviator stress under increasing vertical load corresponding pore water pressure value "u_e."

15.2.5 LOADING CONDITIONS IN TRIAXIAL TESTS

Unlike drainage boundary conditions, a triaxial test is conducted under two types of loading conditions of lateral (or horizontal or radial) and axial loadings. Lateral or radial loading is applied during saturation and consolidation stages and vertical loading (deviator stress) with constant all-around pressure (cell or confined pressure) from consolidation stage in the shearing stage. Drainage boundary condition and application of loading go side-by-side in the triaxial test, and are given in Table 15.3.

Types of triaxial tests, corresponding shear strength parameters, and most suited soils are illustrated in Figure 15.8.

TABLE 15.3
Summary of Drainage Boundary Conditions and Application of Loading in a Triaxial Test

Sr. No.	Test Stage	Drainage Boundary Condition	Application of Loading	Designation Assigned	Test Type
01	Saturation stage	Drainage valve open: water flows into the soil specimen for saturation purposes	Cell or confined pressure (σ_c) applied in increments till full saturation of the specimen is achieved. If necessary, back pressure is also applied into the specimen	$B=\Delta u/\Delta\sigma_c=1$(B=Skempton's pore pressure parameter:	Generally, saturation stage is required for all types of triaxial tests
02	Consolidation stage	Drainage not allowed	Cell or confined or all-around pressure (σ_c) to consolidate the soil specimen until 100% consolidation is achieved	a. Unconsolidated-"U"	Combining designations assigned, we have: Combination of "a & c" yield UU Test Combination of "b & c" yield CU Test Combination of "b & d) yield CD Test
		Drainage allowed	Cell or confined or all-around pressure (σ_c)	b. Consolidated-"C"	
03	Shearing stage	Drainage not allowed	Vertical loading (deviator stress) and cell or confined or all-around pressure (σ_c) has to be constant	c. Undrained-"U"	
		Drainage allowed	Vertical loading (deviator stress) and cell or confined or all-around pressure (σ_c) has to be constant	d. Drained-"D"	

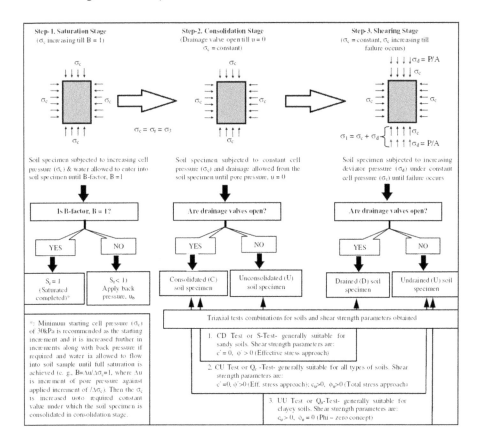

FIGURE 15.8   Types of loading conditions in triaxial tests and shear strength parameters.

15.3   METHOD OF TESTING

The shear strength of soils either on an in situ (undisturbed specimen) or remolded (disturbed) *soil specimen* is determined in a laboratory using a triaxial apparatus by means of a strain controlled mechanism. Triaxial tests of any desired combination of *pre-shear conditions* (e.g. under consolidation stage as consolidated-C or unconsolidated-U) and *during shear conditions* (e.g. under shearing stage as drained-D or undrained-U) result in three combinations of CD, CU, and UU tests, as shown in Figure 15.9.

15.3.1   Pre-Requisite for Triaxial Compression Test

Generally, a triaxial compression test is applicable to all types of soils. A triaxial shear test can be conducted on undisturbed saturated clayey soils or reconstituted soil specimen for determination of undrained shear strength as per requirement. However, neither undisturbed soil specimen can be collected nor compaction test can be conducted in case of cohesionless (sandy soils). Therefore, a relative density

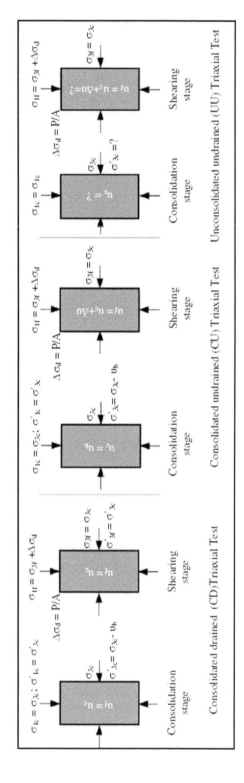

FIGURE 15.9 Three types of triaxial shear tests.

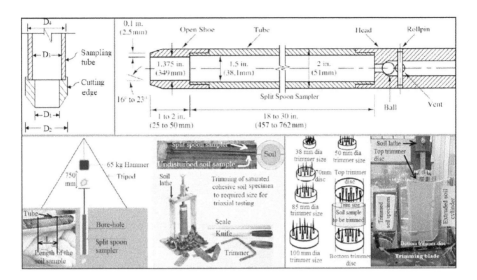

FIGURE 15.10 Preparation of undisturbed soil specimen for triaxial shear testing.

test (Test No. 10) is a pre-requisite for preparation of a remolded specimen for sandy soils for conducting of triaxial shear test, whereas the compaction test (Test No. 9) is a pre-requisite for remolded c-ϕ soils and c-soils.

15.3.2 SIZE OF SOIL PARTICLES IN A SOIL SPECIMEN

While selecting a soil sample for a triaxial test, it may be ensured that the maximum particle size in the soil sample should one-fifth of the height of the specimen (BS 1377).

15.3.3 PREPARATION OF SOIL SPECIMEN

The undisturbed soil specimens of saturated clayey soils at a natural moisture content are collected directly from the field in SPT samplers from desired depth and extracted and trimmed in the laboratory for testing (Figure 15.10). The soil specimen is assumed to be virtually saturated (in in situ condition) NC clay of soft to medium stiff consistency. For a remolded clayey soil specimen of desired size (38 mm diameter, 70 mm diameter, 100 mm diameter), a soil specimen passing through a 0.425 mm sieve may take preparation of the test specimen. For silt-dominated soils, a soil specimen passing through a 2 mm IS sieve may be used. For sand-dominated soils, a soil specimen passing through a 4.75 mm IS sieve may be used. However, sieve size may be chosen by keeping in view the desired size of the soil specimen to be tested. If a reconstituted soil specimen is to be used, then consistency limit tests (LL and PL) are to be conducted on the same soil for preparation of soil slurry. The soil slurry is prepared at a water content of about 1.5 * LL to 2 * LL and thoroughly mixed and de-aired before consolidation to desired consistency.

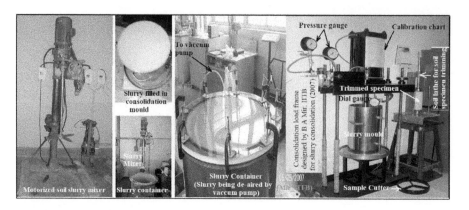

FIGURE 15.11   Preparation of reconstituted clayey specimen for triaxial shear testing.

Special setup is required for preparation of reconstituted clayey specimens from the soil slurry prepared at a water content of twice the liquid limit (e.g. 2 * LL). The soil slurry is de-aired and consolidated on the laboratory floor under the desired stress level in a consolidation load frame (1-D consolidation), as illustrated in Figure 15.11. In this setup (Figure 15.11), the consolidated clay lump is extruded out and cut into desired pieces, which in turn are trimmed using a soil lathe as per the desired soil sample size, maintaining L/D ration of 2 (e.g. 38 mm diameter, 70 mm diameter, 100 mm diameter, whichever is required).

15.4   TESTING EQUIPMENT AND ACCESSORIES

The following equipment and accessories are required for a triaxial shear test (Figure 15.12):

1. Air-dried fine-grained soil passing 0.425 mm (clayey soils), IS sieves 2 mm (silty soils), and 4.75 mm (sandy soils) [**Note:** check sample size to be tested and chose size accordingly]
2. Reconstituted clay sample prepared in lab.
3. Undisturbed soil specimens from field
4. Triaxial testing machine including Load frame (50 kN), triaxial cell, porous stones, filter papers, rubber membranes, O-rings, loading cap, dial gauge or LVDT, proving ring or load cell, data acquisition system, desktop PC, etc. (Figure 15.10)
5. SPT samplers for collection of undisturbed soil samples (for clayey soils) from field
6. Weighing balance accuracy 0.01 g
7. Moisture content cans
8. Squeeze bottle
9. Soil lathe and accessories for undisturbed field specimen trimming
10. Sample extruder

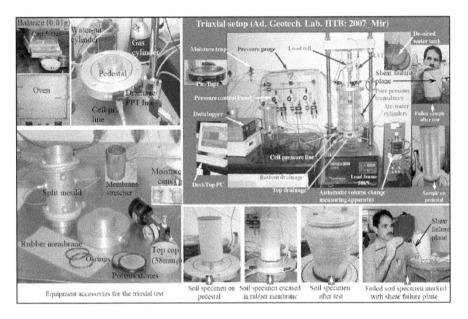

FIGURE 15.12   Triaxial shear testing machine setup and other accessories.

11. Static compaction machine if remolded soil specimen is to be prepared
12. Glass Perspex sheet for mixing soil
13. Relative density apparatus and required accessories for relative density test if required
14. Split mold/membrane stretcher
15. Desiccator
16. Oil or grease
17. De-aired water supply tank
18. Oven
19. Pi-tape
20. Volume change apparatus/burette
21. Vacuum pump
22. Stopwatch

15.5   TESTING PROGRAM

Based on testing requirements and soil type available, the type of triaxial shear test can be chosen and the soil specimen can be prepared accordingly. Generally, there are three types of triaxial compression tests carried out on saturated soil specimens under different drainage and loading conditions.

15.5.1   Unconsolidated Undrained (UU) or Unconsolidated Quick (Q$_U$) Test without Pore Water Pressure Measurement

The unconsolidated undrained (UU) is widely suitable for saturated cohesive soils to determine the undrained cohesion such as s_u or c_u. The "UU" test is conducted in two stages (Saturation and Shear Stages) in the laboratory as described below.

15.5.1.1   Preparation of Soil Specimen

1. The specimen is prepared as explained in Section 15.3.3. It should be assured that the soil sample is trimmed to its final size as required with a L/D ratio of 2. Also, determine the moisture content of trimmed soil as initial water content.
2. After preparing the soil specimen as described in the above step, measure the initial diameter (D_o) and initial length (L_o) of the soil specimen before testing it. Also, measure the thickness of the rubber membrane (t_m) and record modulus (E_m) of rubber membrane.
3. Weigh the soil sample and record its wet weight, W_1 (nearest to 0.01 g).
4. Flush the base pedestal and transducer housing with de-aired water carefully.
5. Apply a thin layer of grease or petroleum jelly to the pedestal base of the confining chamber base and the top loading cap so that air-tight connection is obtained where the O-rings seal the specimen against water leakage either into or out of the soil specimen.
6. Place a saturated porous stone on the pedestal overlaid by a saturated filter paper.
7. Now place the soil sample on the filter paper very carefully and wrap the rubber membrane over it by using a membrane stretcher. Make sure that the rubber membrane covers the bottom pedestal also.
8. Place a saturated filter on the top surface of the soil sample overlaid with a saturated porous stone carefully.
9. Place the top loading cap on the porous stone and seal the soil specimen within the rubber membrane with O-rings at the base pedestal and on the top loading cap by using a membrane stretcher. Make sure that the soil sample is sealed air-tight and there should be no water leakage either into or out of the soil sample.
10. Connect the top drainage line from the base of the triaxial cell with a top loading cap carefully.
11. Now place the triaxial cell cover on the pedestal base so as to cover the soil sample inside the cell and tighten its screws carefully. Make sure that the piston of the load cell is not in touch with the top loading cap and it is well lubricated for its smooth vertical movement without any frictional resistance.
12. Check that all the necessary connections are intact and all the drainage valves are de-aired and closed.
13. Open the air release valve at the top of triaxial cell and allow de-aired distilled water into the triaxial cell surrounding the soil sample. Allow the

water to flow out through the air release valve steadily and make sure that all the air bubbles are removed out of the triaxial cell.

14. The soil sample is ready for the saturation stage followed by the shearing stage.

15.5.1.2 Triaxial Saturation Stage for "UU" Test (application of cell pressure)

1. The various initial parameters of the soil sample before the saturation stage are recorded as:
   - Length $= L_o$ (cm)
   - Diameter, D_o (cm)
   - Cross-sectional area, $A_o = \pi/4 * (D_o)^2$ (cm²)
   - Volume, $V_o = A_o * L_o$ (cm³)
   - Wet mass, M_s (g)
   - Water content, w_o (%)
   - Total soil particle density, $\rho_b = M_s/V_o$ (g/cm³)

   - Dry density, $\rho_d = \dfrac{\rho_s}{1 + w_o/100}$ (g/cc)

   - Voids ratio, $e_o = \dfrac{\rho_s}{\rho_d} - 1 = \dfrac{G * \rho_w}{\rho_d} - 1$

   - Degree of saturation, $S_o = \dfrac{w_o G}{e_o} = \dfrac{w_o \rho_s}{e_o}$ (%)

2. Before shearing the soil sample in the "UU" test, make sure that the soil sample is fully saturated. Therefore, to saturate the soil sample, apply the first increment of cell pressure of about 35 kP and watch/measure the pore water pressure being developed under the applied cell pressure until it remains constant. However, make sure that the piston of the load cell is not in touch with the soil sample.

Note: It may be noted that the cell pressure may be applied in various steps in order to saturate the soil specimen. However, the magnitude of the cell pressure increment is approximately equal to the effective stress (σ'_v) acting on the soil sample in the field.

TABLE 15.4

Cell pressures for saturation stage

Sr. No.	Depth z (m)	Cell pressure increment for saturation stage (kPa)		
		Step-1	Step-2	Step-3
1	less than 5	$2.0\sigma'_v$	$2.5\sigma'_v$	$3.0\sigma'_v$
2	5–10	$1.5\sigma'_v$	$2.0\sigma'_v$	$2.5\sigma'_v$
3	10–40	$1.0\sigma'_v$	$1.50\sigma'_v$	$2.0\sigma'_v$
4	greater than 40	800	1200	1600

Therefore, the value of σ'_v should be specified by the site engineer; however, in the absence of this information, it may be calculated from the equation:

$$\sigma'_v = \sigma_c = \rho g * z = \gamma * z \ (kPa) \tag{15.11}$$

Where: ρ is the soil density (g/cm^3) and z is the relevant depth below the surface (m) at which the soil sample has been collected and $g = 10 \, \text{m/s}^2$ (approximately).

The approximate values of cell pressure increments based on effective stress in the field are given in Table 15.4.

The process for soil sample saturation is also briefly described in sub-section 15.2.4.1 and the B-factor (refer Equation (15.4)) is determined as: $B = \frac{\Delta u_1}{\Delta \sigma_{c)1}}$.

If the B-value is near unity (> 0.98), then the saturation is complete and it is presumed that the soil specimen is fully saturated. However, if the B-value is less than the specified value, then continue the saturation stage by increasing cell pressure in designated increments untl the B-value is about 0.98–1.0. However, if the B-value is too small (< 0.5) against the first designated cell pressure increment, then back pressure (e.g. increased +ve pore pressure) may be applied through the drainage line.

Note: Saturation is continued by the application of alternate back pressure and cell pressure increments. Pressures may be applied either by using a small differential (5 or 10 kPa), or by using a minimum differential pressure equal to the initial effective stress. The latter is considered to be preferable because it tends to reduce the scatter of measured strengths, but it takes longer to complete.

- For example, for applying back pressure (u_b) against constant cell pressure of 100 kPa (as an example), close the drainage valve and increase pore pressure to about 90–95 kPa (check pressure transducer or select using software as the case may be). Make sure that the specified values of cell pressure (100 kPa) and pore pressure (known as back pressure, which will force flow of water into soil specimen for its saturation) are maintained constant.
- Now open the drainage valve and allow the water to flow into the soil sample. Record the response of pore water pressure, which will be increasing. Close the drainage valve when the pore water remains constant and observe the response of pore water pressure for about an hour.
- Once the pore water pressure stabilizes, record its initial value as $\Delta u_{1)i}$.
- Increase the cell pressure by the next designated increment ($\Delta \sigma_{c)2}$) and record the response of pore water pressure. However, drainage valve should be kept closed.
- When the pore water remains constant for about an hour, record the final value of the pore water pressure ($\Delta u_{1)f}$) and determine the net difference between $\Delta u_{1)i}$ and $\Delta u_{1)f}$ as net increase in pore water pressure against cell pressure increment of ($\Delta \sigma_{c)2}$).
- Determine the B-factor to ascertain the degree of saturation: $B = \frac{(\Delta u_{1)f} - \Delta u_{1)i})}{(\Delta \sigma_{c)2})}$.
- If the B-value is near unity (> 0.98), then the saturation is complete and it is

TABLE 15.5
Saturation Stage (Hand Compacted Soil Specimen (Mir 2010))

Sample Depth: 2m, Dry Unit Weight of Soil Solids: 13.6 kN/m³, Target Effective Cell Pressure = 500 kPa

Step No.	Cell press, σ_c (kPa)	Back pressure, u_b (kPa)	Pore pressure, u (kPa)	$\Delta\sigma_c = \sigma_{c2} - \sigma_{c1}$ (kPa)	$\Delta u = u_2 - u_1$ (kPa)	B-factor = $\Delta u/\Delta\sigma_c$	Remarks
–	0	0	0	–	–	0.00	Make sure that all initial values are set to zero
1	35	0	27	35	27	0.77 (<1)	Apply 1st increment of σ_c
	35	30	30	–	–	–	Apply back pressure
2	85	30	77	50	47	0.94	Apply 2nd increment of σ_c and close back pressure *
	85	80	80	–	–	–	Apply back pressure
3	135	80	130	50	50	1.00	Apply 3rd increment of σ_c and close back pressure *
4	235	80	230	100	100	1.00	Apply σ_c
5	335	80	330	100	100	1.00	Apply σ_c
6	435	80	430	100	100	1.00	Apply σ_c
7	535	80	530	100	100	1.00	Apply σ_c
8	650	80	645	115	115	1.00	Apply σ_c
9	650	150	645	–	---	–	Apply back pressure of 150 kPa to get an effective cell pressure of 500 kPa.

Notes

* Maintain the pressure for at least 3–4 hours or until the pore pressure reading has stabilized (i.e. is not changing at a rate greater than 5 kPa/hour). When the pore pressure reading has stabilized under the final pressure increment (cell pressure, σ_c), the specimen is ready for the shear stage in the case of the "UU" test or for the isotropic consolidation stage in the case of the "CU" or "CD" test.

2. Once the saturation stage is complete, check the soil specimen for any volume change as given in Table 15.6. The various initial parameters of the soil sample after saturation stage. In the saturation stage, the water is allowed into soil void pores to remove air so that the soil sample is fully saturated. Therefore, it is presumed that the amount of water entered into the soil sample occupies the void space in the soil sample and as such there is no change in the initial parameters of the soil sample. But it may be noted that the volume of water ($\Delta V_w = \Delta V_v$) entered into the soil void pores increases the wet weight of the soil sample by ΔV_w(cm³), which can be recorded through volume change apparatus after saturation is complete. Therefore, the actual mass of soil sample after the saturation stage is M_s)sat $= M_o + (\rho_w * \Delta V_1)$ (g). Since the amount of this volume of water is negligible as compared to wet mass of the soil sample, therefore, in most of the cases, it is ignored. However, the initial volume (V_o) may not change due to the fact that the water occupies the void space in the soil sample by expulsion of air or by forcing air to dissolve into solution under pressure and the soil sample is not susceptible to swelling when imbibed with water. However, for research purposes, actual measurements of volume change may be recorded and the modified dimensions of the saturated soil specimen may be computed as illustrated in Table 15.5. It may be noted that post-saturation soil specimen dimensions will be taken as initial dimensions for the next triaxial test stage (e.g. for shearing stage in the case of the "UU" test and for consolidation stage in the case of the "CU" or "CD" test).

TABLE 15.6

Test Data for Triaxial Saturation Test (Assuming Soil Specimen Size is 20cm Long having 10cm Diameter, e.g. L/D=2)

Before Saturation Stage			After Saturation Stage		
Initial Length of Sample (cm)	L_o	19.30	Volume Change (ml)	δV_{sat}	-4.05
Initial Diameter of Sample (cm)	D_o	10.058	Length after Saturation (cm)[#]	L_{sat}	19.32
Initial Area of Sample (cm^2)	A_o	79.449	Area after Saturation (cm^2)	A_{sat}	79.59
Initial Volume of Sample (ml)	V_o	1533.361	Diameter after Saturation (cm)	D_{sat}	10.067
L/D Ratio	L/D	>2	Volume after Saturation (ml)	V_{sat}	1537.414

Notes

$L_{sat} = L_o\left(1 - \frac{1}{3}\frac{\varepsilon_v}{100}\right)(cm)$; $\varepsilon_v = \frac{\delta V_{sat}}{V_o}$; $\delta V_{sat} = \delta V_w = $ *Directly recorded from volume change line*

$V_{sat} = V_o - \delta V_{sat}$ (cc); $A_{sat} = A_o\left(1 - \frac{2}{3}\frac{\varepsilon_v}{100}\right)(cm^2)$; $V_{sat} = A_{sat}*L_{sat}$ (cc); $D_{sat} = \sqrt{\frac{4A_{sat}}{\pi}}$ (cm)

presumed that the soil specimen is fully saturated. However, if the B-value is less than the specified value, then repeat the above steps until the required degree of saturation is achieved, as illustrated in Table 15.5.

Note: It may be noted that cell pressure adopted in the above saturation is tentative. But for actual design problems, cell pressure is fixed before starting the saturation stage or any other stage. In that case, the first saturation stage is carried out and it is to be kept in mind that the saturation stage be completed within a fixed value of cell pressure. In case the saturation stage is completed below the fixed value of cell pressure, then cell pressure is increased to a fixed value before starting the consolidation stage or Ssearing stage as the case may be (e.g. no consolidation is allowed in the case of the UU test and shearing stage is directly started after achieving full saturation of soil specimen). If the pore pressure has not stabilized within 8 hours of application of the cell pressure, the engineer should be notified.

15.5.2 TRIAXIAL SHEARING STAGE FOR "UU" TEST (APPLICATION OF DEVIATOR STRESS)

Once the saturation stage of triaxial testing is complete, make sure that all drainage valves are closed and it is ensured that effective cell pressure ($\sigma_c' = \sigma_c - u_b$) remains constant. Now increase the cell pressure (effective confining pressure) to the desired value under which the soil sample is to be sheared in the "UU" test.

The various initial parameters required for an undrained shear test are:

- Constant cell pressure, σ_c (kPa)
- Back pressure, u_b (kPa), if used during saturation stage
- Eff. stress, σ'_c (= $\sigma_c - u_b$) (kPa)

- Post saturation length, $L_{sat} \cong L_o$(mm)
- If presumed that there is no volume change in soil specimen during saturation stage
- Post saturation X-sec $A_{sat} \cong A_o$ (cm^2)
- Strain rate, ε_r (mm/min), depends on data derived from consolidation stage. The maximum strain rate to be applied is:

$$\varepsilon_r = \frac{\varepsilon_f L}{100 t_f} \text{mm/minute} \tag{15.12}$$

Where: ε_f = failure strain (tentatively assumed for undrained test), L = initial is length or height of soil specimen taken during shearing test (e.g. post-saturation in the UU test), and t_f = time to failure.

The rate of strain given in Hight's procedure is 1% per hour. But without pore pressure measurement at the mid-height, to ensure pore pressure equalization the rate of strain should be not greater than that derived from consolidation data on a similar specimen. The rate of strain should not exceed 1% per hour.

Note: The time required to fail the soil sample in undrained tests based on 95% pore pressure equalization within the specimen was given by Blight (1964).

Once the strain rate is fixed, the triaxial shearing stage for the "UU" test for undrained behavior of soil specimen is briefly outlined as follows:

1. Bring the load cell piston in touch with the top loading cap on the soil sample very carefully.
2. Set the zero readings of load cell/proving ring and LVDT/dial gauge and record their initial readings, respectively.
3. Select the designated strain rate; usually 1%/min or 1.27 mm/min or 0.021 mm/s and start the load application.
4. Record the load readings corresponding to 50 divisions of displacement reading until the soil sample fails or at least 20% axial strain is achieved.
5. Once the soil sample fails (e.g. load decreases after peak point), record at least 6 to 8 points after failure to access the residual or ultimate shear strength of the soil sample.
6. Stop the test and release the cell pressure (e.g. after triaxial shear failure of soil specimen), and dislodge the triaxial cell cover from the triaxial cell base carefully.
7. Remove the specimen from the triaxial cell and place it on a white sheet of paper to sketch the mode of shear failure plane.
8. Weigh of the soil specimen again, and put the specimen into the oven for 24 hours for determination of final water content.

Note: In the "UU" test, It is presumed that the water content of the soil sample remains constant during the shearing stage by increasing the deviator stress. Hence

TABLE 15.7
Test Data for the Stress – Strain Plot for "UU" Triaxial Shear Test

Stress-Strain Characteristics of "UU" TEST Sample No.:

Strain L.C. (mm/dvn): = 0.01 mm; PR L.C. (kg/dvn):?; D_o (mm):?; A_o (cm²):?

SITE:	Flood Channel Srinagar;		L_o (mm):?; V_o (cm³):?; 1cc = 10^{-6} m³		
Sample Type	Disturbed & Untreated	BH	I	Depth (m)	1.5

Designated confining stress or cell pressure = $\sigma_c = \sigma_3$ = (kg/cm² or kN/m²) [1(kg/cm²) = 100 kPa = 10t/m²]

Strain Divns	$\Delta L = LC *$ Strain Divns (mm)	Strain, $\varepsilon_a = (\Delta L/L_o) * 100$ (%)	$L = L_o - \Delta L$ (cm)	PR* (Divns)	Load, P = PR * LC (kg)	A_c (cm²)	$\sigma_d = \Delta\sigma = P/A_c$ (kg/cm²)	$\sigma_1 = \sigma_3 + \sigma_d$ (kg/cm²)	$C_u = \sigma_d/2$ (kg/cm²)	C_u (kPa)
0	0									
50	0.5									
100	1.0									
150	1.5									
200	2.0									
250	2.5									
n-1..	–									
n....	ΔL_n									

L_s

$\sigma_{3(1)}$(kPa) known	$\sigma_{d)max}$(kPa) ?	$\sigma_{1(1)}$(kPa) ?	$\varepsilon_{f(1)}$(%) ?	$C_{u)max}$(kPa) ?

* : If automatic triaxial equipment is used, then **PR * (Divns)** can be avoided and the load can be directly recorded by the data logger (this data will plot one Mohr circle)

the test is representative of soils in construction sites where the rate of *construction is very fast* and the pore waters do not have enough time to dissipate.

- Since no drainage is allowed during the test under the application of σ_d, the volume of the soil sample remains and the pore water pressure increases. Therefore, the various parameters in unconsolidated-undrained shear test (without pore pressure measurement) are computed as (Table 15.7):
- Designated confining stress or cell pressure $= \sigma_c = \sigma_r$ (kg/cm^2 or kN/m^2)
- Change in length after triaxial shear test, ΔL (cm) for each time increment
- Calculate axial strain, $\varepsilon_a = (\Delta L/L) * 100$ (%) for each time increment (Note that "L" is length of soil specimen after the Saturation Stage in the case of the "UU" test and length of soil specimen after Consolidation Stage in the case of the "CU" test).
- Corrected cross-sectional area for each value of strain is computed as:

$$A_c = \frac{A_0}{1 - (\varepsilon_a / 100)}$$

- Calculate the load for each time increment, P (from load cell-data logger) (kg or kN)
- Corrected load, P_c (kN) $= P * F$ (where F is a conversion factor if required so)
- Calculate the deviator stress, $\sigma_d = \Delta\sigma = P/A_c$ (kg/cm^2 or kN/m^2) each time increment
- Corrected deviator stress, $\sigma_{d)c} = \sigma_d - \left(\frac{4E_m * (t_m/1000) * (\varepsilon_a/100)}{(D_c/100)} \right)$ after applying membrane correction

Where: E_m (kPa) is the modulus of rubber membrane, t_m (mm) is the thickness of the rubber membrane, ε_a (%) is the axial strain.

D_c (mm) is the corrected post-saturation diameter of the sample ($= \sqrt{4A_c\pi}$), and A_c is the corrected post-saturation average cross-sectional area of the sample.

- Calculate the axial stress, $\sigma_a = \sigma_1 = \sigma_d + \sigma_c$ (kg/cm^2 or kN/m^2) for each time increment

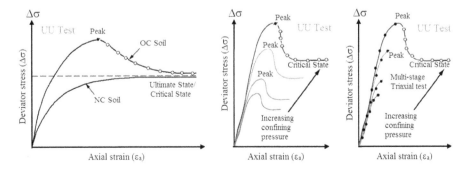

FIGURE 15.13   Stress-strain behavior of soil specimens after triaxial shear test.

- Calculate the axial stress, $\sigma_{a)f} = \sigma_{1)f} = \sigma_{d)f} + \sigma_c$ (kg/cm² or kN/m²) for each time increment

15.5.2.1 Observation Data Sheet and Analysis for "UU" Triaxial Shear Test

1. Record the post–shear test data for each designated chamber stress, as shown in Table 15.7 and plot the stress-strain diagram ($\sigma_d = P/A_c$ versus axial strain $\varepsilon_a = \Delta L/L$) as shown in Figure 15.13 for determination of peak stress and failure strain, respectively. (Note that "L" is length of soil specimen after Saturation Stage in case of "UU" test, refer Table 15.7.)

   Note: It may be assured that at least 6 to 10 readings may be recorded after achieving peak load value or until critical state is obtained as shown in Figure 15.13.

2. Compute corrected Xec. Sec Area post–shear test, $A_{sc} = \dfrac{A_c}{1 - \varepsilon_a/100} * \dfrac{1}{10000}$ (m²), where A_c = Post saturation cross sectional area

3. Compute specimen length after shearing, L_s (cm): $L_o - \Delta L_n = L_{sat} - \Delta L_n$

4. Compute major principal stress, $\sigma_1 = \sigma_{d)c} + \sigma_3 = \sigma_{d)c} + \sigma_c$ (kPa)

5. Compute minor principal stress, $\sigma_3 = \sigma_c$ (kPa)

6. Compute stress ratio, σ_1/σ_3

7. Mean stress, $p = \dfrac{(\sigma_1 + 2\sigma_3)}{3}$ (kPa)

8. Eff. mean stress, $p' = p - u$ (kPa) (if pore water pressure is measured in UU test)

9. Mean stress ratio, $M = \dfrac{q}{p} = \dfrac{\sigma_{d)c}}{p}$

10. Compute ϕ' (Degrees) $= \sin^{-1}(3M/6 + M)$

11. Compute $k_o = 1 - \sin\phi'$ (NC Soils: Jakay, 1944)

12. Undrained strength, $s_u = c_u = \dfrac{q}{2} = \dfrac{\sigma_{d)c}}{2}$ (kPa)

13. Strength ratio, $s_r = \dfrac{s_u}{\sigma_o}$

14. Compute A-factor, $A = \dfrac{\Delta u}{\Delta \sigma_1 - \Delta \sigma_3} = \dfrac{\Delta u}{\Delta \sigma_d}$ (if pore water pressure is measured in UU test)

15. Repeat the test for the second specimen and third specimens for designated confined pressures.

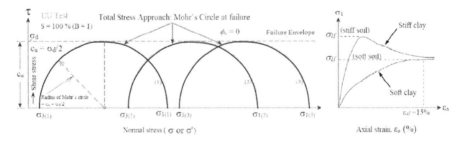

FIGURE 15.14 Mohr-Coulomb failure envelope for UU triaxial shear test.

16. At the end of the shear test and computation of various test parameters, plot the following graphs:

- Axial strain, ε_a (%) versus deviator stress, σ_d (kPa)
- Axial strain, ε_a (%) versus mean stress ratio, $M = \dfrac{q}{p'} = \dfrac{\sigma_{d)c}}{p'}$
- Plot Mohr circle based on major principal stress ($\sigma_1 = \sigma_3 + \sigma_d$) and minor principal stress (σ_3) at failure. They should give the same σ_d value as shown in Figure 15.14. The radius of the Mohr's circle is equal to c_u.
- Plot the Mohr's circle using major and minor principal stresses (σ_1 & σ_3) at failure and the peak value on the Mohr's circle gives the maximum value of the deviator stress. The radius of the Mohr's circle is taken as half of the maximum deviator stress, $R = [\sigma_d/2]$.
- Now draw a straight line, which is tangent to all Mohr's circles. This gives c_u with a horizontal line, i.e. $\phi_u = 0$. Therefore, this test is called $\phi = 0$ test as shown in Figure 15.14 and the undrained shear strength is computed as:

$$c_u = s_u = \frac{\sigma_d}{2}$$

15.5.3 RESULTS AND DISCUSSIONS: UU TEST

The stress-strain behavior of soil samples tested in an unconsolidated-undrained (UU) triaxial test is shown in Figure 15.13. From Figure 15.13, it is seen that OC soils (e.g. dense sands and stiff clays) develop a well-defined peak at smaller failure strain compared to NC soils (e.g. loose sands and soft clays) which exhibit very non-linear behavior without well-defined peak. However, both types of soils posses same ultimate shear strength. The OC soils exhibit brittle behaviour compared to NC soils. Therefore, OC soils or brittle materials can be sheared at higher strain rate compared to NC soils or plastic materials. The maximum or peak compression strength at failure and corresponding failure strain of OC soils is taken corresponding to peak point. However, in the case of NC soils, there is no well-defined peak; therefore, the maximum compression strength at failure is taken at about 15 to 20% failure strain, as shown in Figure 15.13.

Mohr-Coulomb failure envelope for "UU" triaxial shear test based on total stress approach is shown in Figure 15.14. For OC soils, the failure stress is taken as $\sigma_{1)f}$ corresponding to peak point on stress-strain curve. However, in case of NC soils, the failure stress is taken as $\sigma_{1)f}$ corresponding to axial strain of about 15 to 20%. The "UU" is suitable for saturated soft clayey soils, which fail under undrained conditions of rapid loading and the state of stress in the soil sample is plotted in the form of a Mohr's circle. The failure envelope for "UU" shear test is a straight line just tangential to peak points of the Mohr's circles with shear strength parameters as $c_u = \sigma_d/2$ and $\phi = 0$. Further, it may be noted that if an identical soil sample is tested under different state of stress and subjected to rapid loading without pore water pressure dissipation, the Mohr's circle for each such soil sample has the same diameter but the circle shifts towards right side under increasing state of stress, as

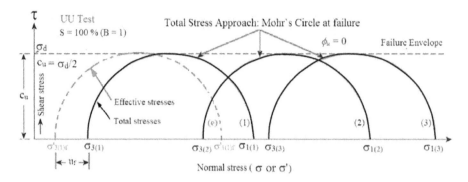

FIGURE 15.15 Mohr's circle for total and effective stresses for "UU" compression tests.

shown in Figure 15.14. Therefore, $\sigma_{1)f}$ and $\sigma_{3)f}$ are major and minor principal stresses at failure.

Furthermore, if pore water pressure is measured for each soil sample tested in the "UU" test, the principal effective stresses σ'_1 ($= \sigma_1-u$) and σ'_3 ($= \sigma_3-u$) can be computed, which yield effective cohesion c'. Since effective stresses (e.g. $\sigma'1$ & σ'_3) are independent of the cell pressure applied in the "UU" test, therefore, only one effective stress circle is plotted as shown in Figure 15.15, which indicates that an effective stress envelope cannot be drawn in the "UU" test. It also seen that the effective stress circle is separated by u_f from the total stress circle and deviator stress is same in both cases, e.g. $(\sigma'_1 - \sigma'_3) = (\sigma_1 - u) - (\sigma_3 - u) = (\sigma_1 - \sigma_3)$.

In the case of 1-D consolidation process, it has been observed that undrained cohesion (c_u) of NC clays increases with increasing depth from ground surface and the ratio of $[c_u/\sigma_{vo}]$ is constant for a given clay type and $[\sigma'_{vo}]$ is vertical effective overburden stress in the ground corresponding to undrained cohesion $[c_u]$. Many researchers have shown that the undrained cohesion $[c_u]$ of saturated NC clays is directly correlated to the plasticity index expressed as:

$$c_u = [(0.11 + 0.0037) * \sigma'_{vo}] \ (Skempton \ 1954) \tag{15.13a}$$

$$c_u = [(0.11 + 0.0037 * \log(PI)) * \sigma'_{vo}] \ (Skempton \ 1957) \tag{15.13b}$$

$$c_u = [(0.45 * (PI/100)^{1/2}) * \sigma'_{vo}] \ (Bjerrum \ \& \ Simons \ 1960) \tag{15.13c}$$

$$c_u = [(0.08 + 0.55 * PI * \sigma'_{vo}] \ (Larsson \ 1977) \tag{15.13d}$$

$$c_u = [(0.129 + 0.00435 * (PI)) * \sigma'_{vo}](Wroth \ \& \ Houlsby \ 1985) \tag{15.13e}$$

$$c_u = [(-0.09 + 0.0092 * PI) * \sigma'_{vo}](Windisch \ \& \ Yong \ 1990) \tag{15.13f}$$

Where: PI = Plasticity index (%)

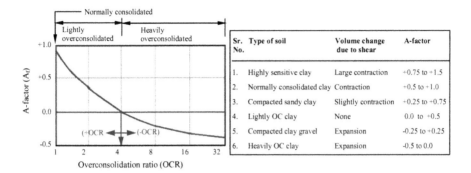

FIGURE 15.16 Typical relationship of A_f to over-consolidation ratio.

Further, in the case of pore pressure measurement, the "A"-parameter at any point may be expressed as:

$$A = \left[\frac{\Delta u}{\Delta \sigma_1 - \Delta \sigma_3} = \frac{\Delta u}{\Delta \sigma_d} \right] \qquad (15.14a)$$

Where: $\Delta \sigma_d$ = Deviator stress

Δu = Pore water pressure change under rapid loading during "UU" test.

Similarly, at failure, the "A"-parameter may be expressed as:

$$A_f = \left[\frac{\Delta u_f}{(\Delta \sigma_1 - \Delta \sigma_3)_f} = \frac{(u_f - u_o)}{(\sigma_1 - \sigma_3)_f} \right] \qquad (15.14b)$$

Where: u_f = Pore water pressure at failure when the soil sample fails

u_o = Initial pore water pressure at the start of shear test, and

$(\sigma_1 - \sigma_3)_f$ = Deviator stress at failure (peak value).

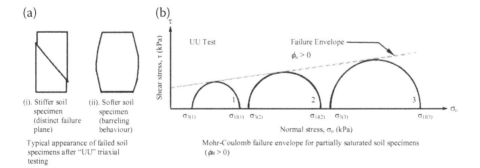

FIGURE 15.17 (a). Schematic representation of soil shear failure planes in "UU" triaxial test; (b). Mohr-Coulomb failure for partially saturated soil specimens in "UU" triaxial test.

A typical relationship between over-consolidation ratio and the "A_f"-parameter at failure is shown in Figure 15.16. Some typical values of the parameter "A_f" from compression tests are also shown in Figure 15.16.

- **Expected Test Results and Observations**

For the academic interest, it is presumed the soil sample being tested in the "UU" test in the soil testing laboratory is fully saturated clayey soil and the failure envelope is horizontal with $c_u = \sigma_d/2$ and $\phi = 0$. However, this is far away from the reality and the soil sample is not fully saturated and soil sample may exist in different classes of consistencies varying from soft to medium stiff consistency, generally referred to as partially saturated soils. Soft soils fail mainly by "bulging/"barreling" without showing distinct failure plane, however, stiffer soils always show a distinct failure plane almost oriented at an angle of 45°, as shown in Figure 15.17(a). Thus, in case of partially saturated soils, the failure envelope will not be a horizontal line and the shear strength parameters are represented by both undrained cohesion and angle of internal friction, as shown in Figure 15.17.

Based on previous discussions, it may be conceded that "UCS" may be the best alternative for testing NC saturated clays and the "UU" test may be adopted for lightly OC soils.

15.5.4 Applications/Role/Significance of and Use of "UU" Test

The test results of a "UU" test are generally used to estimate the shear strength of saturated cohesive soils for field problems where the rate of loading increases more rapidly than the dissipation of pore water pressure. An example of such a case is the stability analysis of an embankment ($\phi_u = 0$) on saturated clayey soils and subjected to dynamic loading.

15.6 CONSOLIDATED UNDRAINED (CU OR Q_C) TRIAXIAL TEST

A consolidated-undrained (CU) test is a novel technique of determination of soil shear parameters inn both total stress approach and effective stress approach. This test is widely suitable for all types of soils. In this test, a soil sample is tested in three separate stages (e.g. saturation stage, consolidation stage, and shearing stage) and drainage and loading conditions are well controlled by the operator or a researcher. The summary of drainage boundary conditions and application of loading in a triaxial compression are given in Table 15.3. This test is also known as a "Consolidated Quick- Undrained" test. This test is conducted in a well-equipped soil testing laboratory and the step-by-step procedure for each stage of this test is given as below:

15.6.1 Triaxial Saturation Stage in "CU" Test

- The process of sample preparation for consolidated undrained "CU" triaxial test is the same as for the **"UU" test** described in Section 15.5.1.1.
- The procedure for conducting triaxial saturation stage for the consolidated

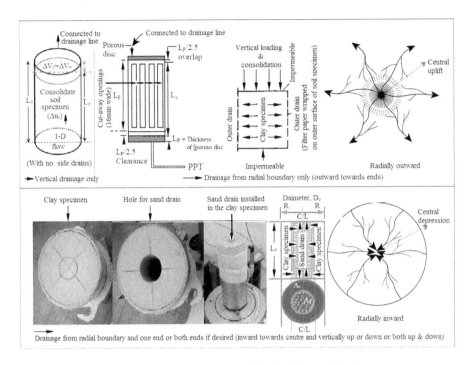

FIGURE 15.18   Drainage boundary conditions for triaxial consolidation test.

undrained "CU" triaxial test is the same as for the "UU" test described in Section 15.5.1.2.

15.6.2   Triaxial Consolidation Stage in "CU" Test

The consolidation method or process is a novel technique in which a saturated soil mass (e.g. saturated clays or clayey soils or generally saturated fine grained soils) is consolidated vertically downward by expulsion of water from soil void pores under gradually applied loading. This process is generally known as consolidation stage in a trixaial compression test in which the pore water pressure is allowed to dissipate out of the soil void pores.

In the consolidation stage, the soil specimen is consolidated with free drainage under the constant confining or cell pressure (σ'_c). Desired drainage has to be chosen for consolidation of the soil specimen as per research program before consolidation stage. Various drainage boundary conditions are (e.g. water can be drained from soil specimen by using any of these boundary conditions) illustrated in Figure 15.18):

a. Drainage from one end only (e.g. 1-D vertical flow preferably from top of soil specimen)

b. Drainage from both ends (e.g. 2-D vertical flow preferably from top & bottom of soil specimen)

c. Drainage from radial boundary only (e.g. either outwards from center of the soil specimen or towards center as desired)

d. Drainage from radial boundary and one end (e.g. combination of above as desired or required)

e. Drainage from radial boundary and two ends

This is very important to choose a drainage boundary condition because time to failure in shearing stage depends on these boundary conditions. Furthermore, drains are effective only if the soil permeability is less than about 10^{-8} m/s. Drains help to equalize pore pressures as well as provide a short radial drainage path. However, it should be ensured that the soil sample has been fully saturated in the saturation stage. When the saturation stage is completed, then the soil specimen is iso-tropically consolidated under the mean effective stress, p', which is computed by the following expression:

$$p' = \left[\frac{\sigma_1' + 2\sigma_3'}{3} = \frac{\sigma_3' + 2\sigma_3'}{3} \right] = [\sigma_3' = \sigma_c'] \qquad (15.15)$$

Where: σ_1' = Effective axial stress (equal to σ_3' in Consolidation stage), and
σ_3' = Effective radial stress or all-around stress equal to confined stress (σ_c').

In the consolidation stage, the hydrostatic pressure remains unchanged and there is volume change due to expulsion of water and decrease in void ratio. However, it should be ensured that the load cell piston does not touch soil specimen during consolidation stage and confining or cell pressure (σ_c') remains constant throughout the consolidation stage.

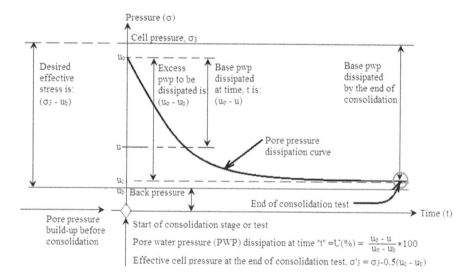

FIGURE 15.19   Representation of pore pressure dissipation under constant cell pressure.

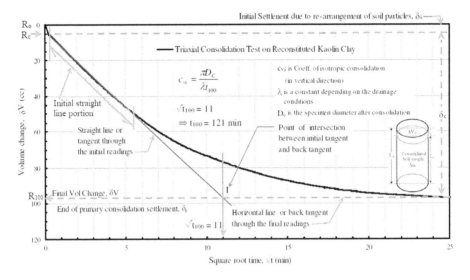

FIGURE 15.20   Time-compression curve for clayey sample during triaxial consolidation by √t method (with radial and one boundary drainage (Mir 2010).

The following steps may be followed for conducting a triaxial consolidation test:

1. Record the post-saturation parameters (these are presumed same as before saturation stage).
2. Check that all drainage valves have been de-aired and are closed before the start of the consolidation test and the loading piston is not in touch with the soil specimen.
3. Apply a cell pressure (σ_c) equal to the desired consolidation pressure and should be maintained constant throughout the consolidation stage or test.
4. Record the initial pore water pressure (u_o) developed against the desired consolidation pressure before the start of the consolidation test.
5. Open the drainage valve and allow water to drain out of the soil specimen under constant cell pressure (σ_c) and record the dissipation of the pore water pressure until the consolidation completes under constant effective cell pressure, as shown in Figure 15.19.
6. Also, record time readings (e.g. volume change readings, δV against time, t) for determination of time for 100% primary consolidation.
7. Plot the volume change readings (δV) against time, ($\sqrt t$) using Taylor's √t Method (or Casagrande's log-t method or Rectangular Hyperbola Method, or whichever method is applicable) for the consolidated soil sample, as shown in Figure 15.20.

Note: The "√t Method" is credited to Taylor (1942, 1948), who found that the method provided reasonably reliable results (determination of coefficient of consolidation and time for 100% consolidation) for clayey soils. The log-t

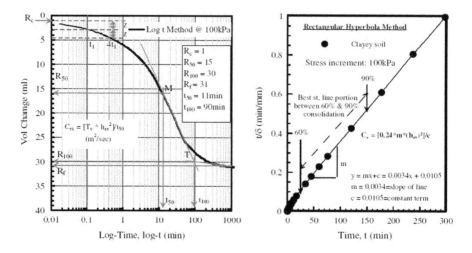

FIGURE 15.21 Time-compression curve for clayey sample during triaxial consolidation by Log-t method and rectangular hyperbola method (Mir 2010).

method is credited to Casagrande (1936) and Casagrande and Fadum (1940) who found that the method provided reasonably reliable results for fine grained soils. However, these two methods are not suitable for silt dominated fine grained soils and fly ash. Therefore, the Rectangular Hyperbola Method is credited to Sridharan et al. (1987) who found that the method (a plot between "t/δ versus t") provided reasonably reliable results which may be suitable for almost all types of fine grained soils and fly ashes.

8. Relationship between the volume change and time (Casagrande method) and rectangular hyperbola method during triaxial consolidation under an all round pressure can also be plotted if desired, as shown in Figure 15.21.

9. The consolidation stage can be terminated when at least 95% dissipation is reached. The method of calculating pore pressure dissipation in terms of degree of consolidation is given by Equation (15.16):

$$U = \left[\frac{(u_o - u_f)}{(u_o - u_b)} * 100 \right] (\%) \tag{15.16}$$

10. Once the degree of consolidation is achieved between 95% and 98%, terminate the consolidation stage by closing the drainage valve and record the end of stage pore pressure measured at the specimen base, denoted by u_c. The effective confining pressure σ'_c can be computed as:

$$\sigma'_c = [\sigma_c - \bar{u}] \tag{15.17}$$

Where: $\bar{u}$ = Mean pore pressure within the soil sample.

TABLE 15.8

Triaxial Consolidation Test Observations and Analysis

At the End of Consolidation (Post-Consolidation Data) in Undrained Triaxial Shear (CU) Test (Mir 2010)

Cell pressure, σ_c (kPa)	Back pressure, u_b (kPa)	Eff. Stress, σ' (= σ_c - u_b) (kPa)	Post-consol length, L_c (cm)	Post-consol X-sec, A_c (cm)	Final consol. void ratio, e_c	Time to failure, t_f (min)	Strain rate, ε_r (mm/min)
550	50	500	19.38	74.52	0.795	212	0.0458

For a saturated soil sample, the pore water pressure is through the bottom drainage valve, which is taken equal to the pore water pressure at the top of the soil sample. Therefore, assuming the pore water pressure distribution within the whole soil sample to be parabolic, the mean pore water pressure within the soil sample is computed as:

$$\bar{u} = \left[\frac{2}{3}u_c + \frac{1}{3}u_b \right] \tag{15.18}$$

Note: If the measured pore pressure u_c is close to applied back pressure, then the mean pore water ($\bar{u}$) can be assumed equal to the arithmetical mean pressure equal to

$\frac{1}{2}(u_c + u_b)$.

11. When the dissipation of pore water pressure remains constant, then close the drainage valve and record the reading of total change in volume (δV_c) during consolidation by volume change apparatus. The consolidated soil specimen is now ready for the compression stage.

15.6.2.1 Observation Data Sheet and Analysis for "CU" Triaxial Consolidation Test

Once the consolidation stage of the triaxial consolidation of the "CU" test is complete, close the drainage valve and check that cell pressure is maintained constant. Triaxial Consolidation Test Observations and analysis are given in Table 15.8. Before the start of the triaxial shear test, the post-consolidation data analysis is worked as follows:

- *Specimen dimensions after consolidation stage*

Since there is about 100% consolidation achieved during consolidation stage and the specimen dimensions change drastically. Therefore, specimen dimensions in the post-consolidation stage are determined, which are used as initial parameters for shearing stage. Specimen dimensions computed after the consolidation stage are:

i. Specimen length after consolidation,

$$L_C = L_o\left(1 - \frac{1}{3}\frac{\varepsilon_{VC}}{100}\right)(mm) \tag{15.19}$$

Where: ε_{VC} = Volumetric strain at the end of consolidation stage given as

$$= \frac{\Delta V_C}{V_o} * 100(\%) \tag{15.20}$$

and ΔV_C = Volume change during consolidation stage = $V_o - V_f$ (cm^3), directly measured at the end of consolidation stage by volume change apparatus.

ii. Specimen Xec-Sec area after consolidation,

$$A_c = A_o\left(1 - \frac{2}{3} * \frac{\varepsilon_{V_c}}{100}\right)(mm^2) \tag{15.21}$$

iii. Specimen diameter after consolidation,

$$D_C = \sqrt{\frac{4A_C}{\pi}}\ (mm) \tag{15.22}$$

iv. As a check, specimen volume after consolidation,

$$V_C = \frac{A_C L_C}{1000} = V_o - \Delta V_C \tag{15.23}$$

Equations (15.19 to 15.23) are based on elastic theory for a small volume change, assuming a Poisson's ratio of 0.5 for saturated clays.

The consolidated length (L_C) and the consolidated area (A_C) are taken as initial dimensions of the soil specimen for triaxial shear test. These parameters are used for calculating axial strain ($\varepsilon_a = \Delta L/L_c$), deviator stress ($\sigma_d = P/A_c$) in the shearing stage, respectively.

- **Coefficient of consolidation (C_{vi})**

In the case of the isotropic triaxial consolidation test, the coefficient of consolidation is denoted by C_{vi}, which can be determined by various methods available such as √t-method, log-t method, and rectangular hyperbola method.

At the end of the consolidation stage, the amount of water drained from the soil sample is taken equal to the volume change of the saturated soil sample. During the consolidation stage, time readings for volume change (δV) in terms of expulsion of water with fixed time intervals (√t) are recorded. At the end of the consolidation test, a consolidation curve is plotted between (δV) and (√t), as

TABLE 15.9

Factors for Calculating C_{vi} (BS 1377: Part 8-1990)

Sr. No.	Drainage boundary conditions during consolidation stage	Values of λ	
		L/D = 2	L/D = r
1	Drainage from one end only	1	$r^2/4$
2	Drainage from both ends	4	r^2
3	Drainage from radial boundary only	64	64
4	Drainage from radial boundary and one end only	80	$3.2(1+2r)^2$
5	Drainage from radial boundary and two ends	100	$4(1+2r)^2$

shown in Figure 15.20. From Figure 15.20, it is seen that about 45 to 50% consolidation curve is almost linear. Therefore, a tangent or a straight line is drawn through this linear portion to bisect the volume change axis at a point "R_c" and the back tangent drawn from the end of the consolidation at the point of intersection "I." The point "R_c" represents the corrected initial or start point of the primary consolidation for the consolidation curve. Similarly, the end point of the consolidation curve represents about 97 to 100% degree of consolidation and the back tangent drawn from the end point of the consolidation curve bisects the volume change axis at "R_{100}," which represents about 100% pore water pressure dissipation. The difference between the compression readings "R_{100} & R_c" represents the primary consolidation (δ_c). Now, draw a vertical line through the point of intersection "I" onto the ($\sqrt{t}$) axis, which represents the required time ($t_{100} = \sqrt{t} * \sqrt{t} = t$) for about 97 to 100 degrees of consolidation. Once the value of t_{100} is known, the value of coefficient of "isotropic consolidation" is calculated as:

$$C_{Vi} = \left[\frac{\pi D_c^2}{\lambda t_{100}} \right] \quad (15.24a)$$

It may be noted if D_c is measured in mm and time (t_{100}) is measured in minutes in Equation [15.24a], then c_{vi} is expressed in m²/year as:

$$C_{vi} = \left[\frac{\pi \left(\frac{D_c}{1000} \right)^2}{\lambda \left(t_{100}/60 * 24 * 365 \right)} \right] = \left[\frac{1.652 D_c^2}{\lambda t_{100}} \right] (m^2/year) \quad (15.24b)$$

Where: D_C = Specimen diameter after consolidation stage

t_{100} = Consolidation against 100% consolidation from the curve (Figure 15.20), and

λ = is a constant depending on the drainage boundary conditions and L/D ratio, which are given in Table 15.9. However, for drainage from radial boundary only, λ is independent of r.

TABLE 15.10

Calculation of C_{vi} from t_{100}

Sr. No.

Initial Specimen Diameter, $C_{vi} = \left[\dfrac{\pi C_{vi} D_c^2}{\lambda t_{100}} = \dfrac{N}{t_{100}}\right] (m^2/year\,approx)\,[15.24a]$

D (mm),	**(Values of** $N = \frac{\pi D_c^2}{\lambda}$ **)**			
	No side drains		**with side drains#**	
	r = 2	**r = 1**	**r = 2**	**r = 1**
1 38	2400	9500	30	83
2 50	4100	16500	52	140
3 100	17000	66000	210	570
4 150	37000	150000	460	1300

t_{100} is in minutes; $r = L/D$; #: "With side drains" applies to drainage from radial boundary and one end.

The value of coefficient of "isotropic consolidation" is also expressed as:

$$C_{vi} = \left[\frac{\pi D_c^2}{\lambda t_{100}}\right] = \left[\frac{N}{t_{100}}\right] (m^2/year\ approx) \qquad (15.25)$$

Where: $N = \frac{\pi D_c^2}{\lambda}$ which relates to typical nominal specimen diameters

Relationships between C_{vi} (m²/year) and t_{100} (minutes) for typical specimen sizes with L/D ratios of 2:1 and 1:1 and for the two most usual drainage conditions (from one end only, and from one end and radial boundary) are summarized in Table 15.10. In this table, the factor N replaces $\pi D_c^2/\lambda$ in Equation (15.24a) and the values of N relates to typical nominal specimen diameters. The calculated value of C_{vi} is reported to two significant Figures.

Note: The value of c_{vi} derived in this way should be used with caution. Calculated values of c_{vi} should be used only for estimating the rate of strain in the triaxial compression test and not in consolidation or permeability calculations. When side drains are used, the value of c_{vi} derived in this way can be grossly in error.

Similarly, the coefficient of consolidation (m²/year) for Radial and one end drainage for "CIU or CU" Test is computed as:

$$C_{vri} = \left[\frac{210}{t_{100}}\right] \qquad (15.26)$$

TABLE 15.11

Coefficient of Consolidation, C_{vi} from Triaxial Consolidation Stage

Sr. No.	Drainage boundary conditions during triaxial consolidation stage	C_{vi} (m^2/year or m^2/Sec)	Remarks
1	Drainage from one end only (1-D: one way vertical flow)	$\dfrac{\pi H_D^2}{t_{100}}$	H_D: Length of drainage path = average or mean soil specimen height = L or H
2	Drainage from both ends (1-D: two way vertical flow from top & bottom)	$\dfrac{\pi H_D^2}{4t_{100}}$	H_D: Length of drainage path = average or mean soil specimen height = L/2 or H/2
3	Drainage from radial* boundary only (Radius will be drainage instead of height)	$\dfrac{\pi R^2 = \pi D_D^2}{16t_{100} = 64t_{100}}$	R: Radius of soil specimen = D/2, and D is diameter
4	Drainage from radial boundary and one end (Radial + 1-way vertical flow)	$\dfrac{\pi H_D^2}{4t_{100}}\left[\dfrac{1}{1 + H_D/R)^2}\right]$	$\dfrac{\pi H_D}{16t_{100}}$ for H_D = R
5	Drainage from radial boundary and both ends (Radial + 2-way vertical flow)	$\dfrac{\pi H_D^2}{4t_{100}}\left[\dfrac{1}{1+2H_D/R)^2}\right]$	$\dfrac{\pi H_D^2}{100t_{100}}$ for H_D = 2R

Note
* "Radial" drainage applies to radial flow outwards to boundary from the center of the soil specimen.

Thus, it is seen that consolidation time (t_{100}) is inversely proportional to the coefficient of consolidation for all drainage boundary conditions (e.g. vertical and radial etc.), as summarized in Table 15.11.

• *Coefficient of volume change or 1-D compressibility (m_{vi})*

The coefficient of volume change or 1-D compressibility is defined as the volumetric strain per unit increase in effective stress in a consolidation test. Thus, for isotropic consolidation test, it is expressed as:

$$m_{vi} = \left[\frac{\Delta V_C}{V_o} * \frac{1000}{\Delta \sigma'}\right] m^2/MN \quad (in\ SI\ units\ and\ \Delta\sigma'\ in\ kPa)\ (15.27a)$$

Where: V_o = Initial volume at the start of consolidation test
ΔV_c = Change in volume during consolidation test, and
$\Delta \sigma'$ = Change in effective stress.

It may be noted that the coefficient of volume change is inverse of the bulk modulus (e.g. $m_v = 1/E$) and varies in the range of $1 * 10^{-3}$ to $1 * 10^{-4}$ (m²/kN). However, the coefficient of volume change or 1-D compressibility derived from the 1-D consolidation test (m_v) is quantitatively the same as the one derived from isotropic triaxial consolidation test (m_{vi}). Therefore, a typical approximate relationship between the two parameters is given as:

$$m_{vi} = [1.5 * m_v] \qquad (15.27b)$$

- **Coefficient of permeability or hydraulic conductivity (k)**

Unlike in a 1-D consolidation test, the coefficient of permeability (k) can also be empirically derived from the isotropic triaxial consolidation parameters "C_{vi} and m_{vi}" as follows:

$$k = \left[C_{vi} * m_{vi} * \rho_{w*}g \right] \qquad (15.28)$$

The Equation (15.28) can be expressed in terms of practical units (SI unit of system) as follows:

$$k = \left[\left(\frac{C_{vi}}{365.25 * 24 * 60 * 60} \right) \left(\frac{m_{vi}}{10^6} \right) (1 * 10^3) * 9.81 \right] (m/sec) \quad (15.29a)$$

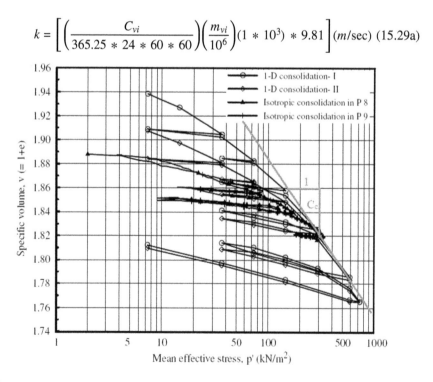

FIGURE 15.22 Isotropic and 1-D consolidation curves for loading & unloading tests (Mir 2010).

Or for practical purposes, if c_{vi} is taken in m^2/year and m_{vi} is taken in m^2/MN, then the coefficient of permeability is calculated as:

$$k_v = [0.31 * 10^{-9} * C_{vi} * m_{vi}] (m/\text{sec}) \tag{15.29b}$$

The accuracy of the coefficient of permeability depends on the reliability of the value of C_{vi}. The result will not be realistic if its value is that of silt or greater (i.e. more than 10^{-8} m/s). If side drains are used, the value of k should not be quoted.

Note: It may be noted that C_v is a function of the stress increment for a given soil and not an constant parameter. However, both coefficient of permeability (k) and coefficient of volume change (m_v) decrease with a decrease in void ratio, C_v which is a function of the ratio $[k\{1+e_o\}/\{m_v * \rho_w g\}]$, remains more or less the same within a considerable pressure range.

- *Compression Index (C_c)*

The slope of the linear portion or normally consolidated line (NCL) of the consolidation curve plotted between the void ratio (on natural scale along y-axis) and the effective stress (on logarithmic scale along x-axis) is generally known as the compression index, which is computed as (Figure 15.22):

$$C_c = \frac{\Delta e}{\ln\left(\frac{p'_o + \Delta p'}{p'_o}\right)} = \frac{(e - e_o)}{\ln\left(\frac{p'_o + \Delta p'}{p'_o}\right)} \tag{15.30}$$

- *Whether soil specimen is NC or OC?*

This is very important to know whether soil specimen being tested is normally consolidated or over-consolidated. This can be recognized simply by knowing the over-consolidation ratio (OCR), which is defined as the ratio of maximum past overburden pressure (or pre-consolidation pressure) to the present effective over-burden pressure, expressed as:

$$OCR = \left[\frac{p'_c}{\sigma'_{vo}}\right]$$

Where: σ'_{vo} = Present effective overburden pressure $[\gamma_s * z]$
Z = Depth of soil sample from ground surface

If OCR = 1 (e.g. $p_c = \sigma_o$), soil specimen is NC
> 1 (e.g. $p_c > \sigma_o$), soil specimen is OC
< 1 (e.g. $p_c < \sigma_o$), soil specimen is under-consolidated, generally ignored, and presumed For example, in case of triaxial testing, if a soil specimen is consolidated under cell pressure of σ_c = 300 kPa and then also sheared under constant cell

pressure of σ_c = 300 kPa, the soil specimen is NC (because-during consolidation stage, σ_c = 300 kPa and during shearing stage, $\sigma_s = \sigma_c$ = 300 kPa, which implies that $p'_c = \sigma_c$ (of consolidation stage) and $\sigma_{vo} = \sigma_c = \sigma_s$ = 300 kPa; hence OCR = 1, σ_s denotes cell pressure during triaxial shearing stage).

However, if the cell pressure at the end of the consolidation stage decreases (e.g. decrease σ_c from 300 kPa to 200 kPa) and the soil specimen is re-consolidated again under cell pressure of 200 kPa, in this case, when the drainage valve is opened, the water will start flowing into the soil specimen (e.g. inward drainage will take place) and the final pore water pressure at the start of consolidation stage (under cell pressure of 200 kPa) will start increasing again. Once consolidation stage under this new cell pressure (e.g. σ_c = 200 kPa) is completed. The soil specimen is sheared under constant cell pressure (e.g. σ_c = 200 kPa). Now the soil specimen will become OC, because the cell pressure, σ_c = p'_c = 300 kPa, was applied in first stage of triaxial consolidation stage and now the soil specimen is sheared under $\sigma_s = \sigma_c$ = 200 kPa of the second stage of triaxial consolidation, which now becomes σ_{vo} (< p_c, hence OCR > 1). In this case, the consolidation curve will have unloading and re-loading curves (e.g. v-logσ plot or v-lnσ plot, where v = specific volume = 1+e). Figure 15.22 shows an illustration of consolidation curves of 1-D consolidation and triaxial consolidation tests.

Also, soil specimen can be made OC by shearing it at lower cell pressure than that of consolidation stage directly. This all depends on the research proposal and testing of specimen is accordingly planned. All the testing data is usually recorded by using data acquisition system, e.g. computerized triaxial testing system.

• **Time to Failure for Undrained Shear Tests (t_f)**

Time required to fail a soil sample during shearing or compression stage is generally known as "time to failure," which is computed from isotropic triaxial consolidation data. Blight (1964) suggested some guidelines for assuming the time to failure for undrained shear tests assuming that the soil sample attain about 95% uniform pressure distribution in the isotropic consolidation test. The approximate correlations for time to failure (t_f) based on the geometry of soil sample and the consolidation data (C_{vi}) are given as follows:

Case-1: *For undrained shear tests without side drains*

For consolidation test with vertical flow, e.g. without providing side drains, Blight (1964) suggested the following expression for computing the time to failure for undrained shear tests:

$$t_f = \left[\frac{1.6 * H^2}{C_{vi}}\right] \qquad (15.31a)$$

Where: H = Length of drainage path and

L_c = Soil specimen average length or height after triaxial consolidation test.

Now, taking $H = L_c/2$, Equation (15.31a) can be modified as:

$$t_f = \left[\frac{0.4L_c^2}{C_{vi}} \right]$$ (15.31b)

Substituting for C_{vi} from Equation (15.24a), we get:

$$t_f = \left[\frac{0.4L_c^2}{\pi D_c^2} * \lambda t_{too} \right]$$ (15.32a)

Now putting $L/D = r$ in above Equation (15.32a), we get:

$$t_f = 0.127r^2\lambda t_{100}$$

From Table 15.9, $\lambda = r^2/4$ for drainage from one end, we get:

$$t_f = [0.0318r^4t_{100}]$$ (15.32b)

However, usually the L/D ratio for the soil specimen is taken as 2 (e.g. $L/D = r = 2$), therefore, time to failure in undrained compression test from Equation (15.32b) is:

TABLE 15.12
Time to Failure (t_f) Corresponding with Different Boundary Conditions (BS 1377: Part 8 – 1990)

Sr. No.	Drainage boundary conditions during consolidation stage	Values of η	Values of λ		Values of F (for r =2)	
			L/D = 2	L/D = r	CU Test[*]	CD Test
A: With no side drains						
1	Drainage from one end	0.75	1	$r^2/4$	0.53	8.5
2	Drainage from both ends	3.0	4	r^2	2.1	8.5
B: With side or radial drains						
1	Drainage from radial boundary only	32.0	64	64	1.43	12.7
2	Drainage from radial boundary and one end	36	80	$3.2(1+2r)^2$	1.8	14.2
3	Drainage from radial boundary and two ends	40.4	100	$4(1+2r)^2$	2.3	15.8

Note

[*] Applies only to plastic deformation of non-sensitive soils. Values correspond to those in Table 1 of BS 1377:Part 8 – 1990.

$$t_f = [0.0318 * (2)^4 * t_{100}] = [0.508 t_{100}] \tag{15.32c}$$

Case-2: *For tests with side drains*

For consolidation tests with side drains and allowing drainage through radial drains only, Blight (1964) suggested the following expression for computing the time to failure for undrained shear tests:

$$t_f = \left[\frac{0.0175 * L_c^2}{C_{vi}} \right] \tag{15.32d}$$

Now substituting the value of C_{vi} in above Equation (15.32d), we get:

$$t_f = \left[\frac{0.0175 * \lambda}{\pi} * \frac{L_c^2}{D_c^2} * t_{100} \right] \tag{15.32e}$$

Further, substitute L/D = r, and λ = 64 (from Table 15.12 for drainage from radial boundary only), we get:

$$t_f = \left[\frac{0.0175 * r^2 * 64}{\pi} * t_{100} \right] \text{ or } t_f = [0.35655 * r^2 * t_{100}] \tag{15.32f}$$

Or if L/D = r = 2, we get:

$$t_f = [0.3565 * (2)^2 * t_{100}] = [1.43 * t_{100}] \tag{15.32g}$$

For the conditions of drainage from the radial boundary with drainage additionally from one end or from both ends, the above factor is increased in proportion to the value of λ, giving the values of "F" shown in the last two columns of Table 15.12. All values of the factor "F" relating (t_f to t_{100}) given in Table 15.12 are for specimens in which the height is equal to twice the diameter (L/D = 2).

It may be noted that the factor "F" depends upon whether or not side drains (radial drains) are fitted and the type of test (e.g. undrained or drained) conducted. For example, from Table 15.12, for drainage boundary condition "from one end" with no side drain for undrained (CU) test, $t_f = 0.53 * t_{100}$. For specimens of L/D ratio of about 2, the factors are independent of the size of the specimen. However, it may be noted that if time to failure (t_f) calculated from above is less that 120 minutes, then the actual time to failure (t_f) should be taken as 120 minutes.

Various factors for calculating C_{vi} and time to failure (t_f) for undrained and drained compression tests are given in Table 15.12.

- **Time to Failure for Drained Shear Tests (t_f)**

Time required to fail a soil sample during shearing or compression stage is generally known as "time to failure," which is computed from isotropic triaxial consolidation data. Theoretically about 95% degree of consolidation or dissipation of excess pore water pressure is acceptable for computing drained shear strength parameters. Henkel and Gilbert (1954) suggested the following correlation between degree of consolidation at failure $\bar{U}_f(\%)$ and time to failure (t_f) for drained shear tests as given below:

$$\frac{\bar{U}_f}{100} = 1 - \frac{L_c^2}{4\eta c_{vi} t_f} = \frac{H_c^2}{\eta c_{vi} t_f} \quad OR \quad t_f = \frac{L_c^2}{4\eta c_{vi}\left(1 - \frac{\bar{U}_f}{100}\right)} \tag{15.33}$$

Where: L_c = Length of specimen post consolidation = $2H_c$ - the height of the soil sample (Bishop and Henkel 1962),

η = Factor depending upon drainage conditions at the specimen boundaries as given in Table 15.12 for various drainage conditions. However, for end drainage, the specimen proportions are immaterial, but for radial drainage the value of η is based on the length being twice the diameter.

Putting $\bar{U}_f$ = 95% in Equation (15.33), and rearranging, "t_f" is given by:

$$t_f = \left[\frac{L_c^2}{0.2\eta C_{vi}}\right] \tag{15.34a}$$

By substituting value of C_{vi} (Equation 15.24a) in Equation (15.34a), the time required for failure, t_f can be calculated as:

$$t_f = \frac{L_c^2}{0.2\eta} * \frac{\lambda t_{100}}{\pi D_c^2} = \left[\frac{5r^2}{\pi} * \frac{\lambda}{\eta}\right] t_{100} \ for \ r = L/D \tag{15.34b}$$

For the usual type of specimen with L/D ratio of 2, the Equation (15.34b) becomes:

$$t_f = \left[\frac{20}{\pi} * \frac{\lambda}{\eta}\right] t_{100} \tag{15.34c}$$

The multiplier $\left[\frac{20}{\pi} * \frac{\lambda}{\eta}\right]$ is denoted by the factor "**F**," values of which are given in Table 15.12. For other values of r, the factor can be computed from Equation (15.34b), taking the relevant values of λ and η from Table 15.12.

However, for brittle or sensitive soils, the value of t_f should be not less than 8.5 $*$ t_{100} minutes for soil specimen without side drains and 14 $*$ t_{100} minutes for the soil specimen with side drains for the drained test.

- **Strain Rate for Triaxial Shear Tests**

Unlike time to failure, the strain rate for triaxial shear tests can be computed using isotropic triaxial consolidation data from the triaxial consolidation stage. For small specimens of soils of low permeability, shearing to failure usually requires a day or two, but for large specimens the time to failure may run into weeks. The permissible rate of deformation that can be applied depends upon the following factors:

- Type of test (undrained or drained)
- Type of soil (drainage characteristics. stiffness)
- Size of specimen
- Whether or not side drains are fitted
- Intervals of strain at which significant readings are to be taken

Strain rate, ε_r (mm/min), depends on data derived from the consolidation stage. However, the maximum strain rate to be applied should be equal to (ε_f/t_f) % *per* minute. The axial compression of the specimen corresponding to a strain of ε_f (%) is $[\varepsilon_f/100] * L$, where L is the specimen length in mm. Therefore, the maximum strain rate to be applied (e.g. should not exceed) is:

TABLE 15.13

Suggested Failure Strains in Triaxial Compression Tests as a General Guideline

Sr. No.	Soil Type	Typical Ranges of Strain at Failure (e.g. @ max^m. Deviator Stress), ε_f (%)[*]	
		CU Test	CD Test
1	In situ clay: NC clay	15–20	5–20
	OC clay	20+	4–15
2	Remolded clay	20–30	20–25
3	Brittle soils (very stiff)	1–5	1–5
4	Compacted boulder clay Dry of O.M.C.	3–10	4–6
	Wet of O.M.C.	15–20	6–10
5	Compacted sandy silt	8–15	10–15
6	Saturated sand dense	25 +	5–7
	Loose	12–18	15–20

Note
[*] In case of any doubt, assume a failure strain less than the tabulated values.

TABLE 15.14

Suggested Failure Strains and Time to Failure in Triaxial Compression Tests

Sr. No.	Clay Type		Liquid Limit (%)	Plastic Limit (%)	Failure Strain, ε_f (%)	Time to Failure, t_f (hours)
1	Undisturbed normally consolidated	Estuarine	103	34	24	46
		Estuarine	116	34	24	50
		Alluvial	28	18	20	10
2	Undisturbed heavily over-consolidated	Lias	56	24	4-8	25
		Weald	43	18	4-8	8
		boulder	30	15	4-6	8
3	Remolded Normally consolidated	London	78	26	22	48
		Weald	43	18	20	30
4	Remolded Over-consolidation ratio of 4	London	78	26	11	48
		Weald	43	18	14	30
5	Remolded Over-consolidation ratio of 24	London	78	26	5	24
		Weald	43	18	7	8

$$\varepsilon_r = \left[\frac{\varepsilon_f L}{100 t_f} \right] (mm/minute) \qquad (15.35)$$

Where: ε_f = Failure strain (tentatively assumed for undrained test and drained test) and

L = Initial is length or height of soil specimen taken during shearing test (e.g. post consolidation), and t_f = time to failure.

The rate of strain given in Hight's procedure is 1% per hour. But without pore pressure measurement at the mid-height, to ensure pore pressure equalization the rate of strain should be not greater than that derived from consolidation data on a similar specimen. The rate of strain should not exceed 1% per hour.

If the "critical" or "ultimate" condition is to be reached, application of strain is continued at the same rate beyond the "peak" point. For failure to occur at maximum value of σ'_1/σ'_3, the corresponding strain at maximum value of σ'_1/σ'_3 should be used for ε_f in the calculation of rate of displacement. However, in case of OC and compacted clays, maximum value of σ'_1/σ'_3 is reached before maximum deviator stress is achieved. For such a case, strain of about $2/3 * \varepsilon_{peak}$ (e.g. strain corresponding to peak point) should be adopted. But, it is recommended that the estimate of strain shall be based on qualitative test results, sound judgment and good experience. Furthermore, estimate of failure strain on lower side will err on the safe side. As a general guideline, the suggested values of strain rate for "CU & CD" tests are provided in Table 15.13 (Head 1998).

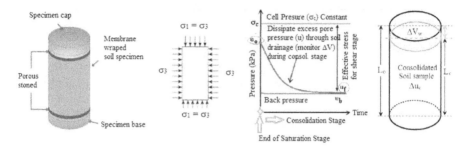

FIGURE 15.23 Schematic of end of triaxial consolidation test (Mir 2010).

Therefore, for desired strain rate to fail a soil sample in the computed time limit (e.g. time to failure, t_f), failure strain (ε_f) must be known, which depends not only on clay type but also on its consolidation theory. Bishop and Henkel (1962) suggested the typical values of failure strain, as given in Table 15.14.

The strain rate for "UU" and "CU" tests can be taken as about that of the unconfined compression test, approx. in the range of 0.5 to 2% per minute with the larger strain rate for softer soils.

15.6.3 SHEARING STAGE FOR "CU" TEST

The triaxial shear test is conducted on a soil specimen, which has been saturated and brought to the required state of effective stress by isotropic triaxial consolidation test. After consolidation is complete, as shown in Figure 15.23, all drainage valves are closed and it is ensured that effective cell pressure ($\sigma'_c = \sigma_c - u_b$) remains constant.

Shearing stage can be conducted either in undrained condition or drained condition as per research program. In this case, shearing stage for undrained behavior of soil specimen is briefly outlined as below:

1. Close all the drainage lines and record final post consolidation data as initial data for shearing stage, e.g. make sure that the following post consolidation parameters required for undrained shear test are computed:
   a. Cell pressure, σ_c (kPa) is maintained constant as from consolidation stage
   b. Back pressure, u_b (kPa), if used during saturation stage
   c. Eff. Stress, σ'_c ($= \sigma_c - u_b$) (kPa)
   d. Post-consol length, L_c (mm)-is considered as initial length for shearing stage
   e. Post-consol x-sec. A_c (cm^2)
   f. Strain rate, ε_r (mm/min), depends on data derived from consolidation stage. The maximum strain rate to be applied is:

$$\varepsilon_r = \left[\frac{\varepsilon_f L_c}{100 t_f} \right] \text{(mm/minute)} \tag{15.36}$$

TABLE 15.15
Undrained Triaxial Shear (CU) Test Data Observation Sheet

Post consolidation data: L_c (cm): 19.38; Dc (cm): ………; Ac (cm²): ………; σ_c (kPa): ………; u_c (kPa): ………; u_b (kPa): ………

Time Hrs	Min	Sec	ΔL (cm)	P (kg)	Corr.P (kN)	L=(L_c−ΔL) (cm)	ε_a=[ΔL/L]*100 (%)	A=A_c/(1−ε_a) (sq.m)	σ_d=P/A(kPa)	Corr.σ_d (kPa)	σ_1=σ_d+σ_3 (kPa)	u (kPa)	σ_1'=σ_1−u (kPa)	σ_3'=σ_3−u (kPa)	σ_1'/σ_3'	A_f	P=(σ_1+2σ_3)/3(kPa)	p'=p−u (kPa)	M=q/p'	S_u=σ_d/2(kPa)	S_u/σ_o
—	—	—	0.00	0	0	19.38	0.00	0.007	0.00	0.00	550.00	74	476.00	476	1.00	0.00	550.00	476.00	0.000	—	—
18	38	16	0.00	1	0.001	19.38	0.00	0.007	1.30	1.30	551.30	74	477.30	476	1.00	0.000	550.43	476.43	0.003	0.65	0.00
18	39	16	0.00	7	0.068	19.38	0.01	0.007	9.11	9.11	559.11	77	482.11	473	1.02	0.38	553.04	476.04	0.019	4.55	0.01
18	40	16	0.00	13	0.126	19.38	0.02	0.007	16.92	16.91	566.92	79	487.92	471	1.04	0.32	555.64	476.64	0.035	8.46	0.02
18	41	16	0.01	17	0.165	19.38	0.04	0.007	22.12	22.11	572.12	82	490.12	468	1.05	0.38	557.37	475.37	0.047	11.06	0.02
18	42	16	0.01	21	0.204	19.37	0.05	0.007	27.32	27.31	577.32	84	493.32	466	1.06	0.38	559.11	475.11	0.057	13.66	0.03
18	43	16	0.01	23	0.223	19.37	0.06	0.007	29.91	29.91	579.92	85	494.92	465	1.06	0.38	559.97	474.97	0.063	14.95	0.03
18	44	16	0.01	25	0.243	19.37	0.06	0.007	32.52	32.51	582.52	87	495.52	463	1.07	0.42	560.84	473.84	0.069	16.25	0.03
18	45	16	0.01	27	0.262	19.37	0.07	0.007	35.12	35.10	585.12	89	496.12	461	1.08	0.44	561.71	472.71	0.074	17.55	0.04
18	46	16	0.02	31	0.301	19.36	0.10	0.007	40.31	40.29	590.31	92	498.31	458	1.09	0.46	563.44	471.44	0.085	20.14	0.04
18	47	16	0.02	35	0.340	19.36	0.11	0.007	45.51	45.48	595.51	95	500.51	455	1.10	0.48	565.17	470.17	0.097	22.74	0.05
18	48	16	0.03	38	0.369	19.36	0.13	0.007	49.40	49.37	599.40	98	501.40	452	1.11	0.50	566.47	468.47	0.105	24.68	0.05
18	49	16	0.03	41	0.398	19.36	0.14	0.007	53.29	53.26	603.29	101	502.29	449	1.12	0.52	567.76	466.76	0.114	26.63	0.06
18	50	16	0.03	45	0.437	19.35	0.15	0.007	58.49	58.45	608.49	103	505.49	447	1.13	0.51	569.50	466.50	0.125	29.22	0.06
n-1………																					
16	42	22	3.53	178	1.7266	15.85	18.22	0.009	189.48	184.76	739.48	418	321.48	132	2.44	1.83	613.16	195.16	0.947	92.38	0.19
16	43	22	3.53	177	1.7169	15.85	18.22	0.009	188.41	183.70	738.41	418	320.41	132	2.43	1.84	612.80	194.80	0.943	91.85	0.19
16	44	22	3.54	178	1.7266	15.85	18.24	0.009	189.44	184.72	739.44	418	321.44	132	2.44	1.83	613.15	195.15	0.947	92.36	0.19
16	45	22	3.54	177	1.7169	15.84	18.26	0.009	188.33	183.61	738.33	417	321.33	133	2.42	1.83	612.78	195.78	0.938	91.80	0.19
16	46	22	3.54	177	1.7169	15.84	18.26	0.009	188.33	183.61	738.33	418	320.33	132	2.43	1.84	612.78	194.78	0.943	91.80	0.19
16	47	22	3.54	177	1.7169	15.84	18.27	0.009	188.31	183.58	738.31	418	320.31	132	2.43	1.84	612.77	194.77	0.943	91.79	0.19
16	48	22	3.54	177	1.7169	15.84	18.27	0.009	190.58	185.85	740.58	418	323.58	133	2.43	1.85	613.53	195.53	0.951	92.93	0.19

Where: ε_f = Failure strain (tentatively assumed for undrained test).
  g. Time to failure, t_f (min)-depends on drainage boundary condition used in consolidation stage.

2. Now bring the cell load piston in touch with the top loading cap of the soil sample carefully.
3. Set initial readings of load cell (or proving ring) and the dial gauge (or LVDT) to zero for measurement of load and vertical compression of the soil sample during shearing stage.
4. Select the designated strain rate as calculated from consolidation data in step 1(g).
5. Check that drainage valve is closed and then start the triaxial shear test.
6. Record the various parameters during undrained shear test at an time interval of about one minute till soil sample fails under applied axial load or deviator stress (Table 15.15):

  • Deformation or change in length, ΔL (mm) to be recorded from strain gauge (using data acquisition system and desktop PC/laptop)
  • Load, P (kN) from load cell to be recorded by using a data logger (using data acquisition system and desktop PC/laptop).
  • Pore water pressure u (kPa) from pressure transducer (using data acquisition system and desktop PC/laptop).

All the testing data is usually recorded by using data acquisition system, e.g. computerized triaxial testing system.

7. Compute the following parameters during undrained shear test against the recorded test parameters till soil sample fails under applied axial load or deviator stress (Table 15.15):

  • Axial strain, ε_a (%) = $\Delta L/L_c$
  • Corrected load, P_c (kN) = P * F (where F is a conversion factor)
  • Actual length during shearing, L_s = L_c-ΔL (mm)
  • Corrected Xec. Sec Area post shear test, $A_{sc} = \dfrac{A_C}{1-\varepsilon_a/100} * \dfrac{1}{10000} (m^2)$, where A_C = Post Consolidation Xex-Sec area
  • Deviator stress, $\sigma_d = P_c/A_c$ (kPa)
  • Corrected deviator stress, $\sigma_{d)c} = \sigma_d - \left(\dfrac{4E_m * (t_m/1000) * (\varepsilon_a/100)}{(D_C/100)} \right)$ after applying membrane correction.

Where: E_m (kPa) is the modulus of rubber membrane, t_m (mm) is the thickness of the rubber membrane, ε_a (%) is the axial strain.
D_c (mm) is the post-consolidation diameter of the sample (= $\sqrt{4A_c\pi}$), and A_c is the post-consolidation average cross-sectional area of the soil sample.

  • Major principal stress, $\sigma_1 = \sigma_{d)c} + \sigma_3 = \sigma_{d)c} + \sigma_c$ (kPa)
  • Minor principal stress, $\sigma_3 = \sigma_c$ (kPa)
  • Effective major principal stress, $\sigma'_1 = \sigma_1 - u$ (kPa)
  • Effective minor principal stress, $\sigma'_3 = \sigma_3 - u$ (kPa)

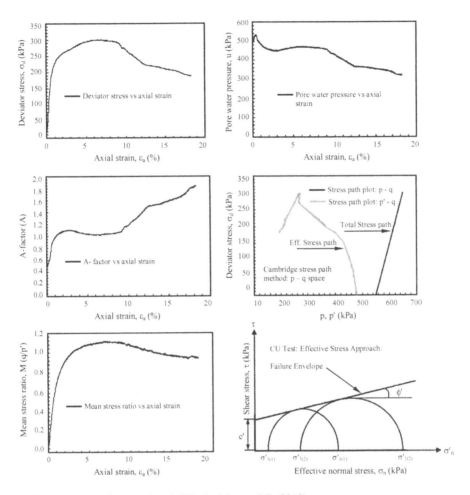

FIGURE 15.24 Test results of CU triaxial test (Mir 2010).

- Stress ratio, σ'_1/σ'_3
- A-factor, $A = \frac{\Delta u}{\Delta\sigma_1 - \Delta\sigma_3} = \frac{\Delta u}{\Delta\sigma_d}$
- Mean stress, $p = \frac{(\sigma_1 + 2\sigma_3)}{3}$ (kPa)
- Eff. mean stress, $p' = p - u$ (kPa)
- Mean stress ratio, $M = \frac{q}{p'} = \frac{\sigma_{d)c}}{p'}$
- Undrained strength, $s_u = c_u = \frac{q}{2} = \frac{\sigma_{d)c}}{2}$ (kPa)
- Strength ratio, $s_r = \frac{s_u}{\sigma_o}$

Where: σ_o is present effective overburden pressure at sample location

8. At the end of shear test and computation of various test parameters (step 7), plot the following graphs (Figure 15.24: Mir 2010):

- Axial strain, ε_a (%) versus deviator stress, σ_d (kPa)
- Axial strain, ε_a (%) versus pore water pressure, u (kPa)

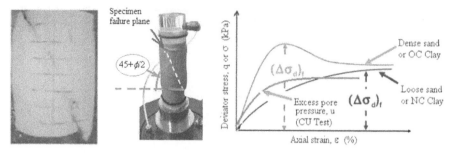

FIGURE 15.25   Schematic of shear failure plane post-shear test.

- Axial strain, ε_a (%) versus A-factor (A)
- Axial strain, ε_a (%) versus mean stress ratio, $M = \dfrac{q}{p'} = \dfrac{\sigma_{d)c}}{p'}$
- Stress path plot between mean effective stress, p' and q (kPa) using Cambridge Stress Path Method
- Mohr`s circle of effective stress representing failure envelope (effective stress approach).

1. When the test is complete, release the cell pressure and dismantle the test assembly carefully.
2. Take out the soil specimen and sketch the shear failure plane at failure, illustrating mode of failure (Figure 15.25).
3. Keep the soil specimen in the oven for 24 hours for determination of final water content.
4. Compute final post-shear parameters (data at failure) as given in Table 15.16.

TABLE 15.16

Post-Shear Parameters in Undrained Triaxial Shear (CU) Test

Max^m. Deviator stress achieved, $q_{max} = \sigma_{dmax}$ (kPa)	299.22	A-factor at failure (at σ_{dmax})	1.02
Mean eff. stress achieved, p' (kPa)	271.29	Pore water pressure at failure, u_f (kPa)	471.88
$M = q_f / p'$	1.10	$k_o = 1 - \sin \phi'$ (NC Soils: Jakay 1944)	0.53
Axial strain at failure, $\varepsilon_{a)f}$ (%)	6.42	$K_o(OC) = K_o(NC) * \sqrt{(OCR)}$: OC Soils	0.53
ϕ' (Degrees) $= \sin^{-1}(3M/6+M)$	27.76	k_o (from plasticity index, PI)	0.57
Minor Eff. Stress σ'_3 (kPa) (at σ_{dmax})	171	Major Eff. Stress σ'_1 (kPa) (at σ_{dmax})	471.88
Max^m. s_u (kPa) (at σ_{dmax})	149.61	Max^m. Stress ratio achieved, σ'_1/σ'_3 (kPa)	2.80
Max^m. s_u/σ_o ratio	0.31	Specimen length after shearing, L_s (cm)	15.84
OCR $(= \sigma_{c)consl}/\sigma_{c)shearing})$	1.00	Max^m. Axial strain achieved, ε_a (%)	18.27

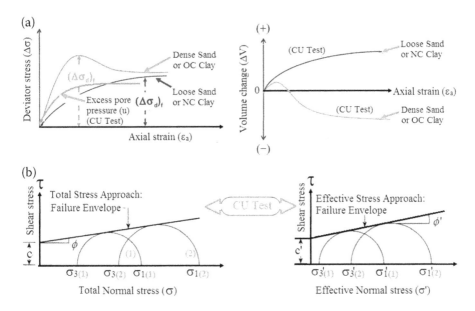

FIGURE 15.26 (a). Typical results (Stress-strain and Vol change) of CU triaxial compression tests; (b). Typical results (Failure envelope) of CU triaxial compression tests.

Note: It should be noted that deviator stress may be corrected taking into consideration thickness and "E" of rubber membrane if required. The rubber membrane used to enclose the soil sample in undrained triaxial compression test results in apparent increase in the measured strength, depending up on the stiffness, thickness, and diameter of the membrane (ASTM D 4767-11). ASTM D 4767-11 clearly suggests the calculation to determine the increase in deviatoric stress due to membrane in a consolidated undrained test, defined as *membrane resistance*, given by the expression:

$$\delta(\sigma_d)_m = (4E_m t_m \varepsilon_a)/(D_c) \tag{15.37}$$

Equation (15.37) is obtained when the rubber membrane is modeled as a hollow cylinder, and the E_m is assumed to be constant along the length (Lade and Hernandez 1977; Baldi and Nova, 1984; Vaid and Negussey 1984; Seed et al. 1989; Bohac and Feda 1992; Greeuw et al. 2001; Noor et al. 2012). The measured deviatoric stress is corrected at each ε_a value; however, the correction is neglected if the error in the deviatoric stress is <5% (ASTMD 4767-11).

15.6.3.1 Typical Results of CU Triaxial Tests

The typical test results of a soil sample tested in isotropic ally consolidated-undrained (CU) triaxial test are illustrated in Figure 15.24, which are self explanatory. From Figure 15.24, it is seen that the soil specimen developed negative pore pressure by the time failure is reached. This may be attributed due to expansion of the soil structure. These pore pressures must be subtracted if positive and added if negative to the confining and vertical stresses to get the effective stresses and effective strength parameters (c' and ϕ'). Further, the results of A-factor and stress path plots show that the soil specimen exhibits as normally consolidated soil. The mean effective stress ratio ($M = q/p'$) is 1.10, which indicates that soil specimen is normally consolidated. Since the consolidation pressure and the effective cell pressure are equal, the over-consolidation ratio (OCR) is 1.0. Typical results for OC and NC soils tested in isotropically "CU" triaxial shear tests are shown in Figure 15.26.

In a consolidated undrained test, pore water pressure increases throughout the test for NC soils as illustrated in Figure 15.26(a). But it may be noted that in case of OC soils, the pore water pressure increases at low stress level (<150 kPa) and decreases below atmospheric pressure (e.g. negative pore water pressure is developed) under high level stresses (> 450 kPa). Mohr's circle based on total and effective stress approaches is also plotted in Figure 15.26(b), which quantifies the difference between the shear parameters for both these approaches. Finally, as word of caution, it may be noted that consolidation of soils in the field occur under different major and minor principal stresses. Therefore, K_o- conditions should be adopted for simulation of actual in situ state of soils in field conditions.

15.6.4 Applications/Role/Significance of and Use of "CU" Test

The results of "CU" test are widely adopted for estimating shear strength of soils in field conditions where the soil has consolidated under foundation loading during construction or under its own weight, and which is then followed by quick increase in loads, causing rapid change in critical stresses during which no further change in water content can take place. The examples of such cases are the stability calculations against failure by shear in consolidated dams, slopes, and other earth structures of cohesive soils (saturated clays) under conditions of rapid drawdown of water, where water has no time to drain out of the voids. Other examples would include cases of consolidated clay soils serving as foundations for grain elevators under conditions of quick unloading.

15.7 CONSOLIDATED DRAINED TEST (CD OR S) TRIAXIAL TEST

A consolidated-drained (CD) test or Slow Test is a novel technique of determination of soil shear parameters in an effective stress approach in which a soil sample is very slowly loaded during shearing stage such that all the pore water present in soil void pores drains out of the soil sample without developing any pore water pressure in the soil sample. This test is widely suitable for sandy soils. Unlike "CU" test, "CD" test is also conducted in three stages in a well-equipped soil testing laboratory. The summary of drainage boundary conditions and application of loading in a triaxial compression are given in Table 15.3. This test is also known as the consolidated drained (CD) test or Slow (S) test. The step-by-step procedure for each stage of this test is given as follows.

TABLE 15.17
Drained Triaxial Shear (CD) Test Data Observation Sheet (Mir 2010)

L (cm)	δL (cm)	e_a (%)	Vol V_o (ml)	Vol Change δV	e_v δV/V_c	P (kN)	A (m²)	σ_d = P/A (σ_1 − σ_3)	σ_1 σ_d + σ_3	u kPa	σ_1' σ_1 − u	σ_3' σ_3 − u	p (σ_1 + 2σ_3)/3	q (σ_1 − σ_3)	p' p − u	M q/p'	Void ratio e	δe δV/V_o (1+e_o)	Sp Vol v = (1 + e_o)
19.18	0	0.00	46.95	0	0.000	0.00	0.01	0.00	225	152	73	73	225	0.00	73	0.00	0.8617	0.00000	1.8617
19.18	0	0.00	47	-0.05	0.003	0.00	0.01	0.00	225	152	73	73	225	0.00	73	0.00	0.8618	-0.00006	1.8618
19.18	0.001	0.01	46.9	0.05	0.003	0.00	0.01	0.00	225	154	71	71	225	0.00	71	0.00	0.8616	0.00006	1.8616
19.18	0	0.00	46.8	0.15	0.010	0.00	0.01	0.00	225	155	70	70	225	0.00	70	0.00	0.8615	0.00019	1.8615
19.18	0.001	0.01	46.8	0.15	0.010	0.00	0.01	0.00	225	154	71	71	225	0.00	71	0.00	0.8615	0.00019	1.8615
19.18	0	0.00	46.75	0.2	0.014	0.02	0.01	2.55	228	153	75	72	226	2.55	73	0.03	0.8615	0.00025	1.8615
19.18	0.001	0.01	46.7	0.25	0.017	0.02	0.01	2.55	228	153	75	72	226	2.55	73	0.03	0.8614	0.00032	1.8614
19.18	0.001	0.01	46.65	0.3	0.021	0.01	0.01	1.27	226	153	73	72	225	1.27	72	0.02	0.8613	0.00038	1.8613
19.18	0.002	0.01	46.6	0.35	0.024	0.02	0.01	2.54	228	153	75	72	226	2.54	73	0.03	0.8613	0.00045	1.8613
n-1.......																			
19.18	0.002	0.01	46.55	0.4	0.027	0.02	0.01	2.54	228	153	75	72	226	2.54	73	0.03	0.8612	0.00051	1.8612
19.18	0.003	0.02	46.55	0.4	0.027	0.03	0.01	3.82	229	153	76	72	226	3.82	73	0.05	0.8612	0.00051	1.8612
19.18	0.002	0.01	46.5	0.45	0.031	0.02	0.01	2.54	228	153	75	72	226	2.54	73	0.03	0.8611	0.00057	1.8611
19.18	0.003	0.02	46.45	0.5	0.034	0.03	0.01	3.82	229	154	75	71	226	3.82	72	0.05	0.8611	0.00064	1.8611
19.18	0.004	0.02	46.45	0.5	0.034	0.03	0.01	3.81	229	153	76	72	226	3.81	73	0.05	0.8611	0.00064	1.8611
19.18	0.003	0.02	46.45	0.5	0.034	0.03	0.01	3.82	229	152	77	73	226	3.82	74	0.05	0.8611	0.00064	1.8611
19.18	0.004	0.02	46.4	0.55	0.038	0.03	0.01	3.81	229	153	76	72	226	3.81	73	0.05	0.8610	0.00070	1.8610
19.17	0.005	0.03	46.45	0.5	0.034	0.04	0.01	5.09	230	153	77	72	227	5.09	74	0.07	0.8611	0.00064	1.8611
19.17	0.005	0.03	46.4	0.55	0.038	0.04	0.01	5.09	230	153	77	72	227	5.09	74	0.07	0.8610	0.00070	1.8610
19.17	0.006	0.03	46.35	0.6	0.041	0.05	0.01	6.36	231	153	78	72	227	6.36	74	0.09	0.8609	0.00076	1.8609

15.7.1 Triaxial Saturation Stage in "CD" Test

- The process of sample preparation for consolidated undrained "CU" triaxial test is the same as for "UU" test described in Section 15.5.1.1
- The procedure for conducting triaxial saturation stage for consolidated undrained "CU" triaxial test is same as for the "UU" test described in Section 15.5.1.2

15.7.2 Triaxial Consolidation Stage in "CD" Test

- The procedure for conducting the triaxial consolidation stage for the consolidated drained "CD" triaxial test is same as for the "CU" test described in Section 15.6.2.

15.7.3 Triaxial Shear Stage in "CD" Test

1. The procedure for the conducting triaxial shear stage for the consolidated drained (CD) test is the same as for the "CU" test except that:
   - The consolidated drained (CD) test is conducted with drainage line open (e.g. pore water pressure is allowed to drain out under gradually applied deviator stress and volume change (δV) measurements are recorded as shown in Table 15.17.
   - In this test, theoretically about 95% degree of consolidation or dissipation of excess pore water pressure is acceptable for computing drained shear strength parameters including time to failure (t_f). However, for brittle or sensitive soils, the value of t_f should be not less than $8.5 * t_{100}$ minutes for the soil specimen for drainage from one end only (without side drains) and $14 * t_{100}$ minutes for the soil specimen with side drains for radial boundary drainage from both ends.
   - The rate of strain ($\varepsilon_r = \frac{\varepsilon_f L}{100 t_f}$ mm/minute) should be not greater than that derived from consolidation data on a similar specimen. The rate of strain should not exceed 1% per hour. The suggested values of strain rate for "CU & CD" tests are provided in Tables 15.13 and 15.14 as a general guide, but the estimate should preferably be based on experience.
   - In a drained test, the peak deviator stress is controlling criteria for determining the strain.
   - The volumetric strain at the end of triaxial shear stage or during triaxial shear stage is given by:

$$\varepsilon_{V_c} = \left[\frac{\Delta V_s}{V_c} * 100 \right] (\%) \tag{15.38}$$

Where: ΔV_s = Volume change during triaxial shear stage (Table 15.17) = $V_c - V_f$ (cm^3)
   = Directly measured at the end of the shear stage by volume change apparatus

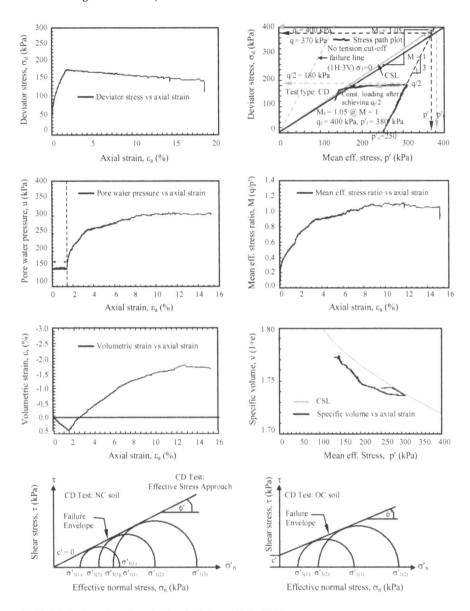

FIGURE 15.27 Results of CD triaxial test (Mir 2010).

V_c = Volume of soil specimen at the end of triaxial consolidation stage (taken as initial volume for triaxial shear stage)

- The corrected X-sec. area of the soil specimen during the triaxial shear stage is computed by the expression:

TABLE 15.18

Post-Shear Parameters in Drained Triaxial Shear (CD) Test (Mir 2010)

Max^m. Deviator stress achieved, $q_{max} = \sigma_{dmax}$ (kPa)	183.90	OCR ($= \sigma_{c)consl}/\sigma_{c)shearing}$)	1.00
A-factor at failure (at max^m. Deviator stress)	0.20	$k_{o(OC)} = k_o(NC) * SQRT(OCR)$: OC Soils	0.70
Mean eff. stress achieved, p' (kPa)	274.95	Pore water pressure, u_f (kPa)	186
$M = q_f/p'$	0.67	Axial strain at failure, $\varepsilon_{a)f}$ (%)	1.62
Max^m. σ_u (kPa)	91.32	Major Eff. Stress σ'_3 (kPa)	584
Max^m. s_u/σ_o ratio	0.23	Minor Eff. Stress σ'_3 (kPa)	215
ϕ' (Degrees) $= \sin^{-1}(3M/6+M)$	17.51	Max^m. Stress ratio achieved, σ'_1/σ'_3 (kPa)	1.85
$k_o = 1 - \sin\phi'$ (NC Soils: Jakay 1944)	0.70	Specimen length after shearing, L_s (cm)	18.56
k_o (from plasticity index, PI)	0.56	Max^m. Axial strain achieved, ε_a (%)	1.62

$$A_s = [A_c * (1 - \varepsilon_v)/(1 - \varepsilon_a)] \quad (sq.\ cm\ or\ sq.\ m)$$

Where: A_c = X-Sec. area at the end of triaxial consolidation stage taken as initial X-Sec. area for triaxial shear stage (Table 15.17), and
$\varepsilon_a = \Delta L/L_c * 100$ (%) = Axial strain during triaxial shear stage (Table 15.17)

The change in void ratio (δe) during the triaxial shear stage is computed as (Table 15.17):

$$\delta_i = [\delta V_s/V_c * (1 + e_c)]$$

Where: e_c = Final void ratio at the end of triaxial consolidation stage
- The void ratio during triaxial shear stage for "CD" test is computed as:
  $e = e_c - \delta e$
- The specific volume during triaxial shear stage for "CD" test is computed as: $v = 1 - e$

Once the shearing stage in "CD" test is complete, other shear parameters are computed in the same manner as for the "CU" test.

2. At the end of "CD" test and computation of various test parameters above (including step 8 of CU test), plot the following graphs (Figure 15.27):

- Axial strain, ε_a (%) versus deviator stress, σ_d (kPa)
- Axial strain, ε_a (%) versus volumetric strain, ε_v (%)
- Axial strain, ε_a (%) versus pore water pressure, u (kPa)
- Axial strain, ε_a (%) versus mean stress ratio, $M = \dfrac{q}{p'} = \dfrac{\sigma_{d)c}}{p'}$
- Stress path plot between mean effective stress, p' and q (kPa) by Cambridge Stress Path method
- Mohr's circle of effective stress representing failure envelope (effective stress approach)

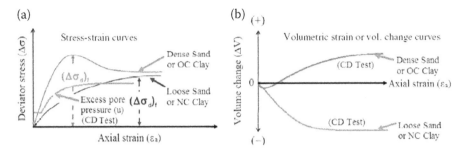

FIGURE 15.28 Typical results (stress-strain curves) of CD triaxial tests.

3. At the end of the drained shear test, dismantle the test assembly and take out the soil specimen and sketch the shear failure plane at failure, illustrating mode of failure (Figure 15.25).

1. Keep the soil specimen in the oven for 24 hours for determination of final water content.
2. Compute final post-shear parameters in CD test (data at failure) as given in Table 15.18.

All the testing data is usually recorded by using a data acquisition system, e.g. computerized triaxial testing system as described in "CU" test.

15.7.3.1   Typical Results of Triaxial "CD" Tests

The typical test results of a soil sample tested in isotropic ally consolidated-drained (CD) triaxial test are illustrated in Figure 15.27, which are self explanatory. Since pore water pressure is not measured in this test, therefore, effective shear strength parameters (c_d' & ϕ_d') are obtained, which also equal to total shear strength parameters (c & ϕ). However, in case of NC clays, the drained cohesion (c_d') is too small and generally ignored. From Figure 15.27, it seen that the pore water pressure remains constant initially under gradually applied loading (e.g. σ_1 varies, however, σ_3 remains constant). However, in this drained shear test, loading was maintained constant after achieving $q_f/2$ as illustrated in Figure 15.27 (stress path plot) beyond which pore water pressure increased indicating that soil specimen behaves as normally consolidated, which is also authenticated by mean effective stress ratio plot. Typical results for OC and NC soils tested in isotropically "CD" triaxial shear tests are shown in Figure 15.28. It seen that the test results are almost same as obtained in "CU" test.

15.7.4   Applications/Role/Significance of and Use of "CD" Test

The results of the "CD" test are widely adopted for estimating strength of soils in connection with field problems where the field stresses develop within the soil mass sufficiently slowly for all changes in moisture content to take place. An example of such a case is the determination of the final bearing capacity of a soil

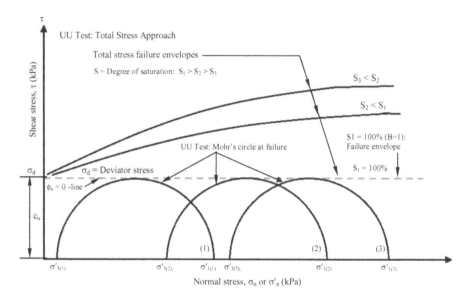

FIGURE 15.29   Typical results (stress-strain curves) of CD triaxial tests.

mass under the foundation of a structure that is erected more slowly than the soil consolidates.

Reliable estimates of the drained shear strength of foundation soils are needed for design of clay embankments in which drainage channels (such as sand drains or PVDs to accelerate the drainage in clayey soils) are embedded. Therefore, the consolidated drained (CD) test is the most common triaxial procedure, as it allows strength parameters to be determined based on the effective stresses (i.e. c'_d and ϕ'_d).

15.8   GENERAL COMMENTS

A triaxial test is a most suitable and widely used soil test in the geotechnical laboratory for evaluation of shear parameters (c, ϕ & c', ϕ') for all types of soils under controlled drainage and loading conditions. However, it may be ensured that the soil samples to be tested are fully saturated. In the case of the "UU" test on saturated cohesive soil with 100% degree of saturation, the failure envelope represented by Mohr's circle is horizontal (e.g. $\phi_u = 0$) as shown in Figure 15.29. However, it also seen from Figure 15.29 that if the degree of saturation decreases, the Mohr's failure envelope is curved (e.g. $\phi_u > 0$).

15.8.1   Some Comments on the Influence of the Type of Test

- An undrained test on non-sensitive plastic clays can be run at a faster strain rate than a drained test on similar material because no movement of water through and out of the specimen is involved. Pore pressure is measured at the base, but it is the value within the middle third that most affects the measured shear strength.

- For soils that fail in a brittle manner (e.g. stiff fissured clays), and for sensitive soils, the rate of strain in an undrained test should be the same as that calculated for a drained test.
- If an automatic data-logging system is used there is no problem because, the machine can be kept running normally and data will accumulate for attention next morning. It is necessary to ensure that the limit of travel will not be reached while unattended (if an automatic limit switch is not fitted), and to check that there is adequate capacity in the recording medium. Prominent signs reading DO NOT SWITCH OFF should be displayed.

15.8.2 EFFECTS OF ACCELERATING TESTS

- Attempts to accelerate the completion of each stage of a test can lead to erroneous results. The most likely effects are summarized as follows:
- Where full saturation is required, incomplete saturation leaves air pockets in the specimen, giving unreliable pore pressure readings.
- The effect on the measured shear strength of running compression tests too quickly was investigated (among others) by Lumb (1968). In genera l the result is to give false values of effective strength parameters, in which the value of c' is too high and ϕ' is too low.

15.9   STATEMENT OF STRENGTH PRINCIPLES

At the end of a triaxial test, the three basic strength principles are concluded below:

Principle I: For NC soils or OC soils with the same maximum preconsolidation pressure p'_c, there is a unique relationship between strength [e.g. maximum stress difference $\Delta\sigma'_{max.} = (\sigma_1 - \sigma_3)_{max.}$] and effective stress at failure.

Principle II: For NC soils or OC soils with the same preconsolidation pressure p'_c, there is a unique relationship among water content, shear stress, and effective stress.

Principle III: For NC soils or OC soils, there is a unique relationship among strength, water content at failure, and effective stress at failure as expressed by the Hvorslev parameters.

15.10   SOURCES OF ERROR

The main sources of error in a triaxial test can be:

1. Soil samples are not fully saturated. This will lead to inaccurate measurement of shear strength parameters.
2. Soil sample selection for a triaxial test plays a vital role. Make sure that proper soil sample is tested in a desired triaxial test (e.g. UU test for clays and CD test for sands). This will save both time and money as a triaxial test is very expansive and time consuming.

3. A soil sample should be saturated by allowing water into the soil sample from bottom drainage line and not from top drainage line.
4. Like wise, water should be allowed to drain out from top drainage line rather that bottom drainage line.
5. The piston of load cell should not be in touch with soil sample during saturation and consolidation; else it will lead to inaccurate measurement of sample size and other allied test parameters.
6. Application of strain rate has direct bearing on shear strength of soils. Any discrepancy in strain rate corresponding to soil type will lead to inaccurate measurement of shear strength parameters.

15.11   PRECAUTIONS

During the triaxial test, the following precautionary measures may be followed:

1. De-air water should be used.
2. Porous stones should be boiled and kept in de-aired water.
3. Apply a thin layer of grease on the pedestal surface of the triaxial cell and on the top cap before placing the soil specimen.
4. Use saturated filter paper and porous stones for the test to reduce saturation time.
5. All the air bubbles should be removed while filling the cell pressure with de-aired water from the top vent on the triaxial cell.
6. Make sure that there is no leakage before the start or during the test.
7. The loading piston should not touch the soil specimen during the saturation and consolidation stage.
8. Check the B-factor carefully for clayey soils (it should be > 0.98).
9. Bring the loading piston in touch with the soil specimen carefully before the start of the shearing stage (e.g. after completion of consolidation stage).
10. Calculate all the test parameters after consolidation and shearing stage carefully (Post consolidation parameters, e.g. diameter and length of soil specimen become initial dimensions for shearing stage).
11. Cell pressure should be maintained constant throughout consolidation and shearing stage.
12. Cell pressure can be reduced in shearing stage if soil sample is to be made OC.
13. Maintain initial density of sand specimens in all the tests using different confining pressures, else there will be appreciable errors in the maximum failure pressure.
14. Take care that sand particles should not penetrate the rubber membrane while compacting sand specimen in the mold.
15. Make sure that the triaxial test is done on saturated soil specimens and no air bubbles are entrapped in the soil specimens.

16. For drained tests, make sure that the rate of strain is such that no excess pore water pressure is developed in the soil sample.

REFERENCES

ASTM D2850-03a. 2007. "Standard Test Method for Unconsolidated-Undrained Triaxial Compression Test on Cohesive Soils." ASTM International, West Conshohocken, PA, 2003. doi:10.1520/D2850-03AR07.

ASTM D 4767-11. 2011. "Standard Test Method for Consolidated Undrained Triaxial Compression Test for Cohesive Soils." ASTM International, West Conshohocken, PA, USA.

ASTM WK3821. 2011. "New Test Method for Consolidated Drained Triaxial Compression Test for Soils (Under Development)." ASTM International, West Conshohocken, PA, 2003.

BS 1377-1. 1990. "Methods of Test for Soils for Civil Engineering Purposes. General Requirements and Sample Preparation." BSI. ISBN 0 580 17692 4.

BS 1377 (Part 7). 1990-clause 8. "Shear Strength Tests (Total Stress): The Unconsoldiated Undrained Triaxial Compression Test, Without Pore Water Pressure Measurement." British Standards, UK.

BS 1377 (Part 8). 1990-clause 7. "Shear Strength Tests (Effective Stress): The Consolidated Undrained Triaxial Compression Test, with Pore Water Pressure Measurement." British Standards, UK.

BS 1377 (Part 8). (1990-clause 8). "Shear Strength Tests (Effective Stress): The Consolidated Undrained Triaxial Compression Test, with Pore Water Pressure Measurement." British Standards, UK.

Baldi, G., and R. Nova. 1984. "Membrane Penetration Effects in Triaxial Testing." *J Geotech Eng* 110(3): 403–420.

Bishop, A. W., and G. Eldin. 1950. "Undrained Triaxial Tests on Saturated Sands and Their Significance in the General Theory of Shear Strength." *Géotechnique* 2(1): 13–32, doi:10.1680/geot.1950.2.1.13.

Bishop, A. W., and D. J. Henkel. 1962. The *Measurement of Soil Properties in the Triaxial Test*. 2nd ed. London: Edward Arnold Ltd.

Blight, G. E. 1964. "The Effect of Non-Uniform Pore Pressures on Laboratory Measurements of the Shear Strength of Soils." *Symposium on laboratory shear testing of Soils. ASTM Special Tech nical Publication No. 361*. American Society for Testing and Materials, Philadelphia, USA, pp. 173–184.

Blight, G. E. 1961. "Strength and Consolidation Characteristics of Compacted Soils." Ph.D. Thesis, University of London.

Blight, G. E. 1960. "Discussion on General Principles and Laboratory Measurements." Proc. Conf. on Pore Pressure and Suction in Soils: 72–73. London: Butterworth.

Bohac, J., and J. Feda. 1992. "Membrane Penetration un Triaxial Tests." *Geotech Test J* 15(3): 288–294.

Casagrande, A. 1936. "Determination of the Pre-consolidation Load and its Practical Significance." Proceedings, 1st International Conference on Soil Mechanics and Foundation Engineering, Cambridge, MA 3: 60–64.

Casagrande, A., and R. E. Fadum. 1940. "Notes on Soil Testing for Engineering Purposes." Harvard University Graduate School of Engineering Publication No. 8.

Casagrande, A., and D. Wilson. 1951. "Effect of Rate of Loading on the Strength of Clays and Shales at Constant Water Content." *Geotechnique* 2: 251–263.

Casagrande, A., and D. Wilson. (1953). "Prestress Induced in Consolidated-Quick Triaxial Tests." *Proc. 3rd Int. Conf. Soil Mech.* 1: 106–110.

Greeuw, G., H. D. Adel, A. L. Schapers, and E. J. D. Haan. 2001. "Reduction of Axial Resistance Due to Membrane and Side Drains." Soft Ground Tech. Geotech. Sp Pub GSP-112, ASCE, pp. 30–42.

Head, K.H. 1998. *Effective Stress Tests, Volume 3, Manual of Soil Laboratory Testing.* 2nd ed. John Wiley & Sons, England. ISBN 978-0471977957.

Henkel, D. J., and G. D. Gilbert. 1954. "The Effect of the Rubber Membrane on the Measured Triaxial Compression Strength of Day Samples." *Geotechnique* 3: 1–20.

Hvorslev, M. J. 1949. "Subsurface Exploration and Sampling of Soils for Civil Engineering Purposes." U.S. Army Corps of Engineers, U.S. Waterways Experiment Station, Vicksburg, MS.

IS 2720 (Part 11). 1993. "Methods of Test for Soils, Part 11: Determination of the Shear Strength Parameters of a Specimen Tested in Unconsolidated, Undrained Triaxial Compression Without the Measurement of Pore Water Pressure." *Bureau of Indian Standards*, New Delhi.

IS 2720 (Part 12). 1981. "Methods of Test for Soils, Part 12: Determination of Shear Strength Parameters of Soil from Consolidated Undrained Triaxial Compression Test with Measurement of Pore Water Pressure." *Bureau of Indian Standards*, New Delhi.

Jakay, J. 1944. "The Coefficient of Earth pressure at Rest." *Journal of the Union of Hungarian Engineers and Architects* 355–358.

Lade, P. V., and S. B. Hernandez. 1977. "Membrane Penetration Effects in Undrained Tests." *J Geotech Eng Div* 103(GT2): 109–125.

Lumb, P. 1968. "Choice of Strain-Rate for Drained Tests on Unsaturated Soils." *Géotechnique* 18(4): 511–514. DOI:10.1680/geot.1968.18.4.511.

Mir, B. A. 2010. "Study of the Influence of Smear Zone Around Sand Compaction Pile on Properties of Composite Ground." Ph.D. Thesis, Deptt. Of Civil Engineering, IIT Bombay, India.

Noor, M. J. M., J. D. Nyuin, and A. Derahman. 2012. "A Graphical Method for Membrane Penetration in Triaxial Tests on Granular Soils." *J Inst Eng, Malaysia* 73(1): 23–30.

Seed, R. B., H. A. Anwar, and P. G. Nicholson. 1989. "Elimination of Membrane Compliance Effects in Undrained Testing." *12th Intl. Conf. Soil Mech Found Eng* 1: 111–114.

Skempton, A. W. (1954). "The Pore-Pressure Coefficients *A* and *B*." *Géotechnique* 4(4): 143–147. doi:10.1680/geot.1954.4.4.143.

Sridharan, A., N.S. Murthy, and K. Prakash. 1987. "Rectangular Hyperbola Method of Consolidation Analysis." *Geotechnique* 37(3): 355–368.

Taylor, D. W. 1942. "Research on Consolidation of Clays." Serial No. 82, Department of Civil and Sanitary Engineering, Massachusetts Institute of Technology, Cambridge, MA.

Taylor, D. W. 1948. *Fundamentals of Soil Mechanics.* New York:Wiley.

Vaid, Y. P., and D. Negussey. 1984. "A Critical Assessment of Membrane Penetration in the Triaxial Test." *Geotech Test J* 7(2): 70–76.

Index

For Product Safety Concerns and Information please contact our EU
representative GPSR@taylorandfrancis.com
Taylor & Francis Verlag GmbH, Kaufingerstraße 24, 80331 München, Germany